Analysis of Free Radicals in Biological Systems

Edited by   A. E. Favier
             J. Cadet
             B. Kalyanaraman
             M. Fontecave
             J.-L. Pierre

Birkhäuser Verlag
Basel · Boston · Berlin

Editors

Prof. Dr. A.E. Favier
Laboratoire de Biochimie C
CHU Albert Michallon
BP 217
F-38043 Grenoble Cédex 9
France

Prof. Dr. B. Kalyanaraman
Biophysics Research Institute
Medical College of Wisconsin
8701 Watertown Plank Road
PO Box 26509
Milwaukee, WI 53226-0509
USA

Prof. Dr. J. Cadet
CEA/Département de Recherche
Fondamentale sur la Matière
Condensée, SESAM/LAN
17, rue des Martyrs
F-38054 Grenoble Cédex 9
France

Prof. Dr. M. Fontecave[1]
Prof. Dr. J.-L. Pierre[2]
[1]Laboratoire de Biochimie Recherche
[2]Laboratoire de Chimie Biomimétique
LEDSS II
Domaine Universitaire
F-38400 Saint Martin d'Hères
France

Library of Congress Cataloging-in-Publication Data

Analysis of free radicals in biological systems
    ISBN 3-7643-5137-3 ISBN 0-8176-5137-3
    1. Free radicals (Chemistry) – Physiological effect – Research –
Methodology. 2. Free radicals (Chemistry) – Analysis. 3. Free
radicals (Chemistry) – Pathophysiology.
RB 170.A53 1995
616.07–dc20

Deutsche Bibliothek Cataloging-in-Publication Data

Favier, Alain E.:
Analysis of free radicals in biological systems / ed. by
A. E. Favier … - Basel ; Boston ; Berlin ; Birkhäuser, 1995
    ISBN 3-7643-5137-3 (Basel…)
    ISBN 0-8176-5137-3 (Boston)

© 1995 Birkhäuser Verlag, PO Box 133, CH-4010 Basel, Switzerland
Printed on acid-free paper produced from chlorine-free pulp.∞
Printed in Germany

ISBN 3-7643-5137-3
ISBN 0-8176-5137-3
9 8 7 6 5 4 3 2 1

Contents

Determination of lipid peroxidation-derivated products

Evaluation of DNA damages

Miscellaneous techniques

Preface

"Oxidative stress" is used as the generic term describing the involvement of reactive oxygen species in various human diseases. The scope of such a topic is becoming increasingly wide. The recent interest in radicals such as nitric oxide and the discovery of new mechanisms such as the effect of free radicals on redox sensitive proteins and genes are enlarging our understanding of the physiological role of free radicals. Oxidative stress is involved in numerous pathological processes such as ageing, respiratory or cardiovascular diseases, cancer, neurological pathologies such as dementia or Parkinson's disease. It still remains difficult, however, to demonstrate by chemical measurement the *in vivo* production of free radicals and even more to realise their speciation. Therefore, the development of new tools and indicators is engrossing many researchers working in this field. Reliable indicators are absolutely necessary not only to monitor the evolution of oxidative stress in patients but also to evaluate the efficiency of new antioxidant treatments.

The French Free radical club of Grenoble, the CERLIB has been involved for many years in the organisation of international training programs on methodology, in order to provide both theoretical and practical help to researchers from various countries. Such training sessions have been highly successful and participants value the opportunity to learn reliable techniques. This positive echo explains why the researchers of CERLIB decided, with the help of Prof. Dr. B. Kalyanaraman, to publish selected techniques on free radical research.

The main aim of this book is to provide a comprehensive survey on recent methodological aspects of the measurement of damage within cellular targets, information which may be used as an indicator of oxidative stress. In this respect, both practical aspects and general considerations including discussions on the applications and limitations of the assays are critically reviewed. One of the major features of the book is the description of high-performance methods, taking into consideration recent achievements. The book should be useful to a large scientific community including biologists, chemists, and clinicians working on the chemical and biological effects of oxidative stress. It will also be pertinent to applied research carried out in the fields of drugs, cosmetics and new foods.

We thank all the authors for their contributions. We hope this book, with excellent reviews covering the field in general and selected methods, will be beneficial to young as well as experienced scientists.

Grenoble, France   *Alain Favier*
February, 1995   *Jean Cadet*
                                         Balaraman Kalyanaraman
                                         Marc Fontecave
                                         Jean Louis Pierre

Analysis of Free Radicals in Biological Systems
Favier et al. (eds)
© 1995 Birkhäuser Verlag Basel/Switzerland

Chemistry of dioxygen and its activated species

J.-L. Pierre

Laboratoire de Chimie Biomimétique, LEDSS (URA CNRS 332), Université Joseph Fourier, BP 53, F-38041 Grenoble Cédex 9, France

Summary. In this article the physico-chemical properties and the chemical reactivities of dioxygen and its activated species are reviewed. The understanding of this chemistry is an essential prerequisite for a comprehensive description of the reactions which are involved in the oxidative stress and, then, for the design of drugs capable of preventing the deleterious effects of active oxygen species.

Introduction

Molecular oxygen is ubiquitous in nature, i.e., it is indispensable for aerobic life. However, once its concentration is higher than normal, it becomes dangerous for living systems. Chemical data for oxygen are necessary for the understanding of oxygen toxicity. In this chapter, some of these aspects will be reviewed.

The predominant role in oxygen toxicity is attributed to free radicals derived from oxygen, formed during cell metabolism, but which may also have an exogenous origin [1–3].

The O_2 molecule

Molecular oxygen or diatomic oxygen O_2, is triplet ($^3O_2$) in its ground state (Fig. 1). It is paramagnetic and cannot react with diamagnetic molecules without removing spin restriction.

The chemistry of O_2 primarily involves reactions with paramagnetic species (reactions not concerned with spin restriction), with "one-electron donors", or with light. O_2 is the substrate of some enzymes (oxygenases, oxidases) and the product of others (superoxide dismutases, catalases).

Analytical chemistry of oxygen using O_2 electrode has been extensively described in earlier works. Due to its relatively non-polar nature, oxygen is much more soluble in organic media than in water.

Singlet oxygen

$^1O_2$ is an excited state (22 kcal/mole for $^1\Delta_g O_2$) which removes spin restriction and is thus more reactive (Fig. 2).

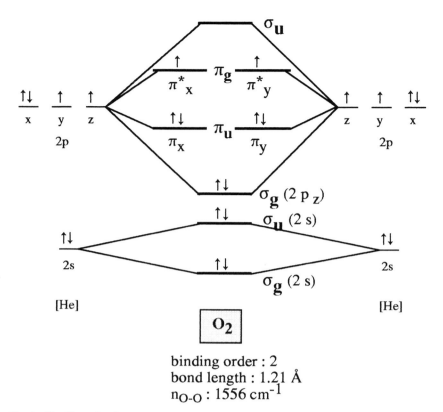

O_2

binding order : 2
bond length : 1.21 Å
n_{O-O} : 1556 cm^{-1}

Fig. 1. The O_2 molecule.

Singlet oxygen is formed in biological systems via photosensitized reactions (the sensitizer in the excited electronic state transfers energy to oxygen), or by chemical excitation reactions which do not involve light excitation. Excitation proceeds either by radical-radical interaction or by the transfer of oxygen using iron(III) of heme, e.g., cytochrome P450):

$$[Fe(III)] + R-O-O-H \longrightarrow [Fe=O] + R-OH$$

$$[Fe=O] + R-O-O-H \longrightarrow [Fe(III)] + R-OH + {}^1O_2$$

In vitro, $^1O_2$ can be obtained by non-photochemical methods using powerful oxidizing agents such as NaOCl, which can react directly with oxidizable species. The best source is an aromatic endoperoxide:

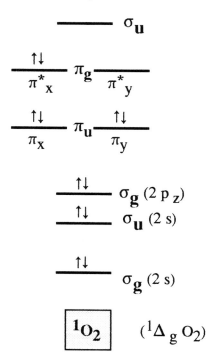

Fig. 2. Singlet oxygen.

$^1O_2$ diffuses across membranes. Its formation in cells is (for example) shown by detecting the photoemission accompanying the reaction:

$$^1O_2 + {}^1O_2 \longrightarrow 2\,^3O_2 + h\nu \ (634 \text{ nm, } 703 \text{ nm})$$

Its biological targets are membranes, nucleic acids and proteins.

A number of compounds can deactivate $^1O_2$, e.g., β-carotene, some amines, N_3^-, phenols, etc.; $^1O_2$ can be trapped and form stable products. The most stable capture products are anthracenes and cholesterol which yield a 5-α hydroperoxide as the only product.

Reductions of dioxygen

O_2 is reduced to H_2O during cellular respiration in mitochondria (two electrons reduction for each oxygen molecule, catalyzed by cytochrome oxidase or other oxidases). One-electron reductions occur in certain cases. Figure 3 shows successive electron transfers, as well as proton transfers which depend on the pH of the medium. Several redox potentials (constituting basic information to predict or justify the direction of electron transfers) are reported [4].

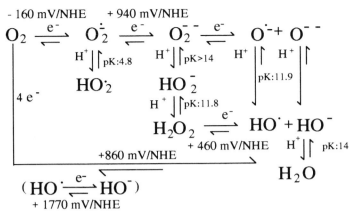

Fig. 3. Chemistry of oxygen: electron and proton transfers.

Metal ions play an important role in these reactions. Thus, the Haber-Weiss reaction: $H_2O_2 + O_2^{\cdot-} \rightarrow O_2 + HO^- + HO^{\cdot}$ is too slow to be taken into consideration in the absence of the catalyst. When catalyzed by Fe^{++} or other ions, it will play an important role.

The superoxide anion radical $O_2^{\cdot-}$

Like O_2, the superoxide is paramagnetic, with a single electron occupying one of the π^* orbitals (Fig. 4). Its chemistry is primary that of reaction with "one-electron donors" or with paramagnetic species. One electron reduction yields O_2^{--} (in practice H_2O_2 in protic media, as a result of its pK), which is diamagnetic (Fig. 4).

$O_2^{\cdot-}$ in aqueous solution is unstable ($2O_2^{\cdot-} + 2H^+ \rightarrow H_2O_2 + O_2$) and stability in organic medium is limited. $HO_2^{\cdot}$ forms in acidic medium (pK = 4.8). $O_2^{\cdot-}$ can be oxidized to $^1O_2$ by strong oxidizing agents such as diacyl peroxides [5–7].

Preparation of $O_2^{\cdot-}$ (in vitro)

$O_2^{\cdot-}$ can be obtained electrochemically in proton-free medium (DMSO, MeCN, etc.). The electrolysis of oxygen at controlled pH furnishes very pure solution of $O_2^{\cdot-}$. $O_2^{\cdot-}$ can also be obtained by dissolving KO_2 in water or in MeCN in the presence of crown ether. Pulsed radiolysis of oxygenated solutions of formate can be used to transiently produce $O_2^{\cdot-}$ in water. $O_2^{\cdot-}$ can be produced biologically (e.g., the reduction of O_2 by the xanthine-xanthine oxidase system or by reduced flavins). The herbicides diquat and paraquat can furnish $O_2^{\cdot-}$:

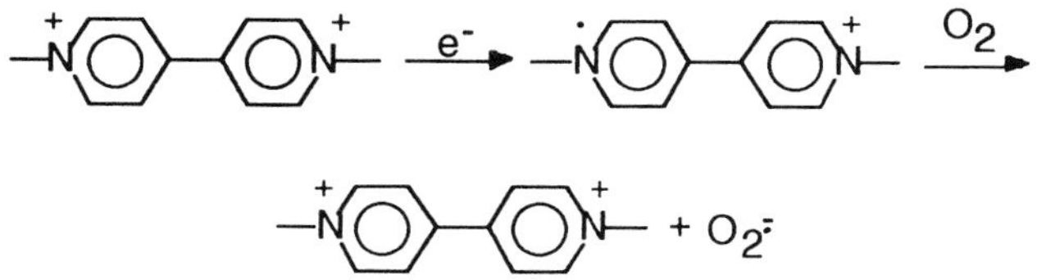

Similarly, quinone drugs such as adriamycin or daunomycin (daunorubicin) also furnish O_2^- by "redox recycling processes", but it is probable that this damage is responsible for its cardiotoxicity:

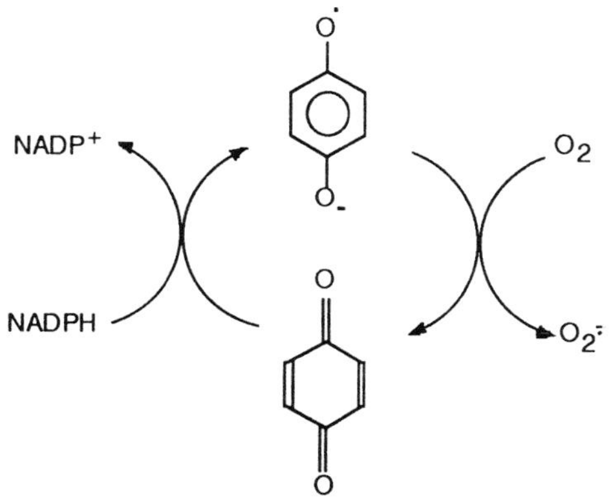

So, vitamin K can produce $O_2^{\cdot-}$, as can all the quinones.

Analytical chemistry

$O_2^{\cdot-}$ can be detected by electron paramagnetic resonance spectroscopy in frozen solution [$g_{//} = 2.1$; $g_{\perp} = 2.00$ in H_2O at $102°K$ and $g_{//} = 2.08$; $g_{\perp} = 2.008$ in MeCN] or by UV spectroscopy [in H_2O: 245 nm ($\epsilon = 2350$) for $O_2^{\cdot-}$ and 225 nm ($\epsilon = 1400$) for $HO_2^{\cdot}$; in MeCN: 255 nm ($\epsilon = 1460$) for $O_2^{\cdot-}$], but these direct methods are not applicable in biological media. $O_2^{\cdot-}$ and $HO_2^{\cdot}$ can be captured and identified by radical scavengers (spin-trapping) such as DMPO:

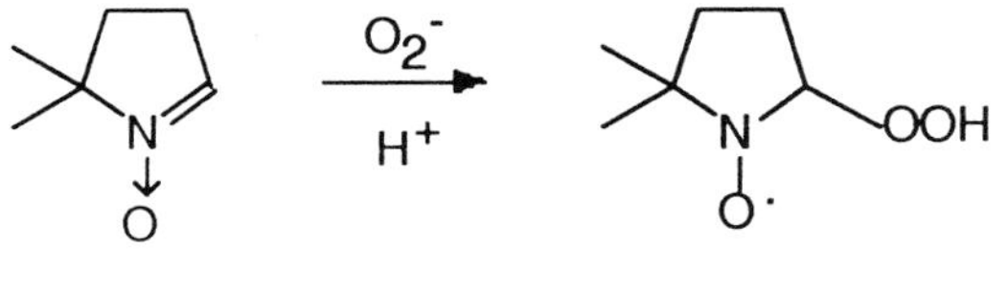

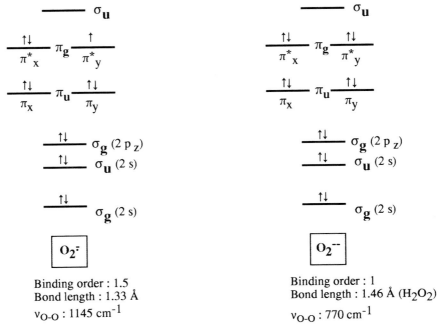

Fig. 4. Superoxide and peroxide anions.

The adduct gives a characteristic EPR spectrum. Cytochrome c and NBT assays are the most frequently used analytical tests; they are well described in specialized works. Chemical methods for measuring $O_2^{\cdot-}$ in water are based on its capture by appropriate indicators and by observing the effect of adding exogenous superoxide dismutase as inhibitor.

Chemical reactivity

– $O_2^{\cdot-}$ is an oxidizing agent or a reducing agent and participates in electron transfers:

$$O_2 \xleftarrow{-e^-} O_2^{\cdot-} \xrightarrow{+e^-} O_2^{-\,-}$$

Examples
– $O_2^{\cdot-} + Cu^{2+} \rightarrow O_2 + Cu^+$ (first step of the catalytic cycle of superoxide dismutase)

– Fe(III)-EDTA, $Fe(CN)_6^{3-}$, $Mo(CN)_8^{3-}$, etc. are reduced by $O_2^{\cdot-}$.

– $O_2^{\cdot-} + Cu^+ + 2H^+ \rightarrow Cu^{2+} + H_2O_2$ (second step of the catalytic cycle of superoxide dismutase)

In water, the dismutation of superoxide is often more rapid than oxidation by $O_2^{\cdot-}$; on the contrary, $HO_2^{\cdot}$ oxidizes $Fe(CN)_6^{4-}$, $Mo(CN)_8^{4-}$, etc.

– $O_2^{\cdot-}$ is a nucleophile

– $O_2^{\cdot-} + RX \rightarrow R-O-O^{\cdot} + X^-$ (SN$_2$ process), then $RO_2^{\cdot}$ gives: $RO_2^{\cdot} + O_2^{\cdot-} \rightarrow RO_2^- + O_2$ and $RO_2^- + RX \rightarrow R-O-O-R + X^-$

– $O_2^{\cdot-}$ is a base

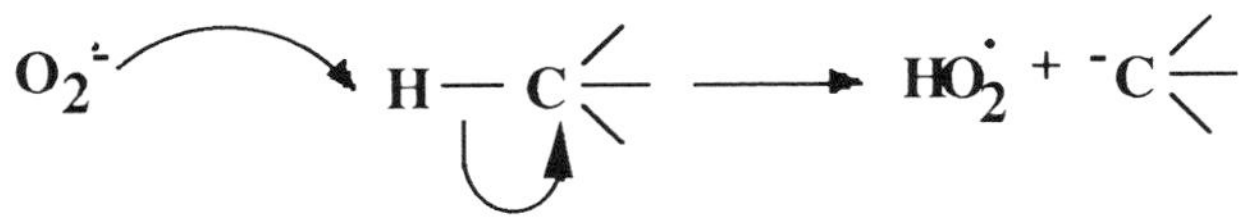

This reaction causes the concomitant dismutation:

$$HO_2 + O_2^{\cdot-} \rightarrow HO_{2-} + O_2$$

– $O_2^{\cdot-}$ is a radical

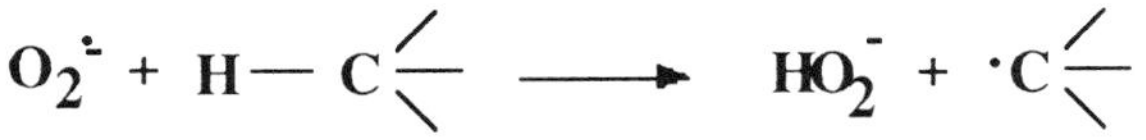

(peroxynitrite)

Peroxynitrite is a strongly toxic species.

$HO_2^{\cdot}$ (perhydroxyl)

Formed in acid medium by the protonation of the superoxide anion radical O_2^- (pK $= 4.8$; at pH: 7, $[O_2^{\cdot-}]/[HO_2] \sim 100$), the radical $HO_2^{\cdot}$ is much more liposoluble than O_2^-. In water, dismutation is much more rapid than that of O_2^-:

$$O_2^{\cdot-} \rightarrow O_2 + O_2^{--}(H_2O_2) \qquad k \leq 0.3 \, \text{mol}^{-1} \, \text{s}^{-1}$$

$$2HO_2^{\cdot} \longrightarrow O_2 + H_2O_2 \qquad k = 8.6 \, 10^5 \, \text{mol}^{-1} \, \text{s}^{-1}$$

The radical $HO_2^•$ adds to C=N double bond and to nitrones. It is not very active as electron donor but is a good acceptor: it is a more powerful oxidizing agent than $O_2^{•-}$ (it is also said, referring to the same result, that H^+ catalyzes the oxidations by $O_2^{•-}$). This results in the HO_2^- anion, a strong base which becomes protonated to form H_2O_2. The $HO_2^•$ radical can abstract $H^•$ from allylic C–H bonds and thus cause lipid peroxidation.

Hydrogen peroxide (H_2O_2)

The formation of hydrogen peroxide from water requires high energy; it is produced by ionizing radiation. Formation from $O_2^{•-}$ by one-electron reduction is catalyzed by metal ions:

Example

$$- \ O_2^{•-} + Fe^{2+} + 2H^+ \longrightarrow H_2O_2 + Fe^{3+}$$

H_2O_2 leads to $HO^•$ radical, either by use of ultra-violet light or by the Fenton reaction, catalyzed by Fe^{2+} or Cu^+ for example:

$$H_2O_2 + Fe^{2+} \longrightarrow HO^- + HO^• + Fe^{3+}$$

Other species can be produced, e.g., ferryl ions:

$$H_2O_2 + Fe^{2+} \longrightarrow (Fe\text{-}OH)^{3+} + HO^- \quad \text{or} \ (FeO)^{2+} + H_2O$$

Biologically, H_2O_2 is normally formed from $O_2^{•-}$ in a reaction considered to be a natural defence against $O_2^{•-}$, catalyzed by superoxide dismutase (SOD):

$$O_2^{•-} + SOD[Cu^{2+} \ state] \longrightarrow O_2 + SOD[CU^+ \ state]$$

$$O_2^{•-} + SOD[Cu^+ \ state] + 2H^+ \rightarrow SOD[Cu^{2+} \ state] + H_2O_2$$

Hydrogen peroxide is toxic as a result of the Fenton reaction, but it is dismuted into H_2O and O_2 by a process catalyzed by catalases, another set of defensive enzymes against the toxicity of oxygen. H_2O_2 can also be produced by peroxidases.

The hydroxyl radical ($HO^•$)

The SOMO (Single Occupied Molecular Orbital) of this paramagnetic species is similar to that of $O_2^{•-}$. Its EPR spectrum can be obtained in frozen solution ($g_{//} = 2.06$ and $g_\perp = 2.01$). The most used spin-trap is DMPO. There exist various chemical methods for detecting $HO^•$.

Formation of HO·

Formation from water requires high energy which can be supplied by ionizing radiation or ultrasound. Photolysis of water at 350 nm also produces HO·. Its chemical formation from H_2O_2 is obtained either by UV light or by catalysis with a complex of Fe(II), Cu(I), Ti(III), etc. A number of pollutants influences its formation directly or indirectly (EtOH, CCl_4, asbestos, paraquat, etc.).

HO· forms in the reactions of numerous reduced metalloproteins; if the reducing agent is O_2^-, the Haber-Weiss reaction occurs, catalyzed by iron:

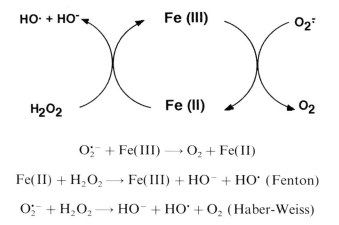

$$O_2^{\cdot -} + Fe(III) \longrightarrow O_2 + Fe(II)$$

$$Fe(II) + H_2O_2 \longrightarrow Fe(III) + HO^- + HO^\cdot \text{ (Fenton)}$$

$$O_2^{\cdot -} + H_2O_2 \longrightarrow HO^- + HO^\cdot + O_2 \text{ (Haber-Weiss)}$$

$O_2^{\cdot -}$ can be replaced by monoelectron donors (e.g., semiquinone).

Reactivity of HO·

The hydroxyl radical is one of the most chemically reactive species known. In biological media, it reacts at the site of formation, i.e., where the catalytic metallic center is located. It attacks lipids, proteins, DNA, sugars, etc., causing a wide variety of damage. The following are examples of chemical reactions:

Hydrogen abstraction and radical formation (first step of a chain reaction)

Hydroxylation of aromatic rings

Electron transfers

HO$^{•}$ behaves as an acceptor (oxidizing agent).

Examples. $HO^{•} + I^{-} \rightarrow HO^{-} + I^{•}$ and also $HO^{•} + I_2 \rightarrow HO^{-} + I_2^{+}$

Conclusion

Oxygen may be the source of toxic species. The design of suitable drugs to prevent or to overcome the deleterious effects of these species (antioxidant therapy) requires the knowledge of the molecular process of oxygen toxicity. The understanding of the chemistry of oxygen and its activated species is an essential prerequisite for defining the reactions involved in the degradation of lipids, nucleic acids, proteins, etc., mediated by active oxygen.

References

1. Bors, W., Saran, M. and Tait, D. (1984) *Oxygen Radicals in Chemistry and Biology*, W. de Gruyter, Berlin, New York.
2. Cadenas, E. (1989) Biochemistry of Oxygen Toxicity. *Ann. Rev. Biochem.* 58: 79–110.
3. Halliwell, B. and Gutteridge, J.M.C. (1989) *Free Radicals in Biology and Medicine*, Clarendon Press.
4. Koppenol, W.H. and Butler, J. (1985) Energetic of interconversion reactions of oxyradicals. *Adv. Free Rad. Biol. Med.* 1: 91–131.
5. Alfans'ev, I.B. (1989) *Superoxide Ion: Chemistry and Biological Implications.* CRC Press, Boca Raton, Vol. I.
6. Alfans'ev, I.B. (1991) *Superoxide Ion: Chemistry and Biological Implications.* CRC Press, Boca Raton, Vol. II.
7. Sawyer, D.T. and Gibian, M.J. (1979) The chemistry of superoxide ion. *Tetrahedron* 35: 1471–1481.

Analysis of Free Radicals in Biological Systems
Favier et al. (eds)
© 1995 Birkhäuser Verlag Basel/Switzerland

Biological sources of reduced oxygen species

V. Nivière and M. Fontecave

Laboratoire de Chimie Bioinorganique, LEDSS 5, Université J. Fourier, URA CNRS 332, BP 53x, F-38041 Grenoble Cédex 9, France

Summary. This paper reviews the various enzymatic sources of superoxide radical, hydrogen peroxide and hydroxyl radical, and their localization within the cell. In the first part, we will consider electron transfer reactions directly from a biological macromolecule (essentially reductases) to oxygen. In the second part, we will consider reactions during which electrons are transferred from the reductase to oxygen via a small molecule which behaves as a substrate of the enzymatic system.

Introduction

Dioxygen is present in all living organisms. Its concentration in the different organs and tissues varies to a great extent, the highest being in the lung, the skin and the heart, and the lowest in the intestine and the bile. However, one can approximately estimate the steady-state concentration of dioxygen to be ca. 10 μM. One can also estimate the concentrations of reduced oxygen species, under non-pathological conditions as follows:

– superoxide: 0.01–0.001 nM
– hydrogen peroxide: 1–100 nM

In general, major contributions of the total cellular production of these oxygen metabolites come from membrane-bound enzymes. In particular, the aerobic life is consistent with the existence of powerful reducing agents (NADPH, NADH) as well as elaborate electron-transfer chains for the reduction of molecular oxygen to water, in respiring cells. This four-electron reduction is accomplished by the *cytochrome oxidase* complex. However, a small but significant amount of intermediate products, superoxide and hydrogen peroxide, is liberated during the reaction. Since cytochrome oxidase is inhibited by cyanide, one can use the cyanide-resistance respiration as an upper limit measure of superoxide production. In *E. coli*, this amounts to 3% of total respiration. Mammalian liver has been estimated to generate 24 nmol of superoxide per min per g. However, the action of superoxide dismutase leads to a very low steady-state concentration of superoxide.

Other reductases, which are not involved in the respiratory chain, may also be responsible for one-electron or two-electron reduction of oxygen. In some cases, these reactions are enzymatically controlled (production of reduced oxygen species by phagocytic cells or production of hydrogen peroxide by oxidases). In other cases, they are due to aberrant reaction pathway, i.e., production of superoxide by the cytochrome P-450 reductase, for example.

Several chemical compounds, from our environment, have been shown to be excellent electron acceptors from biological reductases. The reduced form of these substrates is capable of transferring electrons to oxygen. Thus reducing equivalents, which should normally participate in basic biological processes, are consumed disproportionately to produce reduced oxygen species.

This chapter reviews the various enzymatic sources of superoxide radical, hydrogen peroxide and hydroxyl radical. In the first part, we will consider electron transfer reactions directly from a biological macromolecule (essentially reductases) to oxygen. In the second part, we will consider reactions during which electrons are transferred from the reductase to oxygen via a small molecule, which behaves as a substrate of the enzymatic system (Fig. 1).

Enzymatic activation of molecular oxygen

Formation of superoxide

The cellular sources of superoxide are essentially located in:
Mitochondria. Among enzymes of the electron transfer chain two enzymatic sites of the mitochondrial membrane have been clearly identified as major sources for one-electron reduction of oxygen: *ubiquinone-*

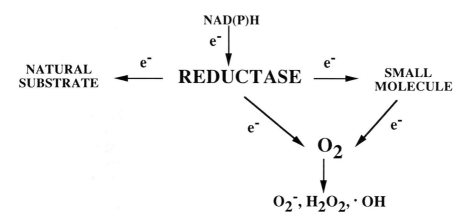

Fig. 1. Scheme of the various enzymatic sources of oxygen toxic radicals in the cell

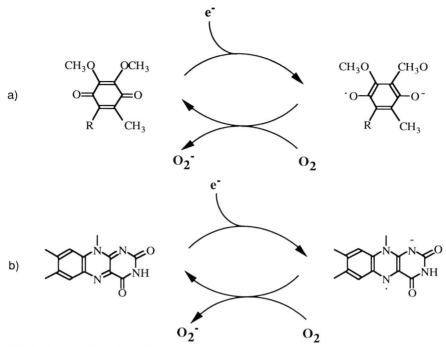

Fig. 2. Autoxidation of the ubisemiquinone (a) and semiflavin (b) in the presence of O_2.

cytochrome c reductase and NADH dehydrogenase (85 and 15% superoxide respectively). This is due to autoxidation of the ubisemiquinone, in the first system, and of the semiflavin cofactor (one-electron reduction of the isoalloxazine moiety), in the second system (Fig. 2). It is generally accepted that mitochondrial hydrogen peroxide (which corresponds to approximately 1% of the mitochondrial reduction of oxygen) is derived from the dismutation of superoxide.

The membrane of the endoplasmic reticulum. Superoxide radicals are produced by the oxy complex of cytochrome P-450 and by the action of NADPH-cytochrome P-450 reductase, a flavoprotein. Hydrogen peroxide is also generated from dismutation of superoxide. It has been shown that this reductase is the major electron source during the activation of exogenous compounds (see below).

The plasmatic membrane of speicalized cells such as leukocytes, from electron transfer chains containing a NADPH oxidase. The latter is a complex system consisting of a flavoprotein dehydrogenase, ubiquinone-50 and a cytochrome b. This system is at the origin of the specific generation of large amounts of superoxide during phagocytic processes achieved by polymorphonuclear leukocytes (PMNL) or other phagocytic cells. Actually, superoxide radicals have been shown to play an important role in the host defense mechanism against microorganisms and they contribute to the phagocytic bactericidal activity. PMNL ingest opsonized particles by encompassing them within phagocytic vesicles formed from the plasma membrane. The NADPH oxidase is rapidly activated followed by a rapid cyanide-insensitive respiratory burst forming superoxide, hydrogen peroxide and hydroxyl radical. The latter is derived from the Haber-Weiss reaction catalyzed by ferric iron:

$$O_2^- + H_2O_2 \xrightarrow{\text{Fe}^{3+}} {}^{\bullet}OH + OH^- + O_2$$

In addition, other oxidants are also produced, as the hypochlorite derived from the activity of myeloperoxidase:

$$H_2O_2 + Cl^- \xrightarrow{\text{myeloperoxidase}} HOCl + OH^-$$

The importance of superoxide and hydrogen peroxide production in PMNL is underscored by the genetic disorder, chronic granulomatous disease, in which the instability to produce reduced oxygen species is associated with a serious impairment of the microbicidal action of the phagocytes and a consequent susceptibility to infection. This disease is due to the abnormally low activity of NADPH oxidase. One should note that part of these radicals is liberated into the extracellular medium. This certainly explains some tissue damage observed during any inflammatory process and the antiinflammatory effect of superoxide dismutase.

The cytoplasm. Various endogenous or exogenous molecules react with oxygen and undergo autoxidation. Virtually all autoxidations result in the formation of toxic reduced oxygen intermediates. Autoxidation of ascorbate, glutathione, pyrogallol, catecholamines such as adrenaline, reduced flavins (generated by NAD(P)H : flavin oxidoreductases [1]) leads to the formation of superoxide. Also, superoxide radicals are released when methemoglobin is formed from oxyhemoglobin.

The following is a summary of sources of superoxide formed in biological systems:

Autoxidation reactions:
– Flavins ($FADH_2$, $FMN\ H_2$)
– Hemoglobin
– Catecholamines
– Tetrahydropteridines
– Aromatic nitro compounds, aromatic hydroxylamines
– Redox dyes (paraquat, adriamycin, antimycin)
– Cu(II) and Fe(II) complexes
– Hydroquinones
– Thiols (glutathione)
– Melanin

Enzymatic reactions:
- Microsomal electron transport chain (Cytochrome P450), NADH-cytochrome b5 reductase, NADPH-cytochrome P450 reductase)
- Mitochondrial respiratory chain (cytochrome oxidase)
- Photosynthetic oxygen reduction (chloroplast photosystem I)
- Leukocytes and macrophages during bactericidal activity (NADPH oxidase in the plasma membrane)
- Oxidases (xanthine oxidase, aldehyde oxidase)
- Ferredoxins
- Metabolism of iron (NADPH : flavin oxidoreductases)

Environmental factors:
- Ultraviolet light, X-rays, metal ions, ultrasound, γ-rays.

Formation of hydrogen peroxide

The major process of generation of hydrogen peroxide is the dismutation of superoxide radicals. However, hydrogen peroxide is also generated directly during two-electron reductions of oxygen by *oxidases*:

- acyl-coA oxidase
- D-aminoacid oxidase
- glycolate or urate oxidase
- xanthine oxidase
- glutathione oxidase
- monoamine oxidase

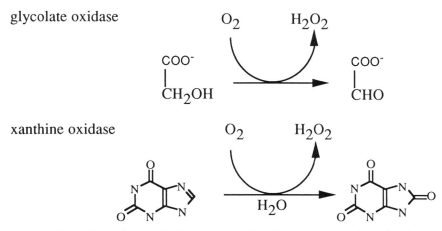

Fig. 3. Various oxidases involved in the formation of steady state concentrations of hydrogen peroxide

Most oxidases are located in peroxisomes, where large concentrations of catalase are generally detected. However, significant amounts of hydrogen peroxide can gain access to the cytosolic compartment. Figure 3 shows the various oxidases involved in the formation of steady-state concentrations of hydrogen peroxide.

Formation of hydroxyl radicals

Oxygen reduction leads to the simultaneous formation of superoxide radicals and hydrogen peroxide. The reaction between O_2^- and H_2O_2, the *Haber-Weiss reaction*, results in the formation of hydroxyl radicals, but this reaction is kinetically too slow to account for the biological generation of these radicals. However, it can be greatly accelerated by catalytic amounts of metal salts (iron or copper).

The mechanism of this reaction is shown below:

$$Fe^{2+} + H_2O_2 \longrightarrow Fe^{3+} + OH^- + {}^{\cdot}OH$$
$$\underline{Fe^{3+} + O_2^- \longrightarrow Fe^{2+} + O_2}$$
$$H_2O_2 + O_2^- \longrightarrow OH^- + {}^{\cdot}OH + O_2$$

This mechanism requires the iron complex to be:

– in the ferric form, reducible by superoxide anions,
– in the ferrous form to be able to transfer electrons to hydrogen peroxide (*Fenton reaction*)

The evidence for the presence of hydroxyl radicals in biological media has led to the hypothesis of the biological significance of a pool of low-molecular weight iron complexes. The nature of this pool has not been elucidated, but iron-citrate or iron-nucleotide complexes have been proposed as possible candidates.

Activation of molecular oxygen catalyzed by small molecules

Endogenous or exogenous low-molecular weight molecules can catalyze the electron transfer to oxygen from various reductases. *NADPH cytochrome P-450 reductase*, within the endoplasmic reticulum, is believed to play a major role in the toxic activation of most of these compounds. Other reductases have also to be considered, including NADH cytochrome b5 reductase, NADH dehydrogenase, xanthine dehydrogenase, aldehyde oxidase, ferredoxin-NADP reductase.

This reaction leads to the formation of superoxide radical which dismutates to form hydrogen peroxide.

Such a cycle requires that:

(a) The redox potential of the redox agent is within the good range to fit with a one-electron reduction of the oxidized form by the reductase system as well as with the oxidation of the reduced form by oxygen.

(b) The hydrophobic or hydrophilic properties of the redox agent favor its penetration into the cellular compartment (membrane or cytosol) containing the reductase. Moreover, the small molecule must have access to the active site of the protein where the electron transfer is achieved.

(c) The intermediate one-electron reduced metabolite has to be stable enough so that no reaction occurs with other electron acceptors, for example amino acid residues of the active site of the protein.
A redox cycle explains the toxic effects of several exogenous compounds:

Nitro-aromatic or nitro-heterocyclic derivatives

This is the case for nitrobenzene as well as for various antimicrobial drugs (nitrofurantoin, responsible for lung fibrosis of the type induced by paraquat; nitrofurazone, a powerful bactericidal agent, mutagen and carcinogen in the rat). Specific reductases are capable of reducing the nitro group to the corresponding anion-radical form, which rapidly returns to the nitro form by transferring one electron to oxygen (see Fig. 4).

Quinone derivatives

This includes antimitotic antibiotics which are metabolized by NADPH : cytochrome P-450 reductase into very reactive semiquinones (adriamycin, daunomycin, streptonigrin).

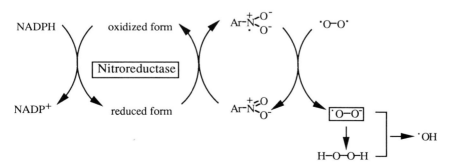

Fig. 4. Toxic effects of nitro-aromatic or nitro-heterocyclic derivatives.

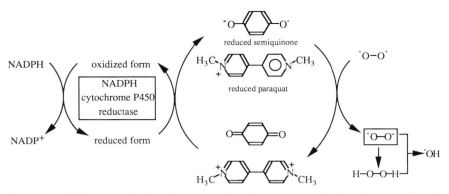

Fig. 5. Reoxidation of reduced semiquinone and reduced paraquat leads to the formation of superoxide.

Reoxidation of the semiquinone leads to the formation of superoxide radicals and hydroxyl radicals (Fig. 5). These toxic metabolites are partly at the origin of chromosome damage observed in cancer cells treated by these antibiotics. However, these drugs may accumulate within other tissues and be responsible for the generation of reduced oxygen species in healthy cells. The cardiotoxicity of adryamycin could be due to such processes.

Pyridinium salts, such as paraquat or diquat

Paraquat is a very toxic herbicide which may induce lung fibrosis and death. It can be reduced by NADPH cytochrome P-450 reductase to a stable radical-cation, very reactive towards oxygen (Fig. 5). Reduced oxygen species then initiate a very strong peroxidation of membrane lipids, during the first step of the fibrosis.

Iron complexes

When iron is bound by phosphate, nucleotides such as ADP or ATP, EDTA, oxalate or malonate, it can catalyze one-electron transfer to oxygen (Fig. 6). The origin of the electron is not clearly identified. Ascorbate, glutathione, cysteine may reduce most of the ferric complexes. However, it is also likely that specific reductases are involved in this reaction.

Iron is stored as ferric oxides and hydroxides within ferritin, the iron storage protein. It is liberated as ferrous iron. The specific physiological reducing system has not been identified. However, it has been shown

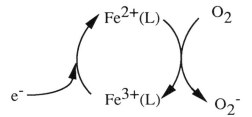

Fig. 6. Formation of superoxide in the presence of iron complexes.

that reduced flavins are excellent reducing agents for mobilization of iron from ferritin. This suggests that NAD(P)H : flavin oxidoreductases might be involved in the reduction of ferritin iron. Thus, it is likely that liberation of ferrous iron under aerobic conditions is accompanied by the formation of significant amounts of reduced oxygen species. Such a process has to be highly controlled by protective enzymes. However these systems may be saturated during situations of iron overload (hemochromatosis, blood transfusions for thalassemia patients) and oxidative stresses catalyzed by iron complexes can be dramatic.

An interesting example of such iron complexes involved in the generation of toxic reduced oxygen species is the *iron-bleomycin complex*. Bleomycin is a glycopeptidic antibiotic used as an anticancer agent. It is able to bind iron both at the ferric and the ferrous state. It is generally accepted that the antimitotic properties of bleomycin are in fact due to the iron complex. Actually, while the ligand behaves as a DNA intercalator, it is the metal ion which catalyzes the formation of superoxide, hydrogen peroxide and hydroxyl radicals. These toxic species are responsible for DNA-strand breaks. In the absence of iron, bleomycin is not able to damage DNA to an appreciable extent.

Reference and further reading

1. Gaudu, P., Touati, D., Nivière, V. and Fontecave, M. (1994) *J. Biol. Chem.* 269: 8182–8188.

Halliwell, B. and Gutteridge, J.M.C. (1990). *In: Methods in Enzymology* 186: 1–85.
Halliwell, B. and Gutteridge, J.M.C. (1989). *In: Free Radicals in Biology and Medicine* 2nd Edition, Oxford University Press (Clarendon), Oxford.
Halliwell, B. and Gutteridge, J.M.C. (1984). *Biochem. J.* 219: 1.
Sies, H. (1991). *In: Oxidative Stress. Oxidants and Antioxidants.* Academic Press, London.
Sies, H. (1986). *Angew. Chem. Int. Ed. Engl.* 25: 1058–1071.

Analysis of Free Radicals in Biological Systems
Favier et al. (eds)
© 1995 Birkhäuser Verlag Basel/Switzerland

Nitric oxide: Chemistry and biology

C. Garrel[1] and M. Fontecave[2]

[1]*Groupe de Recherches des Pathologies Oxydatives, Centre Hospitalier Universitaire, BP 217, F-38043 Grenoble Cédex 9, France*
[2]*Laboratoire d'Etudes Dynamiques et Structurales de la Sélectivité, Université Joseph Fourier, BP 53, F-38041 Grenoble, Cédex 9, France*

Summary. Nitric oxide plays a role in regulation of the vascular tone and platelet aggregation, neuronal transmission and cytostasis. It is generated within cells during oxidation of L-arginine catalyzed by NO synthases. Nitric oxide is a radical species, very reactive towards molecular oxygen, superoxide radical, organic radicals and transition metals. During these reactions, $^{\cdot}NO$ can be converted to NO^+, the nitrosonium ion, NO^-, the nitroxyl anion or to $ONOO^-$, the peroxynitrite anion. Synthetic $^{\cdot}NO$ donors are compounds which can decompose into $^{\cdot}NO$ and may have applications as vasodilators or antiproliferative agents.

Introduction

Nitric oxide ($^{\cdot}NO$) has been implicated in a number of diverse physio-logical processes, including vasodilatation and regulation of vascular tone, inhibition of platelet aggregation, neuronal transmission and cytostasis. Excellent reviews on biological aspects of $^{\cdot}NO$ have recently been discussed [1–6]. The concentration of $^{\cdot}NO$ has to be highly regulated and situations in which $^{\cdot}NO$ levels are either too low or too high may lead to serious adverse effects. In tissues where $^{\cdot}NO$ controls normal vascular tone, low levels can result in pulmonary hypertension, platelet aggregation. It has been suggested that $^{\cdot}NO$ gas or $^{\cdot}NO$ donors such as organic nitrates or sodium nitroprusside are extremely useful in the treatment of this type of hypertension. On the other hand, abnor-mally high levels of $^{\cdot}NO$ are also involved in the hypotension associated with endotoxic shock, inflammatory response-induced tissue injury, mutagenesis and neuronal destruction in vascular stroke. In those cases, inhibitors of NO synthases may provide a strategy for treatment. In this paper, we will describe the basic mechanisms by which $^{\cdot}NO$ regulates blood pressure, neurotransmission and macrophage cytotoxicity. We will also give a general picture of the biochemical process which is responsible for the generation of $^{\cdot}NO$, i.e. the oxidation of L-arginine catalyzed by NO-synthases. It is also important to understand the chemical nature of NO, which is a free radical but may be converted to the reduced NO^- form or the oxidized NO^+ form, both of them displaying chemical reactivities very different from $^{\cdot}NO$ itself. The redox control of the different forms of ˙NO may explain the balance between the toxic and protective effects of ˙NO. Finally, we will review briefly the various synthetic ˙NO donors, which might be useful both for *in vitro* studies or for therapeutic purposes.

The biological effects of ˙NO

˙NO and vasodilatation

A relation between relaxation of smooth muscle present in vascular cell walls and the accumulation of cyclic guanosine monophosphate (cGMP) is established. Nitrovasodilators such as organic nitrates or sodium nitroprusside mediate smooth muscle relaxation after decomposition to ˙NO which activates the enzyme guanylate cyclase and thus stimulates cGMP accumulation. Nitrovasodilators are not, however, the only compounds that effect arterial muscle relaxation. Among others is acetylcholine. The relaxing effect of acetylcholine upon a piece of aorta is completely destroyed if the endothelial cells (on the inner surface of the vessel in contact with the blood) of the aorta were removed. In fact acetylcholine stimulates the endothelial cells to produce a messenger which diffuses out and reaches the vascular smooth muscle where it activates guanylate cyclase to bring about muscle relaxation. This messenger was previously called EDRF (endothelium derived relaxing factor) and later identified as ˙NO. The generation of ˙NO is due to the presence of a constitutive Ca^{2+}/calmodulin-dependent form of NO synthases in vascular endothelial cells, anchored to the inner surface of the cell membrane, by a myristyl side chain [7]. In response to acetylcholine or other drugs, Ca^{2+} channels in the plasma membrane open and the resulting Ca^{2+} influx activates the NO synthase (NOS), responsible for the oxidation of L-arginine to ˙NO and citrulline. The activation is due to the binding of the Ca^{2+}-calmodulin complex to the NOS.

From endothelial cells, ˙NO can also increase cGMP in blood platelets. As a result platelet affinity for both the vascular endothelial surface (platelet adhesion) and for each other (platelet aggregation) are increased.

Maintaining normal blood pressure seems to require continuous synthesis of ˙NO in vascular endothelial cells. However, in some pathological states such as atherosclerosis, this production is not enough to decrease vasoconstriction and hypertension. The vasorelaxant therapy can thus be achieved by treatment with pharmacological agents able to release ˙NO, such as nitroglycerin. ˙NO gas itself seems to be rather efficient to treat vasoconstriction in patients with adult respiratory distress syndrome.

Nitric oxide and neurotransmission

˙NO is one of the main neurotransmitters in the central and peripheral nervous system, even though its precise physiological role is not fully known. It mediates intestinal relaxation in peristalsis, penile erection and the action of glutamate on cGMP levels in the brain. It may also play a role in long-term potentiation and memory. It may mediate major neuronal damage in stroke and neurodegenerative diseases. Most neuronal destruction in stroke seems to result from a massive release, from a stimulated neuron, of glutamate, which, acting through the NMDA receptor of an adjacent neuron, causes neuronal death. Actually, NMDA receptor activation triggers a massive influx of Ca^{2+} into the neuron which, together with calmodulin, activates the constitutive NO synthase. ˙NO diffuses to adjacent cells and kills them. This model is supported by the ability of NOS inhibitors to block the neurotoxicity of glutamate and NMDA in brain cultures. Moreover, drugs which are NMDA receptor antagonists provide marked protection against neuronal damage following vascular occlusion.

However, in some studies ˙NO seems to have neuroprotective effects. The molecular basis of such a paradox will be described later.

Nitric oxide and cytotoxicity

When macrophages are stimulated by interferon-γ (IFNγ), tumor necrosis factor (TNFα), interleukins and lipopolysaccharide (LPS), transcription of the gene of an inducible NOS is increased, and an active enzyme is generated even in the absence of Ca^{2+}. As a consequence, large amounts of ˙NO are produced during several hours following stimulation of NOS synthesis.

˙NO diffuses out and can reach adjacent cells where it reacts with the iron-sulfur centers of several important macromolecules (aconitase, complexes I and II of the mitochondrial electron transport chain) and inhibits ribonucleotide reductase, the enzyme that converts ribonucleotides to the deoxyribonucleotides necessary for DNA synthesis. By this process DNA synthesis is inhibited and cell proliferation ceases. This may be the mechanism by which macrophages inhibit growth of rapidly dividing tumor cells or intracellular parasites.

NO synthases

There are two categories of NO synthases [8]: (i) a constitutive form regulated by Ca^{2+} and calmodulin; (ii) a cytokine-inducible form that is known to be regulated post-transcriptionally [9]. NOS are dimers of

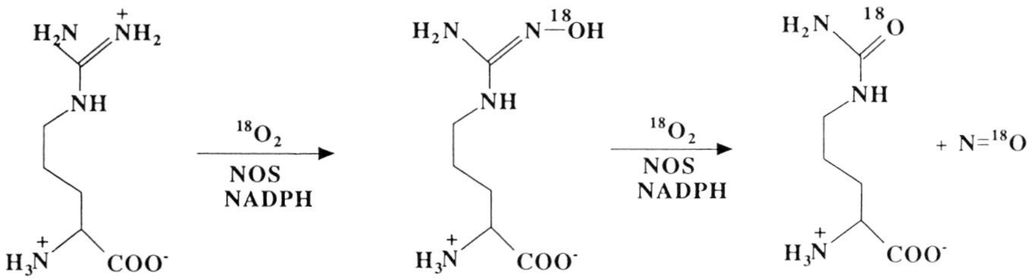

Fig. 1.

$M_r = 130\,000 - 160\,000$ in the native state. They require NADPH and O_2 as co-substrates. As shown in Figure 1, the enzyme catalyzes the oxidation of L-arginine to nitric oxide and citrulline. Until recently all types of NOS were just considered as cytosolic enzymes. However, a constitutive endothelial NOS isoform was found to be membrane associated. There are now several pieces of evidence showing that the fatty myristic acid is covalently bound to the N-terminal part of the protein [7]. Mutation of NH_2-terminal Gly abolished the radiolabeling of endothelial NOS by radioactive myristic acid and rendered the enzyme soluble. The co-translational myristoylation of the enzyme thus allows the association to the membrane. The amino acid sequences derived from the isolated cDNAS for both constitutive and inducible NOS display significant homologies to NADPH-cytochrome P-450 reductase. In particular, the nucleotide binding sequence as well as those sequences associated with FAD and FMN binding were highly conserved when compared with P-450 reductase from rat liver. The sequences from the constitutive NOSs show a consensus for calmodulin recognition. Interestingly, the inducible NOS also has a calmodulin recognition sequence even though this enzyme has not shown a requirement for Ca^{2+} and calmodulin.

As predicted from the sequence comparison, NOS were shown to contain one equivalent each of FAD and FMN per subunit and to require NADPH for activity. It is thus tempting to speculate that electron flow would follow the same path as in the reductase, namely the NADPH would reduce FAD, which in turn reduces FMN.

It is now well established that NOSs also contain a cytochrome P-450 type iron-protoporphyrin IX prosthetic group that functions in the turnover of L-arginine [8]. It thus seems reasonable to make the hypothesis of a unique electron transfer flow, within a single polypeptide chain:

$$\text{NADPH} \longrightarrow \text{FAD} \longrightarrow \text{FMN} \longrightarrow \text{heme.}$$

The need for reduction of the heme moiety is probably related to the function of this prosthetic group during O_2 activation. Thus, this allows the generation of iron-bond reduced oxygen species reactive enough to transfer one oxygen atom to L-arginine and carry out further oxidation of the intermediate N-hydroxyarginine to citrulline and ˙NO. That the oxygen atom in citrulline and ˙NO is derived from O_2 has been determined from isotope studies using $^{18}O_2$ (Fig. 1). In agreement with such a mechanism, it was found that CO, a very good ligand for heme iron, is an inhibitor of NOS.

Only in the case of the constitutive NOS, Ca^{2+} and calmodulin were required for activity. Intracellular Ca^{2+} thus can regulate NOS activity in cells that harbor this isoform. For example, NOS and thus guanylate cyclase are activated in the central nervous system by increases of intracellular Ca^{2+} levels mediated by glutamate and NMDA receptors.

As immunoprecipitated from host cells, all three NOS isoforms are phosphorylated, serine being identified as the phosphate acceptor. However, information is lacking as far as the effect of this phosphorylation on enzyme activity is concerned.

Finally, (6R)-tetrahydro-L-biopterin is also a cofactor of NOS. However, a clear function for this cofactor has yet to emerge. Several hypotheses have been reported: (i) a direct role in the hydroxylation chemistry; (ii) an allosteric effector role; (iii) stabilization of the enzyme. Recently it was shown that NOS exhibits a highly specific site for the tetrahydrobiopterin, which allosterically interacts with the substrate domain and may be located proximal to the prosthetic heme group of the enzyme. Moreover, the cofactor is required for assembly of the active dimer from its subunits *in vitro* [10].

L-arginine analogs are specific inhibitors of NOS. L-NAME and L-NMMA are widely used in *in vitro* studies to determine whether the biological effect is related to the NOS activity and generation of ˙NO. NOS inhibition will also likely be therapeutically important in certain clinical situations. It should be mentioned that L-NMMA was shown to release low amounts of ˙NO during enzymatic oxidation by NOS.

It was observed recently that under conditions of L-arginine or tetrahydrobiopterin deficiency, NOS produces superoxide radicals rather than ˙NO. This univalent reduction of oxygen may arise at the level of the flavins as shown for cytochrome P450 reductase. This now explains why glutamate receptors induce a burst of superoxide in arginine-depleted neurons [11].

The chemical reactivity of ˙NO

˙NO has one unpaired electron and thus can be described as a free radical. Its free radical nature readily explains most of its reactivity. It reacts very efficiently with paramagnetic species such as molecular oxygen, superoxide radicals, transition metal ions and free radicals [1, 5, 12, 13 and references therein].

Reaction with molecular oxygen

In aerobic aqueous solutions, nitrite is the only product of the reaction (no nitrate), according to the following equation:

$$4NO + O_2 + 2H_2O \longrightarrow 4NO_2^- + 4H^+$$

The correct rate law is $-(d[NO]/dt) = k(O_2)(NO)^2$ with $k = 8 - 9 \ 10^6 \ M^{-2} \ s^{-1}$ at 25°C. That the reaction is second order in ['NO] has a great implication as far as the stability and thus availability of 'NO in biological fluids are concerned. Indeed, the known instability of 'NO in the presence of oxygen has raised important doubts regarding the possibility that 'NO could serve as a biological messenger in a variety of physiological processes, travelling from one cell to another, from one intracellular compartment to another and so on. In fact, even in O_2 saturated solutions, at a concentration of 0.1 or 0.01 mM, likely to be found for bioregulatory processes (blood pressure regulation, neurotransmission, etc.), 'NO should be sufficiently long-lived (half-life from 20 min to several hours!). These numbers might even be larger if one considers that O_2 concentration in biological media can be very low (a few μM).

Reaction with superoxide

In deaerated aqueous alkaline solutions (pH 12–13), superoxide reacts with nitric oxide to form the stable peroxonitrite anion.

$$O_2^- + NO \longrightarrow {}^-OONO$$

Under physiological pH condition the rate constant is $k = 6.7 \ 10^9 \ M^{-1} \ s^{-1}$ with $d[{}^-OONO]/dt = k[NO][O_2]$ (faster than the dismutation $(2.10^9 \ M^{-1} \ S^{-1})$. On the other hand, at neutral pH, peroxonitrite is partially protonated to peroxynitrous acid (pKa = 6.8 at 37°C), which rapidly decomposes with a first-order kinetics (rate constant: $0.6 - 1.4 \ s^{-1}$). The half-life of peroxonitrite at pH 7 is about 1–2 s.

The biological relevance of these reactions has to be discussed. Considering the previous kinetic data and the concentrations of reactants likely to occur in normal biological solutions (for example, 10^{-11} M and 10^{-7} M for O_2^- and 'NO, respectively), we end up with very low peroxonitrite concentrations ($< 10^{-9}$ M). This may not significantly contribute to the oxidizing (toxic) properties of the solution.

However, there are situations in which both 'NO and O_2^- are produced in relatively larger amounts. One example is during activation of phagocytic cells. Consequently, under pathological conditions peroxonitrite is expected to be produced in much higher steady-state concentrations (1 μM), which then might be cytotoxic, because of its strong reactivity. Accordingly, peroxonitrite has been detected during activation of macrophages and shown to display bactericidal properties. Another example is during decomposition of SIN 1, the active metabolite of the vasorelaxant drug molsidomine. The spontaneous hydrolysis of SIN 1 is accompanied by an oxygen-dependent release of $^{\cdot}$NO with the concomitant production of superoxide.

When both $^{\cdot}$NO and $O_2^{\cdot-}$ are produced, $^{\cdot}$NO concentration may be underestimated during assays because of such a reaction. It is advisable to use SOD under these conditions.

Reactivity of peroxonitrite

Peroxonitrite is a potent oxidant. However, little is known about its real oxidizing potential and the mechanisms of the reactions. Theoretical values for redox potentials have been found at around 1.2–1.4 V at pH 7. Peroxonitrite is able to hydroxylate benzene and oxidize DMSO at acidic pH. Recently, it was demonstrated that peroxonitrite was a much more efficient oxidant than H_2O_2 during reaction with protein and non protein cysteines. This is an interesting observation since thiols are critical to the active site of many enzymes and for maintaining the native conformation of proteins. Such oxidations may be important mechanisms of NO/OONO dependent toxicity.

In all probability, peroxonitrite has the potential to oxidize a great diversity of important biological molecules including lipids, DNA, etc. This however remains to be established.

There is some uncertainty concerning the mechanisms of the reactions. Several recent data indicate that one cannot rule out a direct oxidizing effect of peroxonitrite with no need for a decomposition step. It has also been suggested that peroxynitrous acid decomposes at physiological pH into hydroxyl radicals, with no requirement for metal ions.

$$HOONO \longrightarrow HO^{\cdot} + NO_2^{\cdot}$$

Recently, it was shown that SIN 1 gives rise to OH$^{\cdot}$ radicals during decomposition, probably because of the intermediate coupling of $^{\cdot}$NO with $O_2^{\cdot-}$. From a practical point of view, this now indicates that SIN 1 may, as a vasodilator, have strong toxic secondary effects. Peroxonitrite can also be a source of nitrating species especially in the presence of transition metal ions such as Cu^{2+} or Fe^{3+}. The Cu^{2+} center of superoxide dismutase catalyzes the nitration of Tyr 108 by peroxonitrite.

Peroxynitrite mediated nitrosation of tyrosines was also observed with Mn- and Fe-containing SOD as well as other Cu-proteins. More generally, Fe(III)-EDTA was also found to catalyze the nitration of phenolic substrates by peroxonitrite. It is clear that such a process, harmful to proteins, might contribute to the toxicity of peroxonitrite *in vivo*.

Hydrogen atom abstraction

Abstraction of hydrogen atoms from neutral molecules is very efficiently performed by hydroxyl radicals. 'NO is much less efficient and very few examples have been reported so far. An interesting one concerns phenols, including α-tocopherol, which can be transformed to phenoxy radicals by 'NO.

Radical coupling

'NO has the ability to react with organic free radicals such as $R_3C^•$, $R-O^•$, $RS^•$. Those radicals are expected to be short-lived intermediates during enzymatic reactions or during degradative processes such as those involved in oxidative stress conditions: lipid, protein, DNA peroxidation.

$$R^• + NO \longrightarrow R-NO \xrightarrow{+R^•} R-N-O^•$$
$$\qquad\qquad\qquad nitroso \qquad\qquad\quad |\ nitroxide$$
$$\qquad\qquad\qquad\qquad\qquad\qquad\qquad\quad R$$

$$RS^• + NO \longrightarrow RS-NO\ thionitrite$$

$$RO^• + NO \longrightarrow RO-NO\ alkylnitrite$$

The importance of all these reactions under biological conditions is very difficult to estimate so far. However, one may speculate that such reactions have to be considered in the case of proteins containing stable amino acid radicals in their active site. One example is ribonucleotide reductase, whose small subunit carries a stable tyrosyl radical absolutely required for enzymatic activity. We have shown by EPR spectroscopy that, during incubation of the enzyme with NO, the tyrosyl radical is scavenged transiently and reappears once 'NO is eliminated from the solution [14]. Such a mechanism may contribute to the observed inhibition of ribonucleotide reductase and DNA synthesis by 'NO and to the cytotoxic effects of activated macrophages. The observed reversibility of the scavenging might be a part of a regulatory mechanism:

Reaction with transition metals and metal nitrosyl complexes

˙NO reacts with a great variety of transition metal complexes (Ni, Mo, Cr, Fe, Ru) generating highly stable and colored nitrosyl complexes.

During binding to the metal ion, internal electron transfer between the metal and ˙NO may take place. The actual electron density of the M–NO bond will be better described as one of the three limit forms depicted below, whose proportion will depend on both the metal and its ligands:

$$[M^{(n-1)+}, NO^+] \longleftrightarrow [M^{n+}, NO] \longleftrightarrow [M^{(n+1)+}, NO^-]$$

As a consequence, NO bound to a metal ion may exhibit a wide range of chemical behaviors due to a change from ˙NO to NO^+ or NO^-. Thus, while free ˙NO is essentially a free radical with poor electrophilic or nucleophilic properties, coordinated ˙NO may display both. This might suggest that, under biological conditions, metals (low-molecular mass complexes or metalloproteins) can activate ˙NO and be a source of NO^+ or NO^-.

One well-known example of biological nitrosyl complexes is deoxy-hemoglobin which binds ˙NO with a much greater affinity than O_2 and CO. Nitrosyl hemoglobin is paramagnetic and can be easily detected by EPR spectroscopy at 77 K, offering a convenient method for NO spin trapping. In the presence of oxygen, the iron center is oxidized. Methemoglobin is formed together with nitrite and nitrate. It should be noted that unlike O_2, ˙NO binds to both iron(II) and iron(III) porphyrins. The great reactivity of ˙NO towards hemoglobin suggests that in red blood cells ˙NO will be converted to NO_3^- rapidly and thus rapidly eliminated.

Another hemoprotein sensitive to ˙NO is guanylate cyclase. This enzyme is responsible for vascular muscle relaxation since, once activated, it catalyzes the conversion of guanosine triphosphate GTP into cyclic guanosine monophosphate cGMP, the accumulation of which, in muscle cells, is accompanied by muscle relaxation. It is assumed that the activation of the enzyme is due to the binding of ˙NO to the iron of the heme moiety.

Non-heme iron centers are also able to make nitrosyl complexes.

Examples

Lipoxygenase
Ferritin. The protein involved in the intracellular storage of iron as polymeric ferric hydroxide and oxide (~ 4500 Fe atoms/molecule). It has been claimed that the reaction between ferritin and ˙NO could lead to iron release from the protein. However, experiments in our laboratory clearly show that this may not be the case (J.P. Laulhère, unpublished results).

Iron-sulfur proteins. There is now accumulating evidence that ˙NO, produced for example during activation of macrophages, modifies the catalytic site of some mitochondrial enzymes by coordinating to iron at their Fe-S clusters, thus inhibiting the whole mitochondrial respiration. This could be one of the mechanisms by which ˙NO is cytotoxic. In the case of mitochondrial aconitase ˙NO could convert the active [4Fe-4S] cluster to the inactive [3Fe-4S] form by binding to the labile iron and removing it out of the cluster. ˙NO might also modulate the post-transcriptional regulation of genes involved in iron homeostasis by interacting with the iron-sulfur center of the iron regulatory factor, a cytoplasmic aconitase, which controls both ferritin mRNA translation and transferrin receptor mRNA stability.

It should be added that iron-sulfur-nitrosyl complexes are paramagnetic species with characteristic EPR signals. This makes EPR spectroscopy a very useful method for observing nitrosation of iron-sulfur proteins in biological solutions.

Some synthetic iron complexes have to be mentioned since they have been used in a biological context. First, iron-nitrosyl complexes may provide a source of ˙NO. The most widely studied complex in this regard is the nitroprusside anion $Fe(CN)_5(NO)^{2-}$ which, under specific conditions, generates fluxes of ˙NO. The complex has been used for inducing hypotension during surgery on the vascular system. Another example is the anion of Roussin's Black Salt $[Fe_4S_3(NO)_7]^-$.

Second, iron complexes have been used as traps of ˙NO yielding EPR-detectable paramagnetic iron-nitrosyl complexes. One example is the diethyldithiocarbamate-Fe complex. However, ˙NO can only be trapped in lipophilic compartments since diethyldithiocarbamate is not water soluble. This problem has been recently solved by the utilization of N-methyl D-glucamine dithiocarbamate which gives water soluble iron-nitrosyl complexes and allows spin-trapping of ˙NO *in vivo* [15].

Nitric oxide reacts with other metal ions, such as copper. Nitrosyl complexes are formed during reaction of ˙NO with copper proteins containing blue sites or binuclear copper centers (tyrosinase, hemocyanin).

NO^+ and NO^-

˙NO can be converted to NO^+, the nitrosonium ion, and NO^-, the nitroxyl anion, during one-electron transfer reactions. Only limited information is available concerning the corresponding redox potentials in water at pH 7.0.

It seems rather easy to generate NO^- from ˙NO even though nothing is known about reducing agents and conditions for the reaction. In aqueous solutions, NO^- generation can be accomplished by decomposition of sodium trioxodinitrate, decomposition of N-hydroxybenzene sulfonamide, or oxidation of cyanamide.

The protonated form of NO^-, HNO, is a weak acid with a pKa value of 4.7, indicating that at physiological pH, NO^- is the predominant form in aqueous solution. NO^- decomposes rapidly to yield nitrous oxide N_2O following dimerization and dehydration.

NO^- chemistry is essentially dominated by its propensity to lose one electron and give back ˙NO. Electron acceptors may be O_2, flavins, metal ions such as Cu^{2+}. NO^- reacts with Fe(III) heme proteins such as methemoglobin and cytochrome c to form NO-adducts (reductive nitrosylation). NO^- has been described to be a potent vasorelaxant.

The conversion of ˙NO to NO^+ is also possible but is achieved at rather high redox potentials. It is possible that NO^+ or a related species is transiently formed during oxidation of ˙NO by O_2. Then this would explain why, only under aerobic conditions, can NO carry out nitrosation of thiols and amines. In fact, ˙NO is not an electrophilic nitrosating agent, unless it is transformed to NO^+.

NO^+ is actually a very strong nitrosating species, during reactions with thiols, alcohols and amines. It thus may modify essential cysteines in enzymes. In particular there is now quite convincing evidence that S-nitroso proteins can accumulate during exposure to ˙NO under aerobic conditions. Even though the S-nitrosocysteines within polypeptide chains are much more stable than the free S-nitrosocysteine, this modification is reversible. Such a process might thus have important implications in terms of regulation of enzyme activities by ˙NO. One example is the nitrosation of essential cysteines of ribonucleotide reductase. This might be a way to use ˙NO for regulating deoxyribonucleotide and DNA synthesis. Inside cells, because of the high concentration of glutathione, S-nitrosoglutathione may accumulate and partly mediate the effects of ˙NO.

Also, nitrosation amine groups of deoxynucleotides and DNA resulting in deamination has been observed, which may contribute to ˙NO-dependent genomic alterations.

The following example shows that ˙NO can exist in distinct redox states which have very different biological actions. ˙NO has been implicated as a mediator of neuronal destruction in vascular stroke. In some studies, however, it seems to have neuroprotective effects. This paradox may be resolved by the observation of Lipton et al. recently reported [16]. The authors find that ˙NO might exert both effects, depending on its redox state. The evidence is that the neurotoxic actions of ˙NO derive from the neutral radical form of the molecule which reacts with superoxide to form peroxonitrite, probably the final neurotoxic agent. On the other hand, when ˙NO is converted to NO^+, it reacts with the thiol group of the NMDA receptor to block neurotransmission. Thus, in cerebral cortical cultures, conditions favoring ˙NO give rise to neurotoxicity, whereas neuroprotective effects occur in the presence of NO$^+$. This may have great therapeutic implications against neuronal stroke damage. A good therapeutic agent should prevent the formation of ·NO and enhance that of NO$^+$. Alternatively, one might seek to develop drugs that are converted only to NO$^+$ and not to ·NO. The electrophilic reactivity of NO$^+$ (nitrosations) has, in that case, to be controlled, to avoid unselective and toxic reactions of such a drug.

Chemical ·NO donors

Different ·NO donors can be used for *in vitro* studies, when ·NO gas is difficult to manipulate. Most of these compounds have, in addition, been used as therapeutic agents in particular because of their vasorelaxant properties.

Organic nitrates

Nitroglycerin has been used as a vasodilator (substance which enlarge blood vessels) for the treatment of angina pectoris. Its vasodilatory effects are related to its metabolism to ·NO within the endothelium.

However, the release of ·NO from organic nitrates is not spontaneous and requires activation through mechanisms (enzymatic?) which have not been unambiguously identified. Thiols might be involved in that activation.

Iron-nitrosyl complexes

Sodium nitroprusside SNP has also been used as a vasodilator. However, *in vitro*, SNP does not spontaneously liberate ·NO. It also requires either reductive (for example with a thiol) or light activation. This is very often not sufficiently appreciated in experiments using SNP as a source of ·NO.

In fact, SNP is rather a source of nitrosonium NO$^+$, and thus behaves as a nitrosating electrophilic species. It has been shown to covert amines to nitrosamines, ketones to oximes.

Sydnonimines

These molecules are heterocyclic compounds derived from morpholine. The liberation of ·NO requires both alkaline pH and the presence of oxygen:

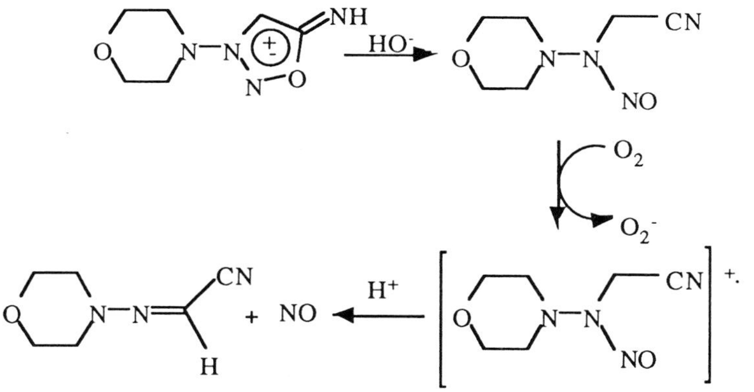

There is one drawback to this ˙NO generation, i.e., the stoichiometric production of superoxide radicals. The reaction between ˙NO and O_2^- generates hydroxyl radicals.

C-nitroso compounds

Upon exposure to light, C-nitroso compounds may undergo a homolytic cleavage of the C–NO bond and generate ˙NO.

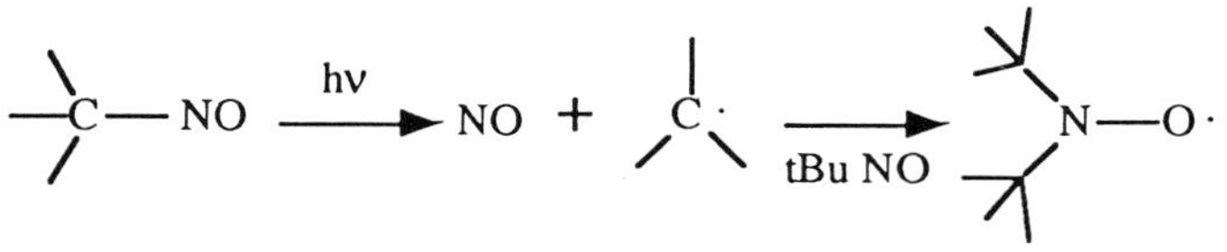

Secondary amine/NO complex ions R₂N[N(O)NO] – (so called NONOates)

These compounds have the following interesting properties:

– They stabilize ˙NO during storage in a solid form.
– They are highly soluble in water.
– They are able to release ˙NO at rates that can be adjusted reliably over a wide range with judicious choice of the carrier nucleophile R_2NH.
– The release of ˙NO is spontaneous and does not require redox or light activation.

$$\underset{(-)}{\overset{\displaystyle \diagdown}{}}N-N\overset{\displaystyle N=O}{\underset{\displaystyle O}{}} \quad \xrightarrow{\ H^+\ } \quad 2\,NO \ + \ \diagup N-H$$

These compounds have both vasorelaxant and antiproliferative (inhibition of DNA synthesis in tumor cells) activities. However, they are powerful mutagens, and therefore caution should be exercised.

Thionitrites (or nitrosothiols)

Thionitrites are red or green colored compounds which can be easily obtained during treatment of thiols with a variety of nitrosating agents [17]. The most commonly synthetic reactions are:

$$RSH \xrightarrow[\text{HCl}]{\text{NaNO}_2} RSNO$$

$$RSH \xrightarrow[\text{tBuONO}]{} RSNO$$

Thionitrites are usually quite unstable in water at physiological pH. They spontaneously release ·NO without redox or photochemical activation, even though light greatly accelerates the reaction. Moreover, the decomposition is strongly catalyzed by Cu^{2+} or Fe^{3+}, even as contaminants of aqueous buffered solutions, and thus inhibited by chelators. The reaction proceeds through a homolytic cleavage of the S–N bond generating a thiyl radical together with ·NO:

$$RSNO \longrightarrow RS^{\cdot} + NO$$

$$2RS^{\cdot} \longrightarrow RSSR$$

These ·NO generators are also very interesting since it is possible to get a wide range of rates of production of ·NO by simple chemical modifications of the R group of RSNO. Highest stabilities are obtained with tertiary thionitrites.

References

1. Butler, A.R. and Williams, D.L.H. (1993) The physiological role of nitric oxide. *Chem. Soc. Rev.* 233–241.
2. Feldman, P.L., Griffith, O.W. and Stuehr, D.J. (1993) The surpsising life of nitric oxide. *C & EN* 20: 26–38.
3. Knowles, R.G. and Moncada, S. (1992) Nitric oxide as a signal in blood vessels. *TIBS* 17: 399–402.
4. Nathan, C. (1992) Nitric oxide as a secretory product of mammalian cells. *FASEB J.* 6: 3051–3064.

5. Stamler, J.S., Singel, D.J. and Loscalzo, J. (1992) Biochemistry of nitric oxide and its redox-activated forms. *Science* 258: 1898–1902.
6. Galla, H.J. (1993) Nitric oxide, NO, an intracellular messenger. *Angew. Chem. Int. Ed. Engl.* 32: 378–380.
7. Liu, J. and Sessa, W.C. (1994) Identification of covalently bound amino-terminal myristic acid in endothelial nitric oxide synthase. *J. Biol. Chem.* 269: 11691–11694.
8. Marletta, M.A. (1994) Nitric oxide synthase structure and mechanism. *J. Biol. Chem.* 268: 12231–12234.
9. Nathan, C. and Zie, Q.W. (1994) Regulation of biosynthesis of nitric oxide. *J. Biol. Chem.* 269: 13725–13728.
10. Klatt, P., Schmid, M., Leopold, E., Schmidt, K., Werner, E.R. and Mayer, B. (1994) The pteridine binding site of brain nitric oxide synthase. *J. Biol. Chem.* 269: 13861–13866.
11. Culcasi, M., Lafond-Cazal, M., Pietri, S. and Bockaert, J. (1994) Glutamate receptors induce a burst of superoxide via activation of nitric oxide synthase in arginine-deploted neurons. *J. Biol. Chem.* 269: 12589–12593.
12. Stamler, J.S. (1994) Redox signaling: nitrosylation and related target interactions of nitric oxide. *Cell* 78: 931–936.
13. Fontecave, M. and Pierre, J.L. (1994) The basic chemistry of nitric oxide and its possible biological reactions. *Bull. Soc. Chim. Fr.* 131: 620–631.
14. Roy, B., Lepoivre, M., Henry, Y. and Fontevave, M. (1995) Inhibition of ribonucleotide reductase by nitric oxide derived from thionitrites: reversible modifications of both subunits. *Biochemistry* 34: 5411–5418.
15. Lai, C.-S. and Komarov, A.M. (1994) Spin trapping of nitric oxide produced *in vivo* in septic-shock mice. *FEBS Lett.* 345: 120–124.
16. Lipton, S.A., Choi, Y-B., Pan, Z-H., Lei, S.Z., Chen, H-S.V., Sucher, N.J., Loscalzo, J., Singel, D.J. and Stamler, J.S. (1993) A redox-based mechanism for the neuroprotective and neurodestructive effects of nitric oxide and related nitroso-compounds. *Nature* 364: 626–631.
17. Roy, B., Du Moulinet D'Hardemare, A. and Fontecave, M. (1994) New thionitrites: synthesis, stability and nitric oxide generation. *J. Org. Chem.* 59: 7019–7026.

Analysis of Free Radicals in Biological Systems
Favier et al. (eds)
© 1995 Birkhäuser Verlag Basel/Switzerland

Pro-oxidant and antioxidant effects of nitric oxide

N. Hogg

Biophysics Research Institute, Medical College of Wisconsin, Milwaukee, WI 53226, USA

Summary. Nitric oxide reacts preferentially with other free radicals. Reaction of nitric oxide with oxygen yields nitrogen dioxide, a potent oxidant. However, this reaction may be too slow to be relevant *in vivo*. Reaction of nitric oxide with superoxide occurs extremely rapidly and generates peroxynitrite. This latter molecule undergoes many deleterious oxidative reactions with biological molecules such as amino acids, sugars and lipids. In an analagous reaction nitric oxide may also react with peroxyl radicals, such as lipid peroxyl, and inhibit free radical chain reactions such as lipid peroxidation.

This chapter discusses experimental sources of nitric oxide and peroxyntirite and the potential pro-oxidant and antioxidant effects of nitric oxide in biological systems.

Introduction

Nitric oxide (NO) is an intriguing and multifaceted molecule in biological systems. The chemical and physical properties of NO place it in a unique class of endogenously synthesized biological molecules – there is really nothing quite like NO. The number of diverse and, at times, contradictory biological activities of NO reported in the literature continues to grow. NO is often refered to as a "reactive free radical" but this qualitative statement denies the essential property of nitric oxide and that is its selective reactivity towards other free radicals. As will be discussed here, this property can lead to both potentially advantageous and profoundly deleterious consequences.

NO reacts with molecular oxygen (Eq. (1)) in third-order process with a rate constant of 6×10^6 $M^{-2}s^{-1}$ [1].

$$2^{\bullet}NO + O_2 \longrightarrow 2^{\bullet}NO_2 \qquad (1)$$

Consequently, the reaction rate decreases sharply as the concentrations of nitric oxide and oxygen are reduced. Thus, at mM concentrations the rate of this reaction is fast, but at lower physiological concentrations (nM) the reaction may be slow enough to be ignored [2]. The use of "half-time" as an index of NO persistence is misleading because of the concentration dependence of this parameter. Perhaps a more useful indicator of the biological persistence of NO is the steady-state concentration that can be sustained by a set rate of NO production in the presence of a constant concentration of oxygen. This is shown in Table

Table 1. The calculated maximum achievable steady-state concentrations of NO when generated at different rates in the presence of 250 μM and 20 μM oxygen

Rate of NO production	$[NO]_{steady\ state}(\mu M)$ in 250 $\mu M\ O_2$	$[NO]_{steady\ state}(\mu M)$ in 20 $\mu M\ O_2$
100 mM/s	7070	25000
10 mM/s	2240	7900
1 mM/s	707	2500
100 μM/s	224	790
10 μM/s	70.7	250
1 μM/s	22.4	79
100 nM/s	7.07	25
10 nM/s	2.24	7.9
1 nM/s	0.707	2.5
100 pM/s	0.224	0.790
10 pM/s	0.0707	0.250
1 pM/s	0.0224	0.079

1 for various rates of NO production in the presence of 250 μM oxygen (acerated water/buffer) and 20 μM oxygen (approximate tissue concentration). Biological rates of NO production are approximately 1–10 nM/s [3] and thus if the reaction with oxygen is the fastest route of NO decomposition, biology could sustain steady-state concentrations of NO in the low μM range. Steady-state concentrations of NO have been measured as approximately 100 nM [4] suggesting that oxygen is not the primary route of NO decomposition.

The question arises as to how NO is destroyed *in vivo*. One potential pathway of NO decomposition is the reaction with oxygen, bound to the heme group of hemoglobin and myoglobin, to generate nitrate and the ferric hemeprotein ($k \approx 4 \times 10^7\ M^{-1}s^{-1}$) [5]. This could be a major pathway of NO destruction if NO diffuses into the blood stream and in tissues rich in myoglobin such as skeletal muscle. The fastest biologically relevant reaction of NO is that with superoxide, as will be discussed later.

In this chapter the advantages and disadvantages of various sources of NO will be discussed, together with the pro-oxidant and antioxidant reactions of NO.

Sources of NO

Nitrosothiols

Nitrosothiols, previously referred to as thionitrites, contain the R–S–NO functional group and are unique among the compounds used to generate nitric oxide as they are formed *in vivo*. Protein nitrosothiols have been detected in human plasma, predominantly as S-nitroso serum albumin [6]. The mechanism of formation of these compounds *in vivo* is poorly understood, but *in vitro* synthesis can be achieved by reaction of most thiols with the nitrosonium cation [7]. This is usually achieved using acidified nitrite (reactions 2–4).

$$NO_2^- + H^+ \longrightarrow HNO_2 \tag{2}$$

$$HNO_2 + H^+ \longrightarrow H_2O + NO^+ \tag{3}$$

$$NO^+ + RSH \longrightarrow RSNO + H^+ \tag{4}$$

The most commonly used nitrosothiols are S-nitrosoglutathione (GSNO), S-nitroso-N-acetyl penicillamine (SNAP), and S-nitrosocysteine (SNOC). Nitrosation of protein thiols has also been achieved with bovine and human serum albumins [8].

All nitrosothiols are intrinsically stable and require the action of a secondary agent before decomposition and NO release occurs. The simplest and best understood mechanism of nitrosothiol decomposition is photolytic homolysis. Ultraviolet light excites the S–N bond resulting in a homolytic cleavage forming NO and the thiyl radical. Visible light of about 550–600 nm will also result in decomposition, but with a much lower efficiency. This is expected as the UV-vis absorbance spectrum of nitrosothiols has a strong band at about 330 nm and a much weaker band ($\varepsilon \approx 10$–$20 \, M^{-1}cm^{-1}$ at 550–600 nm) in the visible region. The thiyl radical, in the absence of competing reactions, will dimerize forming the disulfide [9]. Nitrosothiols are also susceptible to catalytic decomposition by transition metal ions, such as iron and copper, that are (usually) contaminants of water and buffers [10]. This has given rise to the erroneous idea that these compounds spontaneously release NO. Thus measurements of half-life given in the literature are specific to the buffer or the source of water. It is apparent, however, that nitrosothiols have vastly different susceptibilities to this mechanism of release [11]. For this reason SNAP is regarded as a more rapid releaser of NO than GSNO. The mechanism of transition metal ion decomposition is unclear and may involve either redox cycling of the metal ion or transitory metal-thiol bonds. With Hg^{2+}, an agent often used to decompose nitrosothiols, the reaction is not catalytic and a mercury-sulfur bond is formed [10]. The mechanism of NO release from nitrosothiols in biological systems is poorly understood.

$$RSNO + R'H \longrightarrow RSH + R'NO \tag{5}$$

Another important reaction of nitrosothiols is transnitrosylation (Eq. (5)). This represents the donation of NO^+ from a nitrosothiol to a thiol. Thus in a system where a nitrosothiol is added to a mixture of thiols the equilibrium constants for these reactions will determine the eventual location of nitric oxide [12]. It is vitally important to consider the transnitrosylation reaction in biological systems as the effects of nit-rosothiols on receptor/enzyme activity may be due to NO^+ transfer to protein thiols rather than the generation of NO [13].

Sodium nitroprusside (SNP)

SNP has been used for many years as a therapeutic agent for hyperten-sion and angina pectoris. The benefits of such treatment are extremely well documented and are likely due to nitric oxide release. However, the use of this agent to release controlled amounts of nitric oxide is fraught with problems. As with nitrosothiols, SNP is an intrinsically stable molecule and consists of NO^+ bound to ferrous iron. The remaining five ligands are cyanide ions. Photoactivation of SNP causes a charge transfer from the iron to NO^+ resulting in release of NO and oxidation of the iron to the ferric form [14]. However, SNP can undergo many other reactions that make it a particularly problematic source of NO. These reactions include cyanide release, Fe^{n+} release, and superoxide production [15].

NONOates

NONOates are the most recently available generic class of NO releasing compounds. These are stable under basic conditions and decay to release two molecules of NO per molecule of compound at physiological pH. Decomposition is a first-order process that appears to be sponta-neous. These compounds represent the most reliable and unambiguous source of NO to date. Compounds with variable half lives have been reported. Two that are commercially available are spermineNONOate with a half-life of 40 min and diethylamineNONOate with a half-time of 2 min [16]. A major advantage of these compounds is that each molecule of NONOate decays to give two molecules of nitric oxide to leave a non-radical product (e.g., spermine).

NO gas

The techniques for handling NO gas and solutions of NO have been detailed elsewhere [17]. Strict anaerobic conditions are required and the stream of gas must be passed through a solution of sodium hydroxide to remove higher oxides of nitrogen. NO gas dissolves in water to a concentration of approximately 2 mM at room temperature. Unless an adequate delivery system is used, NO gas is not a good model for biological nitric oxide. Bolus addition of NO solution will create a much higher concentration of NO than is ever experienced *in vivo* and under these conditions, the reaction of NO with oxygen is favored, thus generating NO_2, N_2O_3, N_2O_4, and perhaps other oxides of nitrogen. Attempts to solve the problem of delivery of NO solutions have been made. Passage of NO through the bore of gas permeable tubing allows diffusion into the surrounding solution [18]. Slow injection of NO solution using a motor driven gas tight syringe has also been employed [19]. The advantage of these methods over NO-donor compounds is that only pure NO is administered to the system. However, they may be impractical for multi-sample systems and cell culture.

SIN-1

SIN-1 (or 3-morpholino sydnonimine) has been considered as a source of NO, although in fact it would be more correctly termed a peroxynitrite donor. SIN-1 decays spontaneously by a three-step process. The first step is a base-catalyzed ring opening event to generate SIN-1A, which is followed by one electron oxidation by molecular oxygen to give superoxide and the free radical SIN-1B. This latter compound spontaneously releases NO to give the stable end product SIN-1C [20]. NO and superoxide, produced simultaneously will rapidly combine to give peroxynitrite. This compound is an extremely useful model for the cellular generation of nitric oxide and superoxide and for slow continuous peroxynitrite synthesis [21]. SIN-1C is commercially available for use in control experiments.

NO synthase

NO can be generated in cellular systems by the endogenous NO synthase machinery. Constitutive enzyme present in endothelial cells will generate low levels of NO that can be stimulated using vasodilators such as bradykinin and the calcium ionophore A23187. Higher levels of NO can be generated by cytokine/endotoxin stimulated cells (usually macrophages) which will synthesize inducible NO synthase in response to these stimuli. Cellular generation of NO has been extensively discussed elsewhere [22].

Other sources

There are several other sources of NO including organic nitrates and nitrites which appear to generate NO through a thiol-dependent enzymatic process [16]. An interesting new compound is GEA 3162, a SIN-1-like compound that does not generate superoxide [23].

Peroxynitrite synthesis

Peroxynitrite is most usually snthesized by the reactions shown in Equations (6–8).

$$HNO_2 + H^+ \longrightarrow NO^+ + H_2O \tag{6}$$

$$NO^+ + H_2O_2 \longrightarrow ONOOH + H^+ \tag{7}$$

$$ONOOH \longrightarrow ONOO^- + H^+ \tag{8}$$

Sodium nitrite (0.7 M) is mixed with hydrogen peroxide (0.6 M) in sulfuric acid (0.8 M). The NO^+ generated from nitrous acid reacts with hydrogen peroxide to give peroxynitrous acid. This compound is unstable, but rapidly raising the pH by the addition of sodium hydroxide yields the stable peroxynitrite anion. The final solution is bright yellow in color with an absorbance maximum at 302 nm ($\varepsilon = 1670$ $M^{-1}cm^{-1}$). All solutions must be pre-cooled on ice before mixing. The use of a quenched flow reactor gives the greatest yields of peroxynitrite as the time before quenching can be controlled [24]. However, yields of 30–50 mM peroxynitrite can be achieved by successively (and rapidly) emptying acidified hydrogen peroxide and then sodium hydroxide into a beaker containing sodium nitrite [25]. Peroxynitrite solutions can be concentrated by freezing. During this process peroxynitrite partitions to the surface and can be removed. The success of this operation depends upon the type of tube in which the solution is frozen. Excess unreacted hydrogen peroxide can be removed by addition of manganese dioxide granules. It is essential to keep the solution cold during this process as manganese dioxide-catalyzed decomposition of hydrogen peroxide generates heat which can dramatically reduce the yield of peroxynitrite.

Other methods of peroxynitrite production have been described. Potassium superoxide reacts with NO at neutral pH to generate peroxynitrite [26]. The reaction of nitroxyl anion, generated for example for the decomposition of Angeli's salt ($Na_2N_2O_3$), reacts with oxygen to give peroxynitrite [27]. Peroxynitrite can also be generated from the reaction of organic nitrites with hydrogen peroxide in strong base [28]. The reaction of NO_2 with hydroxyl radical forms nitrate through the intermediate formation of peroxynitrite [29] and finally, irradiation of potassium nitrate crystals with UV light results in the formation of peroxynitrite in the solid phase [30]. These methods give valuable insight into the chemical nature of nitrogen oxides, but are less suited as routes of practical synthesis.

Pro-oxidant effects of NO

NO as an oxidant

NO is a poor oxidant. The redox potential ($E^{\circ\prime}$) of the NO/NO$^-$ couple is 0.39 V [29]. This value makes the oxidation of glutathione to the glutathionyl radical thermodynamically unfavorable ($E^{\circ\prime}$ of GS$^\cdot$/GSH is 0.8 V). The oxidation of ascorbate to the ascorbyl radical is thermodynamically favourable ($E^{\circ\prime}$ of ascorbyl radical/ascorbate is 0.3 V), but does not appear to occur at an appreciable rate. NO is also unable to initiate lipid peroxidation by the abstraction of hydrogen from unsaturated fatty acids [31, 41]. Direct oxidative effects of NO gas on protein amino acids have been reported [32]. It is likely that in situations where nitric oxide is observed to be a pro-oxidant higher oxides of nitrogen such as NO$_2$, N$_2$O$_3$, N$_2$O$_4$ and peroxynitrite are being generated. Alternatively, NO may stimulate the release of other pro-oxidant species such as iron [33].

Oxidative effects of peroxynitrite

On combination with superoxide, nitric oxide generates peroxynitrite [26], a powerful biological oxidant [34]. The rate constant for this reaction has been determined as 6.7×10^9 M^{-1}s^{-1} [35]. Peroxynitrite anion is stable and has a pKa of 6.9 at 37°C. Thus at physiological pH, peroxynitrite will be partially protonated. Peroxynitrous acid is unstable and rearranges to form nitric acid. To achieve this intermolecular rearrangement the molecule must pass through a high-energy transition state intermediate that has properties similar to those of the hydroxyl radical [29, 34]. This hydroxyl radical-like activity has also been observed in systems where NO and superoxide are generated simultaneously by either SIN-1 or a combination of xanthine oxidase, acetaldehyde and SNAP [21, 36]. Conventional methods of OH$^\cdot$ detection are unable to distinguish between Fenton reaction derived hydroxyl radical and the peroxynitrite decomposition intermediate. *In vitro*, this activity of peroxynitrite has been used to degrade DNA for footprinting studies [37]. The arguments often used to discredit the involvement of the hydroxyl radical as a biologically relevant oxidant (i.e., short diffusion distance, indiscriminate oxidation) can also be made against this intermediate. However it is likely that most of the pathological potential of peroxynitrite does not involve the OH$^\cdot$ like activity. Peroxynitrite and or peroxynitrous acid will rapidly oxidize thiols [38] and methionine [39] and will initiate lipid peroxidation [40, 41]. Interestingly, peroxynitrite has been shown to release copper from caeriloplasmin [42]. This represents a potential mechanism whereby the short-lived oxidative effects of peroxynitrite are transformed to the persistent oxidative effects of free copper ions. Both peroxynitrite and the simultaneous generation of NO and superoxide will oxidize vitamin E to vitamin E quinone [36, 43, 44] and have also been shown to modify low-density lipoprotein (LDL) to a potentially atherogenic form [41, 45]. The bactericidal activity of peroxynitrite has also been reported [46].

One of the most important and diagnostic reactions of peroxynitrite is the nitration of tyrosine. This reaction is catalyzed by transition metal ions and superoxide dismutase [47]. Perhaps more importantly, protein tyrosine residues are also susceptible to peroxynitrite-dependent nitration. The biological conseqences of this are unclear but are expected to be deleterious. Moreover, this reaction leaves a nitrotyrosine fingerprint that has been successfully detected, using immunochemical techniques, in atherosclerotic lesions [48]. Several cell types including macrophages [49], neutrophills [50] and endothelial cells [51] have been shown to generate peroxynitrite in culture, and it has been suggested that peroxynitrite may play a central role in the pathological mechanisms of cerebral ischemic damage [52] and atherosclerosis [41, 53, 54].

Several methods have been employed to detect peroxynitrite. The first published method used nitration of 4-hydroxyphenyl acetate as a diagnostic marker. The nitrated product, 4-hydroxy-3-nitrophenyl acetate, was detected by UV after HPLC separation [45, 46]. Luminol-dependent chemiluminesence [55] has been used to detect peroxynitrite from activated Kupfer cells [56] and endothelial cells [51]. More recently, dihydroxyrhodamine 123 has been employed as a fluorescent probe for peroxynitrite production [57].

Antioxidant effects of nitric oxide

Inhibition of lipid peroxidation

As mentioned earlier, NO is particularly reactive towards other free radicals. In the case of the reaction between NO and superoxide, peroxynitrite is formed and further oxidative events can occur. However, radical–radical reactions are generally perceived as "chain termination" reactions that result in the formation of stable, non-radical products.

The most important free radical chain reaction that occurs in biological systems is lipid peroxidation. After initiation by an external oxidant, the propagation of lipid peroxidation occurs by reactions 9 and 10, where LH represents an unsaturated lipid; L·, a

$$L^{\cdot} + O_2 \longrightarrow LOO^{\cdot} \tag{9}$$

$$LOO^{\cdot} + LH \longrightarrow LOOH + L^{\cdot} \tag{10}$$

lipid radical; LOO˙, a lipid peroxyl radical and LOOH, a lipid hydroperoxide. The net result of this process is the conversion of lipid into lipid hydroperoxide. The central role of the peroxyl radical in this process can be gleaned from Equations (9) and (10) and compounds that scavenge lipid peroxyl radicals, such as butylated hydroxytoluene and probucol, are potent inhibitors of lipid peroxidation.

Induction of NO synthase was shown to inhibit macrophage-dependent LDL oxidation [31, 58]. NO has recently been shown to be a potent inhibitor of lipid peroxidation in LDL [58] and free fatty acid systems [19]. It is extremely likely that NO inhibits lipid peroxidation by scavenging lipid peroxyl radicals (reaction 11). The rate constant for the reaction between NO and organic peroxyl radicals has been measured as $1-3 \times 10^9$ $M^{-1}s^{-1}$ [60]. Reaction 11 leads to the formation of nitrosylated lipid products and

$$LOO˙ + NO \longrightarrow LOONO \tag{11}$$

such compounds have been detected during the oxidation of linolenic acid in the presence of NO [19]. The reactivity and pharmacology of these NO-lipids adducts have yet to be established.

Other potential antioxidant mechanisms of NO

Nitric oxide has been shown to inhibit the cytotoxic effects of superoxide and hydrogen peroxide to fibroblasts and neuronal cells [61]. The mechanism for this protection is unknown, but the possibility arises that the reaction between NO and superoxide can, in some circumstances, be protective [62]. It can be envisaged that in situations in which either NO or superoxide are particularly toxic, the production of peroxynitrite may redirect the oxidative event/damage to a site that is more easily repaired (e.g., glutathione). The inhibitory effects of NO on hydrogen peroxide-induced toxicity are less well established and appear to be system dependent as other reports in the literature show enhanced cell killing by hydrogen peroxide in the presence of NO [63].

NO, in high concentrations, has also been shown to inhibit hydroxyl radical formation resulting from either UV irradiation of hydrogen peroxide or the Fenton reaction [64]. In the case of UV irradiation, direct scavenging of hydroxyl radical by nitric oxide was envisaged. It was suggested that inhibition of the Fenton reaction occurred by the binding of nitric oxide to a ferrous iron complex preventing redox cycling. *In vivo* it is conceivable that redox active iron is sequestered in a thiol-NO complex, however, such complexes remain to be identified. Complexes of iron (II) and thiols have been used to detect NO formation due to their characteristic electron spin resonance signature [65].

The effects of NO on potential pro-oxidant enzymes have also been studied. It has been reported that NO inhibits the lipid oxidation reactions of lipoxygenase and cyclo-oxygenase by a mechanism that involves conversion of the catalytic iron to the ferrous form [66, 67], thus inactivating the enzyme. Another report indicates activation of cyclo-oxygenase by NO in cellular systems [68]. NO has also been reported to reversibly inhibit xanthine oxidase activity by a direct action on the flavin component of the enzyme [69]. However, xanthine oxidase was not inhibited by slow generation of NO from NO donor compounds or by lower, more physiological concentrations of NO [19]. Further studies are required to clarify which reactions of NO are relevant *in vivo*.

Conclusion

Nitric oxide, since the discovery of its relevance in biological systems, has repeatedly been shown to be a two-faced character. The task of uncovering the mechanistic basis for this often contradictory behavior is complex. The process of lipid peroxidation is unique in that the two faces of NO can be explained at the molecular level. It is clear that the combined effects of NO and superoxide depend critically on the rates of generation of these two free radicals. At stoichiometric, or sub-stoichiometric NO, stimulation of lipid peroxidation occurs [19, 41]. However, as the rate of NO generation surpasses that of superoxide, the antioxidant effects of NO become dominant [19]. The extent to which these mechanisms contribute to the effects of NO *in vivo* remain to be determined. However, it is compelling to believe that NO is a vital regulator of oxidative pathologies.

References

1. Wink, D.A., Darbyshire, J.F., Mims, R.W., Saavedra, J.E. and Ford, P.C. (1993) Reactions of the bioregulatory agent nitric oxide in oxygenated media: Determination of the kinetics of oxidation and nitrosation by intermediates generated in the NO/O$_2$ reaction. *Chem. Res. Toxicol.* 6: 23–27.
2. Kharitonov, V.G., Sundquist, A.R. and Sharma, V.J. (1994). Kinetics of nitric oxide autoxidation in aqueous solution. *J. Biol. Chem.* 269: 5881–5883.
3. Kelm, M. and Schrader, J. (1988). Production of nitric oxide by the isolated guinea pig heart. *Eur. J. Pharm.* 155: 317–321.
4. Beckman, J.S. and Crow, J.P. (1993). Pathological implications of nitric oxide superoxide and peroxynitrite formation. *Biochem. Soc. Trans.* 21: 330–334.
5. Doyle, M.P. and Heokstra, J.W. (1981) Oxidation of nitrogen oxides by bound dioxygen in hemoproteins. *J. Inorg. Biochem.* 14: 351–358.
6. Stamler, J.S., Jaraki, O., Osbourne, J., Simon, J., Keaney, J., Vita, J., Singel, D., Valeri, C.R. and Loscalzo, J. (1992) Nitric oxide circulates in mammalian plasma primarily as an S-nitroso adduct of serum albumin. *Proc. Natl. Acad. Sci. USA* 89: 7674–7677.
7. Field, L., Dilts, R.V., Ravichandran, R., Lenhert, P.G. and Carnahan, G.E. (1978) An unusually stable thionitrite from N-acetyl-D,L,-penicillamine; X-ray crystal and molecular structure of 2-(acetylamino)-2-carboxy-1,1-dimethyethyl thionitrite. *J. Chem. Soc. Chem. Com.* 249–250.

8. Keaney, J.F., Jr., Simon, D.I., Stampler, J.S., Jaraki, O., Scharfstein, J., Vita, J.A., Loscalzo, J. (1993) NO forms an adduct with serum albumins that has endothelium-derived relaxing factor properties. *J. Clin. Invest.* 91: 1582–1589.

9. Singh, R.J., Hogg, N., Joseph, J. and Kalyanaraman, B. (1995) Photosensitized decomposition of S-nitrosothiols and 2-methyl-2-nitrosopropane: possible use for site-directed nitric oxide production. FEBS Lett. 360: 47–51.

10. McAninly, J., Williams, D.L.H., Askew, S.C., Butler, A.R. and Russel, C. (1993) Metal ion catalysis in nitrosothiol (RSNO) decomposition. *J. Chem. Soc. Chem. Commun.* 1758–1759.

11. Mathews, W.R. and Kerr, S.W. (1993) Biological activity of s-nitrosothiols: The role of nitric oxide. *J. Pharm. Exp. Therap.* 267: 1529–1537.

12. Mayer, D.J., Kramer, H., Özer, N., Coles, B. and Ketterer, B. (1994) Kinetics and equilibria of S-nitrosothiol-thiol exchange between glutathione, cystine, penicillamines and serum albumin. *FEBS Lett.* 345: 177–180.

13. Park, J.-W., Billman, G.E. and Means, G.E. (1993) Transnitrosation as a predominant mechanism in the hypotensive effects of S-nitrosoglutathine. *Biochem. Mol. Biol. Int.* 30: 885–891.

14. Wolf, S.K. and Swinehart, J.H. (1975) Photochemistry of pentacyanonitrosyl ferrate (2-), nitroprusside. *Inorg. Chem.* 14: 1049–1053.

15. Rao, D.N.R., Elguindi, S. and O'Brien, P.J. (1991) Reductive metabolism of nitroprusside in rat hapatocytes and human erythrocytes. *Arch. Biochem. Biophys.* 266: 30–37.

16. Maragos, C.M., Morley, D., Wink, D.A., Dunams, T.M., Saavedra, J.E., Hoffman, A., Bove, A.A., Isaac, L., Hrabie, J.A. and Keefer, L.K. (1991) Complexes of NO with nucleaphiles as agents for the controlled biological release of nitric oxide. Vasorelaxant effects. *J. Med. Chem.* 34: 3242–3247.

17. Feelisch, M. (1991) The biochemical pathways of nitric oxide formation from nitrovasodilators: Appropriate choice of exogenous NO donors and aspects of preparation and handling of aqueous NO solutions. *J. Cardiovasc. Pharm.* 17 (suppl. 3): S25–S33.

18. Tamir, S., Lewis, R.S., de Rojas Walker, T., Deen, W.M., Wishook, J.S. and Tannenbaum, S.R. (1993) *Chem. Res. Toxicol.* 6: 895–899.

19. Rubbo, H., Radi, R., Trujillo, M., Telleri, R., Kalyanaraman, B., Barnes, S., Kirk, M. and Freeman, B.A. (1994) Nitric oxide regulation of superoxide and peroxynitrite-dependent lipid peroxidation. *J. Biol. Chem.* 269: 26066–26075.

20. Feelisch, O., Ostrowski, J., and Noak, E. (1989) On the mechanism of NO release from sydnonimines. *J. Cardiovasc. Pharm.* 14 (Suppl. 11): 513–522.

21. Hogg, N., Darley-Usmar, V.M., Wilson, M.T. and Moncada, S. (1992) Production of hydroxyl radicals from the simultaneous generation of superoxide and nitric oxide. *Biochem. J.* 281: 419–424.

22. Förstermann, U., Closs, E.I., Pollock, J.S., Nakane, M., Schwarz, P., Gath, I. and Kleinert, H. (1994) Nitric oxide synthase isoenzymes. Characterisation, purification, molecular cloning and functions. *Hypertension* 23: 1121–1131.

23. Malo-Ranta, U., Ylä-Herttuala, S., Metsä-Ketelä, T., Jaakkola, O., Moilanen, E., Vuorinen, P. and Nikkari, T. (1994) Nitric oxide donor GEA 3162 inhibits endothelial cell-mediated oxidation of low-density lipoprotein. *FEBS Lett.* 337: 179–183.

24. Reed, J.W., Ho, H.H. and Jolly, W.L. (1974). Chemical synthesis with a quenched flow reactor. Hydroxytrihydroborate and peroxynitrous acid. *J. Am. Chem. Soc.* 96: 1248–1249.

25. Hughes, M.N. and Nicklin, H.G. (1968) The chemistry of pernitrites. Part 1. Kinetics of decomposition of pernitrous acid. *J. Chem. Soc.* (A) 450–452.

26. Blough, N.V. and Zafiriou, O.C. (1985) Reaction of superoxide with nitric oxide to form peroxynitrite in alkaline aueous solution. *Inorg. Chem.* 24: 3502–3504.

27. Donald, C.E., Hughes, M.N., Thompson, J.M. and Bonner, F.T. (1986) Photolysis of the N=N bond in trioxodinitrate: Reaction between triplet NO^- and O_2 to form peroxonitrite. *Inorg. Chem.* 25: 2676–2677.

28. Ramón Leis, J., Peña, M.E. and Ríos, A. (1993) A novel route to peroxynitrite synthesis. *J. Chem. Soc. Chem. Comms.* 1298–1299.

29. Koppenol, W.H., Moreno, J.J., Pryor, W.A., Ischiropoulos, H. and Beckman, J.S. (1992) Peroxynitrite, a cloaked oxidant formed by nitric oxide and suproxide. *Chem. Res. Toxicol.* 5: 834–842.

30. King, P.A., Anderson, V.E., Edwards, J.O., Gustafoson, G., Plumb, R.C. and Suggs, J.W. (1992) A stable solid that generates hydroxyl radical upon dissolution in aqueous solutions: Reactions with proteins and nucleic acid. *J. Am. Chem. Soc.* 114: 5430–5432.
31. Jessup, W., Mohr, D., Gieseg, S.P., Dean, R.T. and Stocker, R. (1992) The participation of nitric oxide in cell free- and its restriction of macrophage-mediated oxidation of low-density lipoprotein. *Biochim. Biophys. Res. Commun.* 183: 598–604.
33. Reif, D.W. and Simmons, R.D. (1990) Nitric oxide mediates iron release from ferritin. *Arch. Biochem. Biophys.* 283: 537–541.
34. Beckman, J.S., Beckman, T.W., Chen, J., Marshall, P.A. and Freeman, B.A. (1990) Apparent hydroxyl radical production by peroxynitrite: Implications for endothelial injury from nitric oxide and superoxide. *Proc. Natl. Acad. Sci. USA* 87: 1620–1624.
35. Huie, R.E. and Padamaja, S. (1993) The reaction of NO with superoxide. *Free Rad. Res. Comm.* 18: 195–203.
36. Hogg, N., Darley-Usmar, V.M., Wilson, M.T. and Moncada, S. (1993) The oxidation of α-tocopherol in human low density lipoprotein by the simultaneous generation of superoxide and nitric oxide. *FEBS Lett.* 326: 199–203.
37. King, P.A., Jamison, E., Stahs, D., Anderson, V.E. and Brenowitz, M. (1993) 'Footprinting' proteins on DNA with peroxynitrous acid. *Nucleic Acids Res.* 21: 2473–2478.
38. Radi, R., Beckman, J.S., Bush, K.M. and Freeman, B.A. (1991) Peroxynitrite oxidation of sulfhydryls. *J. Biol. Chem.* 266: 4244–4250.
39. Pryor, W.A., Jin, X. and Squadrito, G.L. (1994) One- and two-electron oxidations of methionine by peroxynitrite. *Proc. Natl. Acad. Sci. USA* 91: 11173–11177.
40. Radi, R., Beckman, J.S., Bush, K.M. and Freeman, B.A. (1993) Peroxynitrite-induced membranelipid peroxidation: The cytotoxic potential of superoxide and nitric oxide. *Arch. Biochem. Biophys.* 288: 481–487.
41. Darley-Usmar, V.M., Hogg, N., O'Leary, V.J., Wilson, M.T. and Moncada, S. (1992). The simultaneous generation of superoxide and nitric oxide can initiate lipid peroxidation in human low-density lipoprotein. *Free Rad. Res. Comms.* 17: 9–20.
42. Swain, J.A., Darley-Usmar, V.M. and Gutteridge, J.M.C. (1994) Peroxynitrite releases copper from caeruloplasmin: implications for atherosclerosis. *FEBS Lett.* 342: 49–52.
43. Hogg, N., Joseph, J. and Kalyanaraman, B. (1994) The oxidation of α-tocopherol and trolox by peroxynitrite. *Arch. Biochem. Biophys.* 314: 153–158.
44. De Groot, H., Hegi, U. and Seis, H. (1993) Loss of α-tocopherol upon exposure to nitric oxide or the syndnonimine SIN-1. *FEBS Lett.* 315: 139–142.
45. Graham, A., Hogg, N., Kalyanaraman, B., O'Leary, V.J., Darley-Usmar, V. and Moncada, S. (1993) Peroxynitrite modification of low-density lipoprotein leads to recognition by the macrophage scavenger receptor. *FEBS Lett.* 330: 181–185.
46. Zhu, L., Gunn, C. and Beckman, J.S. (1992) Bactericidal activity of peroxynitrite. *Arch. Biochem. Biophys.* 298: 452–457.
47. Beckman, J.S., Ischiropoulos, H., Zhu, L., van der Woerd, M., Smith, C., Chen, J., Harrison, J., Martin, J.C. and Tsai, M. (1992) Kinetics of superoxide dismutase- and iron-catalysed nitration of phenolics by peroxynitrite. *Arch. Biochem. Biophys.* 298: 438–445.
48. Ischiropoulos, H., Zhu, L. and Beckman, J.S. (1992) Peroxynitrite formation from macrophage-derived nitric oxide. *Arch. Biochem. Biophys.* 298: 446–451.
49. Beckman, J.S., Ye, Y.Z., Anderson, P.G., Chen, J., Accavitti, M.A., Tarpey, M.M. and White, C.R. (1994). Extensive nitration of protein tyrosines in human atherosclerosis detected by immunohistochemistry. *Biol. Chem. Hoppe-Seyler* 375: 81–88.
50. Carreras, M.C., Pargament, G.A., Catz, S.D., Poderoso, J.J. and Boveris, A. (1994). Kinetics of nitric oxide and hydrogen peroxide production and formation of peroxynitrite during the respiratory burst of human neutrophils. *FEBS Lett* 341: 65–68.
51. Kooy, N.W. and Royall, J.A. (1994) Agonist-induced peroxynitrite production from endothelial cells. *Arch. Biochem. Biophys.* 310: 352–359.
52. Beckman, J.S. The double-edged role of nitric oxide in brain function and superoxide-mediated injury. *J. Dev. Physiol.* 15: 53–59.
53. Hogg, N., Darley-Usmar, V.M., Graham, A. and Moncada, S. (1993) Peroxynitrite and atherosclerosis. *Biochemical Society Transactions* 21: 358–362.
54. White, C.R., Brock, T.A., Chang, L.-Y., Crapo, J., Brisco, P., Ku, D., Bradley, W.A., Gianturco, S.H., Gore, J., Freeman, B.A. and Tarpey, M. (1994). Superoxide, and peroxynitrite in atherosclerosis. *Proc. Natl. Acad. Sci. USA* 91: 1044–1048.

55. Radi, R., Cosgrove, T.P., Beckman, J.S. and Freeman, B.A. (1993) Peroxynitrite-induced liminol chemiluminescence. *Biochem. J.* 290: 51–57.
56. De Groot, H., Hegi, U. and Sies, H. (1993) Loss of α-tocopherol upon exposure to nitric oxide or the sydnonimine SIN-1. *FEBS Lett.* 315: 139–142.
57. Kooy, N.W., Royall, J.A., Ischiropoulos, H. and Beckman, J.S. (1993) Peroxynitrite-mediated oxidation of dihydrorhodamine 123. *Free Rad. Biol. Med.* 16: 149–156.
58. Jessup, W. and Dean, R.T. (1993) Autoinhibition of murine macrophage-mediated oxidation of low-density lipoprotein by nitric oxide. *Atherosclerosis* 101: 145–155.
59. Hogg, N., Kalyanaraman, B., Joseph, J., Struck, A. and Parthesarathy, S. (1993) Inhibition of low-density lipoprotein oxidation by nitric oxide. Potential role in atherogenesis. *FEBS Lett.* 334: 170–174.
60. Padmaja, S. and Huie, R.E. (1993). The reaction of nitric oxide with organic peroxyl radicals. *Biochem. Biophys. Res. Commun.* 195: 539–544.
61. Wink, D.A., Hanbauer, I., Krishna, M.C., DeGraff, W., Gamson, J. and Mitchell, J.B. (1993). Nitric oxide protects against cellular damage and cytotoxicty from reactive oxygen species. *Proc. Natl. Acad. Sci. USA* 90: 9813–9817.
62. Rubanyi, G.M., Ho, E.H., Cantor, E.H., Lumma, W.C. and Parker Botelho, L.H. (1991) Cytoprotective function of nitric-oxide: Inactivation of superoxide radicals produced by human leukocytes. *Biochem. Biophys. Res. Commun.* 181: 1392–1397.
63. Ioannidis, I. and de Groot, H. (1993) Cytotoxicity of nitric oxide in Fu5 hepatoma cells: Evidence for co-operative action with hydrogen peroxide. *Biochem. J.* 296: 341–345.
64. Kanner, J., Harel, S. and Granit, R. (1991) Nitric oxide as an antioxidant. *Arch. Biochem. Biophys.* 289: 130–136.
65. Komarov, A., Mattson, D., Jones, M.M., Singh, P.K. and Lai, C.-S. (1993) *In vivo* spin trapping of nitric oxide in mice. *Biochem. Biophys. Res. Commun.* 195: 1191–1198.
66. Kanner, J., Harel, S. and Granit, R. (1992) Nitric oxide, an inhibitor of lipid oxidation by lipoxygenase, cyclooxygenase and hemoglobin. *Lipids* 27: 46–49.
67. Floris, R., Piersma, S.R., Yang, G., Jones, P. and Wever, R. (1993) Interaction of myeloproxidease with peroxynitrite. A comparison with lactoperoxidase, horseraddish peroxidase and catalase. *Eur. J. Biochem.* 215: 767–775.
68. Salvemini, D., Misko, T.P., Masferrer, J.L., Seibert, K., Currie, M.G. and Needleman, P. (1993). Nitric oxide activates cyclooxygenase enzymes. *Proc. Natl. Acad. Sci. USA* 90: 7240–7244.
69. Fukahori, M., Ichimori, K., Ishida, H., Nakagawa, H. and Okino, H. (1994) Nitric oxide reversibly suppresses xanthine oxidase activity. *Free Rad. Res.* 21: 203–212.

Analysis of Free Radicals in Biological Systems
Favier et al. (eds)

Oxidative damage to DNA

J. Cadet, M. Berger, B. Morin, S. Raoul and J.R. Wagner[1]

CEA/Département de Recherche Fondamentale sur la Matière Condensée – SESAM/LAN, F-38054 Grenoble Cédex 9, France
[1]*Département de Médecine Nucléaire et de Radiobiologie, Université de Sherbrooke, Sherbrooke, Québec, T6G 1Z2, Canada*

Summary. The main oxidation reactions of the four major purine (adenine, guanine) and pyrimidine (cytosine, thymine) bases of DNA and related nucleosides are critically reviewed. These include the reactions mediated by hydroxyl radical, singlet oxygen, hydrogen peroxide, together with one-electron processes involving the transient formation of radical cations. In addition, the main available assays for monitoring the formation of oxidized bases within cellular DNA are presented (mostly chromatographic methods associated with various detection techniques).

Introduction

Oxidative reactions of DNA, a critical cellular target, are ubiquitous and are involved in mutagenesis, carcinogenesis, aging and cellular lethality [1]. These deleterious processes may arise under various conditions of oxidative stress associated in particular with intracellular endogenous oxidants [for recent reviews, see 2–4] (i.e., the leakage of reactive oxygen species from mitochondria and endoplasmic reticulum). Oxidative processes have been shown to be induced by several environmental carcinogens [5] and diet. Several oxidized DNA bases and nucleosides and, in particular, 8-oxo-7,8-dihydro-2′-deoxyguanosine (8-oxodGuo) have been used as bioindicators of oxidative stress within DNA of isolated cells and tissues. In addition, the measurement of the above compounds in biological fluids such as urine may be used for assessing oxidative damage to DNA [6]. Other important sources of oxidation processes are provided by physical agents such as ionizing radiation and near-ultraviolet/visible light [7, 8]. In the latter case, oxidation of DNA would require the presence of endogenous or exogenous photosensitizers including flavins and porphyrins.

Base lesions, abasic sites, DNA strand breaks and DNA-protein crosslinks represent the four main classes of oxidative DNA damage [9]. Emphasis has been placed in this brief survey on the hydroxyl radical and one-electron oxidation mediated decomposition products of the base moieties of both DNA and model compounds. In addition, the main oxidation reactions of highly specific singlet oxygen with the guanine base of DNA are described. Finally, a survey of the main approaches (HPLC separations associated with various spectroscopic detections, gas chromatography-mass spectrometry, postlabeling techniques, immunoassays, etc.) involving either initial acid hydrolysis or enzymatic digestion of DNA which were recently developed for monitoring the formation of oxidative DNA base damage in cells, tissues and biological fluids are also presented.

Oxidative reactions of the purine and pyrimidine base moieties

The main reactive oxygen species involved in oxidation of DNA are listed in Table 1. Their reactivities with DNA components and the chemical reactions of base radical cations that are produced by one-electron reactions are also indicated. It appears that $OH^{•}$ radical and related reactive species including ferryl ions, both of which may be generated through the Haber-Weiss reaction inside the cell, are the most reactive agents. The complex nature of the oxidation reactions of the base moieties of DNA is evident for guanine, a nucleobase which presents the lowest ionization potential among the DNA components. On the other hand, the initial events of the radical oxidation of the thymine base are much simpler. However, the final product distribution consists of more than 20 compounds due to the occurrence of rearrangement processes and the presence of diastereoisomers.

Oxidative reactions of the DNA guanine base

Radical processes involving OH radical and one-electron oxidation
The two overwhelming oxidation products of the purine moiety of 2'-deoxyguanosine (1; Fig. 1) resulting from either the reaction with $OH^{•}$ radical or the transformation of the guanine radical cation (one-

Table 1. Reactive species and radicals involved in oxidative stress

Species-radicals	Reactivity with DNA
Ferryl ion	Oxidizes bases and sugar moieties
Hydrogen peroxide (H_2O_2)	Oxidizes adenine
Hydroperoxide radical (HO_2)	Not detectable reactivity
Hydroxyl radical ($OH^{•}$)	Oxidizes bases and sugar moieties
Peroxinitrite ($ONOO^{-}$)	Oxidizes bases and sugar moieties
Oxyl ($RO^{•}$) and peroxyl ($RO^{•}$) radicals	Oxidized sugar moieties
Ozone (O_3)	Oxidizes pyrimidine and purine bases
Purine and pyrimidine radical cations	Hydration and deprotonation
Singlet oxygen ($^1O_2$)	Oxidizes guanine
Superoxide radical (O_2^{-})	Not detectable (reduction of $ROO^{•}$)

electron oxidation) have been isolated and identified as 2,2-diamino-4-[(2-deoxy-β-D-*erythro*-pentofuranosyl)amino]-5(2*H*)-oxazolone(6; Fig. 1) and its precursor 2-amino-5-[(2-deoxy-β-D-*erythro*-pentofurano-syl)amino]-4*H*-imidazol-4-one (5; Fig. 1) [9, 10]. The mechanism of their production as shown in Figure 1 may be interpreted in terms of a transient formation of the oxidizing guanilyl radical (3; Fig. 1) which may arise either from dehydration of the OH• adduct at C-4 (4; Fig. 1) or deprotonation of the guanine radical cation (2; Fig. 1). The neutral radical (3; Fig. 1) which may exist in several tautomeric forms is implicated in a rather complicated decomposition pathway. This in-volves the opening of the pyrimidine ring at the C5–C6 bond, followed by the transient formation of a peroxyl radical arising from the addition of molecular oxygen to tautomeric C(5) carbon centred radical and subsequent nucleophilic addition of a water molecule across the 7,8-ethylenic bond. Following rearrangement, this leads to the formation of the unstable imidazolone (5; Fig. 1) (half-life = 10 h in aqueous solution at 20°C) which is then quantitatively converted into the oxazolone (6; Fig. 1) [9, 10].

It should be noted that the nucleophilic addition of a water molecule in the sequence of events giving rise to (6; Fig. 1) represents an interesting model system for investigating the mechanism of the genera-tion of DNA-protein cross-links under radical-mediated oxidative con-ditions [11]. The formation of 8-oxo-7,8-dihydro-2'-deoxyguanosine (8-oxodGuo) (8; Fig. 1) is a minor process when 2'-deoxyguanosine (1; Fig. 1) is exposed to OH radicals in aqueous aerated solution. However, the yield of formation of 8-oxodGuo (8; Fig. 1) increases at the expense of the oxazolone derivative (6; Fig. 1) in double-stranded DNA [12]. This is even more striking, given the transformation reactions of the guanine radical cation (2; Fig. 1) within native DNA. Under the latter conditions a significant hydration reaction which was not observed within the free nucleoside (1; Fig. 1) and short oligonucleotides was found to give rise to the formation of 8-oxodGuo (8; Fig. 1) through the transient 8-hydroxy-7,8-dihydro-2'-deoxyguanosyl radical (7; Fig. 1) [12]. This illustrates the complexity of the radical oxidation reactions of the guanine moiety of DNA and also shows the similarity in the decomposition pathways mediated by hydroxyl radical and one-electron oxidation. It should be noted that the formamidopyrimidine derivative (9; Fig. 1) is not formed in the presence of oxygen, at least at the nucleoside level.

Oxidation of 2'-deoxyguanosine by Fenton reagents
The use of Fenton reagents to oxidize (1; Fig. 1) leads to a complete change in the quantitative product distribution since the formation of 8-oxo-7,8-dihydro-2'-deoxyguanosine (8; Fig. 1) was considerably en-

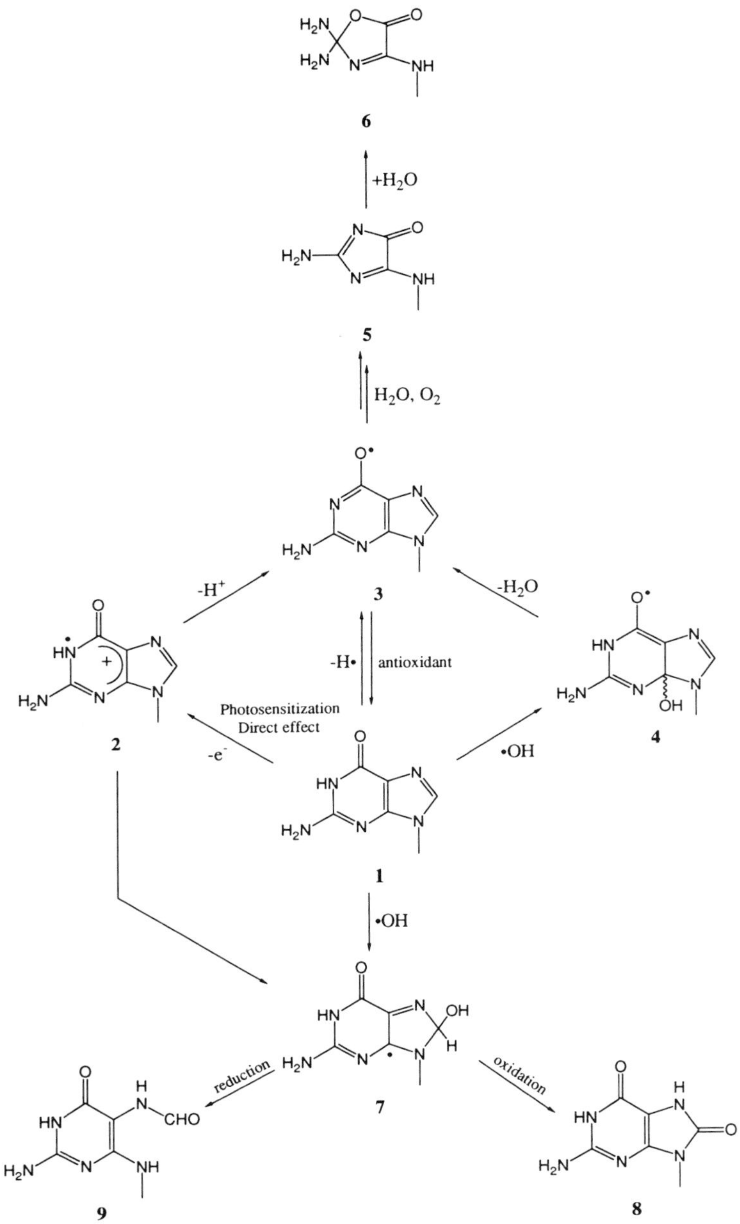

Fig. 1. Main radical oxidation pathways of the guanine moiety within nucleosides and DNA.

hanced, at the expense of (5; Fig. 1) and (6; Fig. 1) which were almost abolished. This may be explained in terms of a competition between the reduction of the oxidizing guanilyl radical (3; Fig. 1) and its reaction with molecular oxygen [13, 14].

Singlet oxygen oxidation of the guanine moiety
The two main products of $^1O_2$-mediated oxidation of 2′-deoxyguanosine (1; Fig. 1) were identified as the 4R* and 4S* diastereoisomers of 4-hydroxy-8-oxo-4,8-dihydro-2′-deoxyguanosine. A likely mechanism for the formation of the latter specific oxidation products involves a [4 + 2] cycloaddition of $^1O_2$ to the guanine moiety according to the Diels-Alder mechanism producing unstable 4,8-endoperoxides [15, 16]. It should be added that 8-oxodGuo (8; Fig. 1) is also produced by the reaction of $^1O_2$ with 2′-deoxyguanosine (1; Fig. 1) but only in a ratio of 1 to 7 with respect to 4-hydroxy-8-oxo-4,8-dihydro-2′-deoxyguanosine. However, in double-stranded DNA, the formation of 8-oxodGuo (8; Fig. 1) becomes predominant at the expense of 4-hydroxy-8-oxo-4,8-dihydro-2′-deoxyguanosine [17].

Oxidation reactions of the adenine base

Radical processes involving OH˙ radical and one-electron oxidation
The major oxidation product of the base moiety of 2′-deoxyadenosine (dAdo) with OH˙ radical in aerated aqueous solution has been identified as 8-oxo-7,8-dihydro-2′-deoxyadenosine (8-oxodAdo) [13]. A reasonable mechanism for the formation of 8-oxodAdo involves oxidation of the initially generated 8-hydroxy-7,8-dihydro-2′-deoxyadenosyl radical. It should be noted that under these conditions the formation of the formamidopyrimidine derivative which requires the reduction of the latter purine radical [18] is formed in a low yield (Raoul and Cadet, unpublished data). In addition, 2′-deoxyinosine (dIno) which derives from dehydration of the OH˙ radical adduct at C-4 of the adenine moiety and subsequent conversion of the 6-aminyl radical was not observed. On the other hand, dIno is the predominant decomposition product of the one-electron oxidation reaction of dAdo [8]. This may occur *via* the initial formation of adenyl radical cation followed by fast deprotonation to give rise to the 6-aminyl radical.

Non-radical hydrogen peroxide-mediated oxidation
In the absence of reduced transition metals, hydrogen peroxide is able to react specifically with adenine to generate adenine N^1-oxide [19]. This oxidized base has been detected using a sensitive high performance liquid chromatographic (HPLC)-^{32}P-postlabeling assay in bacterial DNA upon exposure to hydrogen peroxide [20].

Oxidation reactions of the thymine base

Hydroxyl radical oxidation
The main hydroxyl radical-mediated oxidation product of thymidine (10; Fig. 2), a DNA model compound, in aerated aqueous solution has been isolated and characterized on the basis of extensive NMR and mass spectrometric measurements. About 50% of the overall decomposition products are represented by nine hydroperoxides that have been assigned as the *cis* and *trans* diastereoisomers of 6-hydroperoxy-5-hydroxy-5,6-dihydrothymidine (16; Fig. 2) and 5-hydroperoxy-6-hydroxy-5,6-dihydrothymidine (15; Fig. 2) [21] together with 5-hydroperoxymethyl-2'-deoxyuridine (17; Fig. 2). Interestingly, the structure of the thymidine hydroperoxides was recently confirmed by chemical synthesis [22]. It should be stressed that the mixture of the hydroperoxides may be completely resolved by reverse phase high performance liquid chromatography (RP-HPLC) and each of the peroxides is individually detected using a sensitive post-column reaction method [23]. The bulk of the stable decomposition products has been also identified. This includes the following nucleosides that are ranged in their decreasing order of quantitative importance: N-(2-deoxy-β-D-*erythro*-pentofuranosyl) formamide (20; Fig. 2) > the four *cis* and *trans* diastereoisomers of 5,6-dihydroxy-5,6-dihydrothymidine (19; Fig. 2) > the 5R* and 5S* forms of 1-(2-deoxy-β-D-*erythro*-pentofuranosyl)-5-hydroxy-5-methylhydantoin (21; Fig. 2) > the 5R* and 5S* diastereoisomers of 1-(2-deoxy-β-D-*erythro*-pentofuranosyl)-5-hydroxy-5-methylbarbituric acid (18; Fig. 2) > 5-hydroxymethyl-2'-deoxyuridine (22; Fig. 2) > 5-formyl-2'-deoxyuridine (23; Fig. 2) [24–27]. Interesting information on the structure and the redox properties of the transient pyrimidine radicals has been otained using pulse radiolysis methods combined with the redox titration technique [28, 29]. Based on the product analysis (*vide supra*), we propose a likely mechanism for the OH⁻-mediated oxidation of the pyrimidine moiety of thymidine (10; Fig. 2). The predominant reaction (60%) within the thymine residue is the OH⁻ addition at carbon C-5 [7, 28, 30] giving rise to the C-6 centered reducing radical (13; Fig. 2). The formation of the isomeric C-5 oxidizing radical (12; Fig. 2) which arises from the addition of the OH⁻ radical at C-6 occurs with a lower yield (35%). The third reaction, which is a minor process (5%), involves the abstraction of a hydrogen atom from the methyl group, generating the aromatic radical (14; Fig. 2). Then, the pyrimidine radicals (12–14; Fig. 2) are converted into the corresponding peroxyl intermediates through a fast reaction with oxygen which is controlled by diffusion [31]. It is likely that about half of the latter peroxyl radicals are transformed into the hydroperoxides (15–17; Fig. 2), subsequently to a reduction step involving superoxide radicals [32]. The hydrolytic decomposition of the hydroperoxides

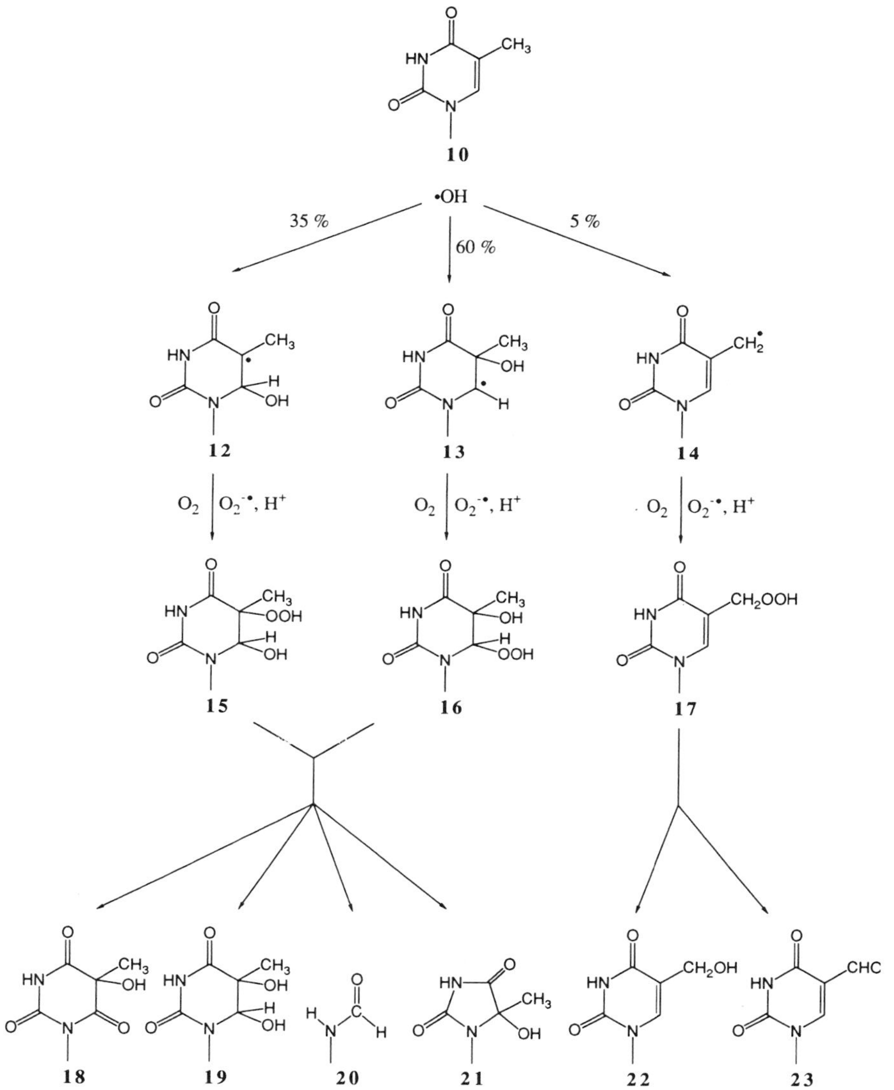

Fig. 2. Hydroxyl radical-mediated oxidation of the base moiety of thymidine (10) in aerated aqueous solution.

whose lifetimes vary from several days to 1 week at 37°C is quite specific. It was shown that the *trans* and *cis* diastereoisomers of the 6-hydroperoxide (16; Fig. 2) are predominantly degraded into (18; Fig. 2). On the other hand, the major decomposition products of (15; Fig. 2) have been identified as the 5R* and 5S* diastereoisomers of (21; Fig. 2)

[22]. The rest of the hydroperoxyl pyrimidine radicals (~ 50%) may undergo a competitive dismutation reaction giving rise to highly reactive oxyl radicals [7]. In particular, the oxyl radicals originated from (15; Fig. 2) and (16; Fig. 2) are expected to be involved in a hydrogen abstraction reaction leading to the formation of the diol (19; Fig. 2).

Another reaction to be considered for an oxyl radical is the β scission process. This may lead to the cleavage of the 5,6-pyrimidine bond with subsequent loss of a pyruvyl group and ring contraction to provide rearrangement products such as 5R* and 5S* diastereoisomers of 1-(2-deoxy-β-D-*erythro*-pentofuranosyl)-5-hydroxy-5-methylhydantoin (21; Fig. 2). This decomposition pathway applies to the OH radical oxidation to the thymine moiety in double-stranded DNA. This is inferred from several studies aimed at isolating and characterizing oxidized bases or nucleosides after suitable hydrolysis of oxidized DNA. 5,6-Dihydroxy-5,6-dihydrothymine and 5-hydroxy-5-methylhydantoin have been found to be generated within [$^{14}CH_3$-thymine] DNA upon exposure to OH· radicals [33]. A HPLC-chemical postlabeling assay has been used for assessing the radiation-induced induction of 5-hydroxymethyl-2'-deoxyuridine (13; Fig. 2) and the four *cis* and *trans* diastereoisomers of (19; Fig. 2) in aerated aqueous solution [34, 35]. In addition, the hydroxyl radical-mediated formation of 5-formyl-2'-deoxyuridine (12; Fig. 2) was also shown to occur in DNA [36]. Other assays involving the combined use of either gas chromatography or HPLC separation methods with mass spectrometric detection have been used for monitoring the formation of OH· radical-induced decomposition products of the thymine moiety of DNA after appropriate hydrolysis [37, 38].

One-electron oxidation reaction
Photoexcited 2-methyl-1,4-naphthoquinone (menadione) has been used to efficiently generate the pyrimidine radical cation of thymidine (10; Fig. 2) through one-electron oxidation. Two main reactions were inferred from the isolation and characterization of the main oxidized nucleosides. In this respect the unique formation of the four *cis* and *trans* diastereoisomers of 5-hydroperoxy-6-hydroxy-5,6-dihydrothymidine (15; Fig. 2) may be rationalized in terms of the specific formation of the oxidizing pyrimidine radical (12; Fig. 2) through hydration of the thymidine radical cation [32, 39]. The competitive deprotonation reaction of the latter intermediate is likely to explain the formation of 5-hydroperoxymethyl-2'-deoxyuridine (17; Fig. 2) which represents about 40% of the overall production of thymidine hydroperoxides.

Oxidation reactions of the cytosine base

Hydroxyl radicals are able to add to the 5,6-ethylenic bond of cytosine with a preference for the C-5 carbon [7] in a manner analogous to that observed for thymidine. The formation of the four *cis* and *trans* diastereoisomers of 5,6-dihydroxy-5,6-dihydro-2'-deoxyuridine is likely to result from the fast addition of molecular oxygen to the hydroxyl radical adduct at either the C-5 or the C-6 position [40]. The resulting peroxyl radicals are expected to behave like the analogous thymine hydroxyperoxyl radicals by undergoing a dismutation reaction and/or being involved in reduction processes. However, it should be noted that the corresponding cytosine hydroxyhydroperoxides have not yet been isolated, probably because they are too unstable. In additon, specific oxidation products of cytosine have been isolated and characterized. These include 5-hydroxy-2'-deoxycytidine, the two *trans* diastereoisomers of N-(2-deoxy-β-D-*erythro*-pentofuranosyl)-1-carbamoyl-4,5-dihydroxy-imidazolidin-2-one, N^1-(2-deoxy-β-D-*erythro*-pentofuranosyl)-N^4-ureidocarboxylic acid and the α and β anomers of N-(2-deoxy-D-*erythro*-pentosyl)biuret [40]. A reasonable mechanism for the formation of the latter three classes of modified nucleosides involves an intramolecular cyclization of 6-hydroperoxy-5-hydroxy-5,6-dihydro-2'-deoxycytidines as a common initial pathway [41]. Such transpositions of the pyrimidine ring have been demonstrated by isotopic labeling experiments involving the incorporation of a ^{18}O atom in the carbamoyl group of the diastereoisomers of the 4,5-dihydroxy-imidazolidin-2-one derivatives. The same modified nucleosides but with a different relative distribution were shown to be generated through one-electron oxidation of the cytosine moiety [40]. In addition, the deprotonation of the transient pyrimidine radical cation gives rise to the formation of 2'-deoxyuridine and the release of cytosine as the result of oxidation of the sugar residue at C-1 [42]. It should be added that the formation of 5-hydroxy-2'-deoxycytidine and 5-hydroxy-2-deoxyuridine has been monitored by HPLC-electrochemical detection within DNA exposed to oxidative stress [43].

Measurement of oxidative DNA base damage

Assays from monitoring oxidized bases within DNA

The measurement of oxidative base damage in tissue and cellular DNA remains a challenging analytical problem [44]. This may be due, in part, to the high level of sensitivity that is required (the threshold of detection is close to one single lesion per 10^5-10^6 normal bases in a sample size of DNA lower than 30 μg) together with the multiplicity and the lability of the lesion. In addition, another limiting factor deals with the occurrence of autoxidation reactions during the work-up of DNA, which may induce a significant level of oxidized base damage. This is particularly the case for 8-oxodGuo (8; Fig. 1), whose formation is significantly

Table 2. Methodologies for measuring oxidative DNA damage (adapted from [44])

Methods	DNA	Sensitivity[a]	Amount of DNA (μg)
HPLC/electrochemistry	Hydrolyzed	1×10^{-5}	25–50
HPLC/MS (thermospray)	Hydrolyzed	$10^{-4} - 10^{-5}$	30–40
HPLC-fluorescence	Hydrolyzed	5×10^{-5}	4–8
CG/SIM-MS	Hydrolyzed	1×10^{-5}	50–100
HPLC-^{32}P-postlabeling	Hydrolyzed	1×10^{-6}	1–5
Immunology (RIA-ELISA)	Intact	$10^{-5} - 10^{-6}$	2–10
DNA-glycosylases (alkaline elution)	Intact	5×10^{-7}	10

[a]Sensitivity is indicated with respect to normal bases

enhanced under Fenton reaction conditions (*vide supra*). In addition to the measurement of damage within isolated cells, two main approaches are currently developed for targeting specific oxidative DNA damage. The available assays for monitoring oxidative base damage within the whole DNA or after hydrolysis of the biopolymer have been critically reviewed in two recent surveys [38, 44]. In most cases, the methodology requires either enzymatic digestion or chemical hydrolysis of DNA (Tab. 2).

The most sensitive assays currently available involve radioactive postlabelling of either nucleoside 3'-monophosphates and dinucleoside monophosphates which are substrates for polynucleotide kinases. The sensitivity of the method is about one adenine N^1-oxide per 10^7 normal bases in a sample size of 1 μg of DNA [20, 45]. Similar assays have been developed for the detection of 5-hydroxymethyluracil, 8-oxo-7,8-dihydroguanine, 5,6-dihydroxy-5,6-dihydrothymine, phosphoglycolate residues and abasic sites (for a recent review, see [45]). It should be added that chemical postderivatization methods involving radioactive and fluorescence detection are also available [45]. However, the most widely used assay deals with the measurement of 8-oxodGuo (8; Fig. 1) by high performance liquid chromatography-electrochemical detection. Another interesting assay involved the combined use of gas chromatography (GC) with mass spectrometry [37, 46]. However, this method, despite its high intrinsic sensitivity, requires about 200 μg of DNA [44]. It should be also mentioned that derivatization process has been shown to induce significant levels of 8-oxodGuo (8; Fig. 1) when dGuo is present in the reaction mixture [47]. This may explain the relatively high value of 8-oxodGuo when measured by GC-MS assay [48] by comparison with the more accurate HPLC-electrochemical detection method [49]. It is a requisite that the damage to be measured has to be prepurified from the related normal constituent prior to the GC-MS analysis.

Non-invasive assays

Evaluation of the effects of oxidative stress on DNA in humans requires the development of non-invasive assays. Attempts are being currently made to use the release of oxidized bases and nucleosides in urine as an index of DNA damage. For this purpose, HPLC-EC and GC-MS assays have been applied to the measurement of several oxidized DNA compounds including 8-oxo-7,8-dihydroguanine, 5-hydroxymethyluracil and their corresponding nucleosides [50–52]. A significant increase in the release of 8-oxodGuo (8; Fig. 1) was observed in human urine and in the leukocytes of cigarette smokers (for a review, see [53]). These few examples illustrate the potential use of oxidized bases and related nucleosides as indicators in epidemiological studies designed to correlate dietary factors and lifestyles with cancer risk.

Conclusion

This review clearly shows the complexity of the oxidation reactions with DNA. There is still an important need for development of accurate and sensitive methods of detection of oxidative base damage in cellular and tissue DNA. Non-invasive methods are particularly relevant for epidemiological studies. Further developments would involve both molecular biology techniques (polymerase chain reaction) and modern analytical methods such as the capillary electrophoresis associated with the sensitive and versatile electrospray mass spectrometry detection technique. Efforts are currently underway to determine the biological role of oxidized base damage. One major aspect deals with repair studies which have already shown the major role played by glycosylase proteins in the enzymatic removal of damaged bases (for recent papers and reviews, see [54–56]). The mutagenic assessment of oxidized DNA bases is based on site-specific incorporation of a single modification in sequence defined oligonucleotides [57]. Then, the resulting oligonucleotides can either be used as templates for DNA replication or elongated further prior to incorporation in eukaryotic cells for mutagenic evaluation (for a recent review, see [58]).

Acknowledgement
This work was partly supported by a grant from Centre Jacques Cartier, Lyon, France (JC-JRW).

References

1. Marnett, L.J. and Burcham, P.C. (1993) Endogenous DNA adducts: Potential and paradox. *Chem. Res. Toxicol.* 6: 771–785.

2. Ames, B.N. and Gold, L.S. (1991) Endogenous mutagens and the cause of aging and cancer. *Mutat. Res.* 250: 3–16.

3. Sies, H. (1991) *Oxidative stress, oxidants and antioxidants.* Academic Press, Inc. New York.

4. Lindahl, T. (1993) Instability and decay of the primary structure of DNA. *Nature* 362: 709–715.

5. Srinivasan, S. and Glauert, H.P. (1990) Formation of 5-hydroxymethyl-2'-deoxyuridine in hepatic DNA of rats treated with gamma-irradiation, diethylnitrosamine, 2-acetylaminofluorene or the peroxisome proliferator ciprofibrate. *Carcinogenesis* 11: 2012–2024.

6. Shinenaga, M.K., Gimeno, C.L. and Ames, B.N. (1989) Urinary 8-hydroxy-2'-deoxyguanosine as a biological marker of *in vivo* oxidative DNA damage. *Proc. Natl. Acad. Sci. USA* 86: 9697–9701.

7. von Sonntag, C. (1987) The chemical basis of radiation biology. Taylor Francis, London.

8. Cadet, J. and Vigny, P. (1990) The photochemistry of nucleic acids. *In*: H. Morrison (ed.): *Bioorganic Photochemistry*, Vol. 1, Wiley and Sons, New York, pp 1–272.

9. Cadet, J. (1994) DNA damage caused by oxidation, deamination, ultraviolet radiation and photoexcited psoralens. *In*: K. Hemminki, A. Dipple, D.G.E. Shuker, F.F. Kadlubar, D. Segerbäck and H. Bartsch (eds: *DNA adducts: Identification and biological significance.* Lyon International Agency for Research on Cancer, IARC Scientific Publications, No. 125, pp 245–276.

10. Cadet, J., Berger, M., Buchko, G.W., Joshi, P.C., Raoul, S. and Ravanat, J.-L. (1994a) 2,2-Diamino-4-[(3,5-di-O-acetyl-2-deoxy-β-D-*erythro*-pentofuranosyl)amino]-5-(2*H*)-oxazolone: A novel and predominant radical oxidation product of 3',5'-di-O-acetyl-2'-deoxyguanosine. *J. Am. Chem. Soc.* 116: 7403–7404.

11. Morin, B. and Cadet, J. (1994) Benzophenone photosensitisation of 2'-deoxyguanosine: Characterization of the 2R and 2S diastereoisomers of 1-(2-deoxy-β-D-*erythro*-pentofuranosyl)-2-methoxy-4,5-imidazolidinedione. A model system for the investigation of photosensitized formation of DNA-protein crosslinks. *Photochem. Photobiol.* 60: 102–109.

12. Kasai, H., Yamaizumi, Z., Berger, M. and Cadet, J. (1992) Photosensitized formation of 7,8-dihydro-8-oxo-2'-deoxyguanosine (8-hydroxy-2'-deoxyguanosine) in DNA by riboflavin: a non singlet oxygen mediated reaction. *J. Am. Chem. Soc.* 114: 9692–9694.

13. Berger, M., de Hazen, M., Nejjari, A., Fournier, J., Guignard, J., Pezerat, H. and Cadet, J. (1993) Radical oxidation reactions of the purine moiety of 2'-deoxyribonucleosides and DNA by iron-containing minerals. *Carcinogenesis* 14: 41–46.

14. Cadet, J., Berger, M., Buchko, G.W., Incardona, M.-F., Morin, B., Raoul, S. and Ravanat, J.-L. (1994b) DNA oxidation: Characterization of the damage and mechanistic aspects. *In*: R. Paoletti (ed.): *Oxidative processes and antioxidants.* Raven Press, New York, pp 97–115.

15. Buchko, G.W., Cadet, J., Berger, M. and Ravanat, J.-L. (1992) Photooxidation of d(TpG) by phthalocyanines and riboflavin. Isolation and characterization of dinucleoside monophosphates containing the 4R* and the 4S* diastereoisomers of 4,8-dihydro-4-hydroxy-8-oxo-2'-deoxyguanosine. *Nucleic Acids Res.* 20: 4847–4851.

16. Cadet, J., Ravanat, J.-L., Buchko, G.W., Yeo, H.C. and Ames, B.N. (1994c) Singlet oxygen DNA damage: Chromatographic and mass spectrometric analysis of damage products. *Methods Enzym.* 234: 79–88.

17. Ravanat, J.-L. and Cadet, J. (1995) Reaction of singlet oxygen with 2'-deoxyguanosine and DNA. Identification and characterization of the main oxidation products. *Chem. Res. Toxicol.* 8: 379–388.

18. Steenken, S. (1989) Purine bases, nucleosides, and nucleotides: Aqueous solution redox chemistry and transformation reactions of their radical cations and e⁻ and OH adducts. *Chem. Rev.* 89: 503–520.

19. Mouret, J.-F., Polverelli, M., Sarrazin, F. and Cadet, J. (1991) Ionic and radical oxidations of DNA by hydrogen peroxide. *Chem. Biol. Interactions* 77: 187–201.

20. Mouret, J. F., Odin, F., Polverelli, M. and Cadet, J. (1990) ³²P-postlabelling measurement of adenine N-1-oxide in cellular DNA exposed to hydrogen peroxide. *Chem. Res. Toxicol.* 3: 102–110.

21. Cadet, J. and Téoule, R. (1975) Radiolyse gamma de la thymidine en solution aqueuse aérée. 1.-Identification des hydroperoxydes. *Bull. Soc. Chim. Fr.* 879–884.

22. Wagner, J.R., van Lier, J.E., Berger, M. and Cadet, J. (1994) Thymidine hydroperoxides. Structural assignment, conformational features, and thermal decomposition in water. *J. Am. Chem. Soc.* 116: 2235–2242.

23. Wagner, J.R., Berger, M., Cadet, J. and van Lier, J.E. (1990) Analysis of thymidine hydroperoxides by post-column reaction high-performance liquid chromatography. *J. Chromatogr.* 504: 191–196.
24. Cadet, J., Ulrich, J. and Téoule, R. (1975) Isomerization and new specific synthesis of thymine glycol. *Tetrahedron* 31: 2057–2061.
25. Cadet, J., Ducolomb, R. and Hruska, F.E. (1979) Proton magnetic resonance studies of 5,6-saturated thymidine produced by ionizing radiation. *Biochim. Biophys. Acta.* 563: 206–215.
26. Cadet, J., Nardin, R., Voituriez, L., Remin, M. and Hruska, F.E. (1981) A ^{1}H and ^{13}C nmr study of the radiation-induced degradation products of 2'-deoxythymidine derivatives: N-(2-deoxy-β-D-*erythro*-pentofuranosyl) formamide. *Can. J. Chem.* 59: 3313–3318.
27. Lutsig, M.J., Cadet, J., Boorstein, R.J. and Teebor, G.W. (1992) Synthesis of the diastereomers of thymidine glycol, determination of concentrations and rates of interconversion of their *cis-trans* epimers at equilibrium and demonstration of differential alkali lability within DNA. *Nucleic Acids Res.* 20: 4839–4845.
28. Fujita, S. and Steenken, S. (1981) Pattern of OH radical addition to uracil and methyl- and carboxyl-substituted uracils. Electron transfer of OH adducts with N,N,N',N'-tetramethyl-p-phenylenediamine and tetranitromethane. *J. Am. Chem. Soc.* 103: 2540–2545.
29. Jovanovic, S.V. and Simic, M.G. (1986) Mechanism of OH radical reactions with thymine and uracil derivatives. *J. Am. Chem. Soc.* 108: 5968–5972.
30. von Sonntag, C. (1992) Some aspects of radiation-induced free-radical chemistry of biologically important molecules. *Radiat. Phys. Chem.* 39: 477–483.
31. Isildar, M., Schuchmann, M.N., Schulte-Frohlinde, D. and von Sonntag, C. (1982) Oxygen uptake in the radiolysis of aqueous solutions of nucleic acids and their constituents. *Int. J. Radiat. Biol.* 41: 525–533.
32. Wagner, J.R., van Lier, J.E. and Johnston, L.J. (1990) Quinone sensitized electron transfer photoxidation of nucleic acids: chemistry of thymine and thymidine radical cations in aqueous solution. *Photochem. Photobiol.* 52: 333–343.
33. Téoule, R., Bert, C. and Bonicel, A. (1977) Thymine fragment damage retained in the DNA polynucleotide chain after gamma irradiation in aerated solutions. *Radiat. Res.* 72: 190–200.
34. Teebor, G., Cummings, A., Frenkel, K., Shaw, A., Voituriez, L. and Cadet, J. (1987) Quantitative measurement of the diastereoisomers of *cis* thymidine glycol in gamma-irradiated DNA. *Free Rad. Res. Commun.* 2: 303–309.
35. Frenkel, K., Zhong, Z., Wei, H., Karkoszka, J., Patel, U., Rashid, K., Georgescu, M. and Solomon, J.J. (1991) Quantitative high-performance liquid chromatography analysis of DNA oxidized *in vitro* and *in vivo*. *Anal. Biochem.* 196: 126–136.
36. Kasai, H., Iida, A., Yamaizumi, Z., Nishimura, S. and Tanooka, H. (1990) 5-Formyldeoxyuridine: a new type of DNA damage induced by ionizing radiation and its mutagenicity to Salmonella strain TA102. *Mutat. Res.* 243: 249–253.
37. Dizdaroglu, M. (1993) Quantitative determination of oxidative base damage in DNA by stable isotope-dilution mass spectrometry. *FEBS Lett.* 315: 1–6.
38. Frenkel, K. and Klein, C.B. (1993) Methods used for analyses of "environmentally" damaged nucleic acids. *J. Chromatogr.* 618: 289–314.
39. Decarroz, C., Wagner, J.R., van Lier, J.E., Krishna Murali, C., Riesz, P. and Cadet, J. (1986) Sensitized photo-oxidation of thymidine by 2-methyl-1,4-naphthoquinone. Characterization of the stable photoproducts. *Int. J. Radiat. Biol.* 50: 491–505.
40. Wagner, J.R., van Lier, J.E., Decarroz, C., Berger, M. and Cadet, J. (1990) Photodynamic methods for oxy radical-induced DNA damage. *Methods Enzym.* 186: 502–511.
41. Polverelli, M. (1983) *Modifications radio-induites de la cytosine en solution aqueuse aérée et après irradiation gamma du DNA d*'Escherichia coli. PhD Thesis, University of Grenoble.
42. Decarroz, C., Wagner, J.R. and Cadet, J. (1987) Specific deprotonation reactions of the pyrimidine radical cation resulting from the menadione mediated photosensitization of 2'-deoxycytidine. *Free Rad. Res. Commun.* 2: 295–301.
43. Wagner, J.R., Hu, C.-C. and Ames, B.N. (1992) Endogenous damage of deoxycytidine in DNA. *Proc. Natl. Acad. Sci. USA* 89: 3380–3384.
44. Cadet, J. and Weinfeld, M. (1993) Detecting DNA damage. *Anal. Chem.* 65: 675A–682A.

45. Cadet, J., Odin, F., Mouret, J.-F., Polverelli, M., Audic, A., Giacomoni, P., Favier, A. and Richard, M.-J. (1992) Chemical and biochemical postlabeling methods for singling out specific oxidative DNA lesions. *Mutat. Res.* 275: 343–354.
46. Fuciarelli, A.F., Wegher, B.J., Blakely, W.F. and Dizdaroglu, M. (1990) Quantitative measurement of radiation-induced base products in DNA using gas-chromatography-mass spectrometry. *Int. J. Radiat. Biol.* 58: 397–415.
47. Ravanat, J.-L., Stadler, R.H. and Turesky, R.J. (1995) GC-MS and HPLC-EC detection of 8-oxoguanine. 2nd Winter Research Conference on Oxidative DNA damage: Bioindicators and repair. Les-2-Alpes, France, January 24–25, 1995.
48. Halliwell, B. and Dizdaroglu, M. (1992) The measurement of oxidative damage to DNA by HPLC and GC/MS techniques. *Free Rad. Res. Commun.* 16: 75–88.
49. Floyd, R.A., Watson, J.J., Wong, P.K. Altmiller, D.H. and Rickard, R.C. (1986) Hydroxyl free radical adduct of deoxyguanosine: sensitive detection and mechanism of formation. *Free Rad. Res. Commun.* 1: 163–172.
50. Faure, H., Incardona, M.-F., Boujet, C., Cadet, J., Ducros, V. and Favier, A. (1993) Gas chromatographic-mass spectrometric determination of 5-hydroxymethyluracil in human urine by stable isotope dilution. *J. Chromatogr. Biom. Appl.* 616: 1–7.
51. Shigenaga, M.K., Aboujaoude, E.N., Chen, Q. and Ames, B.N. (1994) Assays of oxidative DNA damage biomarkers 8-oxo-2'-deoxyguanosine and 8-oxoguanine in nuclear DNA and biological fluids by high performance liquid chromatography with electrochemical detection. *Methods Enzym.* 234: 16–33.
52. Incardona, M.-F., Bianchini, F., Favier, A. and Cadet, J. (1995) Measurement of oxidized nucleobases and nucleosides in human urine by using a GC/MS assay in the selective ion monitoring mode. This volume, pp.
53. Cadet, J., Berger, M., Girault, I., Incardona, M.-F., Molko, D., Polverelli, M., Raoul, S. and Ravanat, J.-L. (1995) DNA modifications due to oxidative damage. *In*: C. Bardinet, C. Glaeser and J.L. Royer (eds): *CODATA 94 – Environmental Aspects; in press.*
54. Boiteux, S. (1993) Properties and biological functions of the NTH and FPG proteins of *Escherichia coli*: two glycosylases that repair oxidative damage in DNA. *J. Photochem. Photobiol. B. Biol.* 19: 87–96.
55. Hatahet, Z., Kow, Y.W., Purmal, A.A., Cunningham, R.P. and Wallace, S.S. (1994) New substrates for old enzymes. *J. Biol. Chem.* 269: 11814–11820.
56. Tchou, J., Bodepudi, V., Shibutani, S., Antoshechkin, I., Miller, J., Grollman, A.P. and Johnson, F. (1994) Substrate specificity of Fpg protein. *J. Biol. Chem.* 269: 15318–15324.
57. Guy, A., Duplaa, A.M., Ulrich, J. and Téoule, J.R. (1991) Incorporation by chemical synthesis and characterization of deoxyribosylformylamine in DNA. *Nucleic Acids Res.* 19: 5815–5820.
58. Grollman, A.P. and Moriya, M. (1993) Mutagenesis by 8-oxoguanine: an enemy within. *Trends Genet.* 9: 246–249.

Analysis of Free Radicals in Biological Systems
Favier et al. (eds)

Oxidation of low density lipoprotein (LDL) by cells

S. Parthasarathy

*Department of Gynecology and Obstetrics, Emory University School of Medicine,
Box 21246, Atlanta, GA 30322, USA*

Cellular mechanism of oxidation of LDL

Is there a need to study the cellular mechanism(s) of oxidation of LDL? If antioxidants are safe and effective even at relatively high doses in retarding the progression of experimental atherosclerosis, why is it important to study the specific mechanism(s) by which oxidative process may be initiated? True, vitamin E may be safe at quite high doses without evidence of major toxicity [1], however, antioxidants such as BHT, BHA and probucol do have toxicity and side-effects associated with their intake. Also, very little else is known about their normal physiological functions or their potential interactions with other drugs.

Current evidence suggests that if the oxidation of LDL is important in atherosclerosis, it is likely to occur in the subendothelial intima rather than the plasma compartment [2]. In fact, oxidation, when it occurs in the plasma may be actually beneficial. Oxidized LDL is rapidly cleared from circulation by liver [3, 4] and such oxidations may even lower plasma cholesterol. Information on the availability of antioxidants in the normal or atherosclerotic intima is meager.

An understanding of the oxidation process and the molecular mechanism(s) involved would have the following advantages. First, such an understanding might lead to the development of more specific inhibitors that would not interfere with physiological oxidations. Second, we can begin to understand the factors leading to the cellular malfunction of the oxidative pathway and can then devise preventive strategies. And third, site-specific inhibitor therapy can be initiated if abnormalities of metabolic functions are confined, for instance, to just the artery.

The mechanism(s) by which cells may oxidize LDL has been one of the most discussed topics in the study of modified lipoproteins. A number of mechanisms has been proposed for the oxidation of LDL by cells. Evidence presented in Table 1 would suggest that cells do play a role in the oxidation of LDL or other lipoproteins.

There does not appear to be a specific interaction between the cell and the LDL. Cells appear to be capable of oxidizing lipoproteins other

Table 1. Evidence for the involvement of cells in the oxidation of LDL

(1) LDLs incubated with cells are more extensively modified than LDLs incubated in the absence of cells.
(2) Cell-conditioned media alone could not sustain oxidation when subsequently incubated with LDL. Studies by Heinecke et al. [5] showed that the continuous presence of cells is essential in the oxidation of the lipoprotein. If lipoproteins were removed from the cells and the incubation was continued in their absence, the oxidation did not proceed to completion and was lower than in the undisturbed incubations continued for the same duration.
(3) Studies by Parthasarathy et al. also showed that LDL separated from cells by a dialysis membrane was not oxidized [6].

than LDL, such as β-VLDL and HDL [7, 8]. In fact, the oxidation of β-VLDL by cells results in the generation of even greater amounts of TBARS than does the oxidation of comparable amounts of LDL. These findings, at least, do not seem to suggest the involvement of any specific receptor.

A number of cell types (at least, *in vitro*) have been reported to oxidize LDL [2, 9]. However, use of inappropriate cell culture media, ^{125}I-labeled lipoproteins and older lipoprotein preparations might have led to the erroneous conclusion that many different divergent cell types are capable of initiating massive lipid peroxidation in the medium. It is more likely that most cell types contribute very little to the oxidation of LDL. It is more likely that cells generate oxygen radicals or H_2O_2 in the medium in meager quantities and more specialized cells such as monocyte/macrophages are better suited to provide the oxidative function.

The cell type that may be responsible for the initiation of the oxidation of LDL in the artery remains a matter of conjecture. If the endothelial cells are the important cell type involved in the oxidative

Table 2. Role of cells in the oxidation of LDL

(1) Cells may play an active role in the oxidation of LDL. Cells may generate reactive oxygen species or other oxidants that initiate the oxidation of LDL in the presence of metal ions or other reactions that may have peroxidase activity. These may include superoxide radicals, hydrogen peroxide, lipid peroxides and nitric oxide. The latter may combine with superoxide radicals to generate the potent oxidant, peroxynitrite.
(2) Cells may also play a passive role. Cells do not themselves participate in the oxidative process but instead provide substrates that generate extra-cellular oxidants. This may result from the uptake of oxidized thiols such as cystine followed by its release as cysteine or glutathione which in turn may generate a variety of oxidants in the presence of redox metals. The specific radicals that are generated in these reactions that may participate in the oxidation of LDL have not been characterized.
(3) Cells, in addition to providing the pro-oxidant environment, also participate in the propagation of lipid peroxidation in the medium. Cells may contribute to this process by peroxidase reactions. Extracellular heme and secreted cellular peroxidases may participate in these reactions. The question of whether in early atherosclerotic lesions damaged red blood cells or heme abounds has not been satisfactorily addressed. For such reactions to occur, the generation of H_2O_2 or LOOH in the medium is essential.
(4) Cells possibly deplete the antioxidants from the lipoprotein, thus increasing the spontaneous rate of lipoprotein oxidation.

process, then the intactness of the endothelium, the predisposition of certain areas of the artery to the development of the lesion, the role of hemodynamic factors, the effects of plasma components and a host of other factors should be taken into consideration. If monocytes/ macrophages are the predominant cell types involved in the oxidation, then the specific nature of these cells, the oxidative enzymes that are present and that can be induced in these cells, and above all, the constituents of the sub-endothelial milieu should be considered. If these cells play a dominant role in the oxidation of LDL, then one has to also consider the redox factors that lead to their recruitment in the artery wall. Table 2 describes the ways by which cells may contribute to the oxidation of LDL.

Active participation of cells

Superoxide-dependent mechanisms

Superoxide radical was one of the first radicals to be suspected in the oxidation of LDL because of the inhibition of oxidation by superoxide dismutase (SOD). Endothelial cells, smooth muscle cells, and mono- cytes have been suggested to oxidize LDL by a superoxide-dependent mechanism. Heinecke et al. [5] using smooth muscle cells in DMEM medium showed that these cells were incapable of oxidizing LDL unless micromolar quantities of iron or copper were added to the medium. The results of this study and those of Steinbrecher et al. [10] also showed substantial modification of LDL in the absence of cells when 10 μM iron or copper was present in the medium. In a subsequent study, Heinecke et al. [11] observed no oxidation of LDL without micromolar additions of iron or copper. The addition of superoxide dismutase inhibited the oxidation; the inhibition was attributed to the catalytic activity of the protein.

The same study showed the time-dependent generation of SOD- inhibitable release of superoxide radicals from cells in the absence of added iron or copper. If cells are capable of releasing superoxide radicals in the unsupplemented medium, its incapacity to oxidize LDL in the absence of added iron or copper would indicate that these radicals are not involved in the oxidation process – not, at least, as an initiator of LDL oxidation. Furthermore, most of the oxidative process thus could be attributed to the presence of metals in the medium. The ability of superoxide dismutase to inhibit the oxidative process might then reflect the ability of the enzyme to inhibit the metal-dependent oxidation process, as reported by Parthasarathy et al. [6].

Thus, (i) cellular generation of superoxide radicals was independent of the presence of copper, (ii) cell-free oxidation was poor even in the presence of added copper, and (iii) the combined presence of cells, copper, and LDL was needed for oxidation. The following questions need to be addressed. Are superoxide radicals generated by the cells necessary to initiate oxidation of LDL? Or do superoxide radicals generated by cells only propagate the oxidation of LDL by metals? The question of the effect of superoxide radicals on copper-mediated oxidation of LDL also needs to be addressed.

Jessup's [12] studies have shown that when LDL was subjected to oxidation by superoxide radicals generated by radiolysis, vitamin E appeared to be depleted despite lack of any evidence of lipid peroxidation. The possibility that there is an accelerated antioxidant depletion that would enhance the rate of lipid peroxidation by metals needs to be considered. In retrospect, a determination of LOOH and the rate of subsequent oxidation by copper of LDL exposed to these cells in the absence of added copper would have contributed a great deal to our understanding of the mechanism by which these cells oxidize LDL.

The oxidation of LDL by monocytes has been described by several investigators. Studies by Cathcart et al. [13, 14] first demonstrated that both monocytes and PMN are capable of oxidizing LDL and that activation of these cells enhanced their capacity to oxidize LDL. These studies and those of Hiramatsu et al. studied the oxidation of LDL by monocytes and concluded that the generation of superoxide by these cells was important in the oxidation of LDL [15]. Superoxide dismutase added soon after the stimulation of superoxide production by phorbol myristate acetate (PMA) in monocytes completely inhibited the oxidation of LDL. In contrast, SOD added several hours later had no effect. Thus superoxide by itself was suggested to be the initiator of lipid peroxidation. The NADPH oxidase that generates massive quantities of superoxide on stimulation by PMA was implicated. Supportive evidence was provided by Hiramatsu et al. [15], who observed that cells from a patient suffering from chronic granulomatous disease were unable to oxidize LDL even in the presence of PMA, whereas cells from control subjects were. Because these studies were carried out using just one subject, however, more detailed studies are warranted.

Similar studies by Cathcart et al. [14] confirmed the inhibitory effect of SOD on the oxidation by zymosan-activated monocytes. Their results indicated that the presence of LDL during the intense release of superoxide radical was not required and that oxidation of LDL could occur long after the respiratory burst had subsided. These results would suggest that a product, possibly H_2O_2, derived from superoxide could be an important component.

The importance of superoxide radicals in the oxidation of LDL by endothelial cells has been studied by several investigators. Early studies by van Hinsberg et al. [16] showed that SOD, even at very high concentrations, had no effect on the oxidation of LDL by these cells. On the other hand, the studies by Steinbrecher [17] suggested that superoxide radicals might, after all, be involved in the oxidation of LDL because of the ability of SOD to inhibit the oxidation of LDL and the ability of these cells to generate SOD-inhibitable superoxide radicals. However, the cytochrome reduction that measured the release of superoxide radicals by the cells was only partially inhibitable by SOD, whereas the modification was completely inhibitable by SOD. The results suggest that cells release components that are only partially sensitive to inactivation by SOD, whereas propagation of LDL oxidation is subject to considerable inhibition by SOD. Steinbrecher's studies showed that cells that oxidized LDL readily generated more superoxide radicals that did the cells that oxidized LDL poorly [17].

The role of superoxide radicals in the initiation of the oxidation of LDL needs to be critically examined. A number of factors, such as cell type and cell numbers used, concentration of peroxides, EDTA and LDL, and components such as iron, copper, thiols, and phenol-red in the medium, may be important in defining a role for superoxide anion in the oxidation of LDL. More important, what is measured as evidence of oxidation may directly affect the interpretation of the effect of SOD on oxidation. For example, if superoxide radicals are able to generate a few molecules of lipid peroxides in the absence of metal ions, then rapid propagation of lipid peroxidation is very unlikely.

The roles of superoxide radicals in LDL oxidation are summarized in Table 3.

Our recent results have indicated that LOOH itself has the capacity to generate superoxide radical and enhance the oxidation of lipids [18]. When we incubated oxidized phosphatidyl ethanolamine with unoxidized phosphatidyl ethanolamine, oxidation of the latter was inhibited by inclusion of SOD. No added metals or enzymes were present in the system. McNally and coworkers have speculated that superoxide radicals may be derived from a lipoxygenase pathway [19]. Others have provided evidence that LOOH may generate superoxide radicals and hydroxy radicals in the presence of lipoxygenase activity or amino

Table 3. Roles of superoxide radicals in the oxidation of LDL by cells

(1) The superoxide radical (inefficient in catalyzing lipid peroxidation) may be generated by cells as an important means of providing extracellular hydrogen peroxide which, in the presence of redox metal, generates more potent oxidants. Alternatively, the protonated form of superoxide radical is more reactive and is capable of oxidizing lipids, and it is this species which may be involved.

(2) The origin of the superoxide radical needs to be established. NADPH oxidase ahs been implicated, at least in monocytes, but unstimulated cells also appear to oxidize LDL. The role or the requirement of cysteine in these cells is not understood, and the role of superoxide anion in other cell types that generate cysteine should be established.

(3) There have been conflicting reports regarding the effect of added superoxide dismutase on the oxidation of LDL. Studies with smooth muscle cells, stimulated monocytes, and endothelial cells have suggested that SOD inhibits oxidation effectively. Evidence to the contrary has also been presented, however, using endothelial cells and macrophages.

Table 4. When superoxide radicals may participate in the oxidation of LDL

(1) When the cells have a poor capacity to oxidize LDL, i.e., when the rate of oxidation of
    LDL or the rate of peroxide generation in LDL is slow (e.g., smooth muscle mediated
    oxidation of LDL).
(2) When high concentration of LDL is present in the incubation medium (200 μg/ml or
    more). When there is low rate of peroxide generation and high amounts of unoxidized
    lipids are present in the medium, the superoxide radicals generated from LOOH may be
    quenched by SOD.
(3) When the media do not adequately support peroxidation (high phenol red, low metal
    concentration).
(4) When fresh or unlabeled LDL is used (lower amounts of lipid peroxides and slower rate
    of propagation).

compounds. These findings suggest that when there are more peroxides associated with LDL and the cell has a limited capacity to initiate further oxidation, the rate of propagation may depend on the
ability of the peroxides associated with LDL to generate superoxide
radicals and induce lipid peroxidation. Under these conditions, SOD
might be an effective inhibitor. Interestingly, the amount of SOD
needed to inhibit such reactions is rather large (10–100 μg/ml), similar to the amount required for inhibiting the oxidation of LDL by
cells. In contrast, even very large amounts of the xanthine-xanthine
oxidase generated superoxide production can be quenched by as little
as 5 μg of SOD.

A thorough analysis of the literature suggests that superoxide radicals may be involved in the oxidation of LDL under the conditions
shown in Table 4.

Role of nitric oxide in the oxidation of LDL
The role of nitric oxide (NO) in the oxidation of LDL has received
considerable attention in recent years. NO, by virtue of its ability to
combine with superoxide radicals, would be expected to oxidize LDL
by generating peroxynitrite, a potent oxidizer [20]. NO may also play
a protective role by inactivating radicals that would otherwise oxidize
LDL.

Evidence for a pro-oxidant role for NO was provided by Darley
Usmar et al. who used peroxynitrite and other compounds which would
simultaneously generate superoxide radicals and nitric oxide [21, 22].
The LDL was readily modified in the presence of peroxynitrite to a
form recognized by macrophages. Loss of tocopherol with the formation of the corresponding quinone was concentration- and time-dependent. The interesting feature of this modification is that relatively fewer
TBARS are generated and no added metals ions are required. A further
distinguishing feature of this modification is that no LDL associated,

pre-formed peroxide was needed because peroxynitrite is itself capable of hydrogen abstraction from the fatty acid and could initiate oxidation. An extensive loss of lysine was observed, which was interpreted to suggest that lipid peroxidation products were involved. In contrast, very little loss of lysine was observed when free amino acid lysine, at comparable concentrations, was similarly treated.

Jessup et al. showed that NO alone was incapable of oxidizing LDL to a form recognizable by macrophages [23, 24]. When LDL was subjected to oxidation by the decomposition of SIN-1, a compound that simultaneously generates superoxide radicals and NO and forms peroxynitrite, lipid peroxides were formed, but without any increase in macrophage uptake. The discrepancy between these results and those of Darley Usmar et al. could be due to differences in the time of incubation and the concentrations of the peroxynitrite generator. Micromolar quantities of SIN-1 and 3-h incubations were used in the studies of Jessup et al., whereas studies by Darley Usmar et al. [21, 22] used 24-h incubations and millimolar quantities of peroxynitrite. The increased quantities of LOOH generated and longer incubations undoubtedly caused the greater degree of lysine modification observed in the latter study.

Evidence for an antioxidant role for NO is also suggested by several studies. The studies of Jessup et al. showed a remarkable feature of the modification of LDL by cells [23]. When the production of NO by cells was inhibited using NO synthesis inhibitors, the oxidation of LDL was actually increased. We also have observed that the N-monomethyl arginine, the inhibitor of NO synthesis, actually increased the oxidation of LDL by cells (unpublished results). In a subsequent study, Jessup and Dean [24] induced the release of nitric oxide by macrophages by treatment with IFN-γ or LPS and demonstrated reduced oxidation of LDL by treated cells. This was shown to be due to an actual induction of the enzyme activity. These results suggested not only that NO synthesis was not necessary for the oxidation of LDL by cells, but also that NO may actually inhibit the oxidation of LDL. The inhibition of LDL oxidation by nitric oxide was not due to its effect on the cell. Hogg et al. [25] have shown in the cell-free system that the oxidation of LDL by copper or ABAP is inhibited by nitric oxide generated by sodium nitroprusside by a light-dependent mechanism. The generation of NO during the oxidation of LDL by copper inhibited the formation of conjugated dienes. The generation of TBARS and the subsequent macrophage degradation of the incubated LDL was also affected. Similar results were obtained with another NO generator called SNAP. Recently, the inhibition of oxidation of LDL by cells in the presence of reagent-generated NO was described [26]. The mechanism of such an inhibition is currently under intense investigation.

Role of 15-lipoxygenase in the oxidation of LDL

There are basically three kinds of evidence suggesting a role for 15-lipoxygenase in atheroslerosis. This review will not focus on 5-lipoxygenase, as such an enzyme has never been proposed to play a role in the oxidation of LDL, although several publications have mistakenly assumed such a role and went to lengths to disprove the participation of such an enzyme activity.

Studies of in vitro *cell culture systems.* Most such evidence came from the use of inhibitors and the measurement of specific products. Most lipoxygenase inhibitors are antioxidants. The acetylinic acid analog of arachidonic acid, Eicosa tetraynoic acid (ETYA) has long been used as an inhibitor of lipoxygenases as well as other oxidases. The use of this compound in the presence of lipoprotein deficient serum caused effective inhibition of the oxidation of LDL. There was no cytotoxicity in these studies in the presence of lipoprotein deficient serum. Based on this evidence, together with the findings that LDL inside a dialysis bag was protected against oxidation and that several recombinant superoxide dismutase preparations were ineffective in preventing the oxidation of LDL, we proposed a role for lipoxygenase [6]. The unlikeliness that lipoxygenase would act directly on the lipoprotein led us to suggest that perhaps a cell-derived peroxide is transferred to the lipoprotein in the medium. This study was subseqently extended to the macrophage system with additional demonstration of actual formation of 15-lipoxygenase products and their inhibition during incubation with ETYA [27]. Unfortunately, these studies have been misunderstood, misquoted, and misrepresented by several investigators, particularly by Jessup et al. [28] and by Sparrow and Olszewski [29], who failed to note that there was no suggestion of a direct reaction between lipoxygenase and LDL and there was no suggestion that it served as the sole oxidizing system capable of propagation reaction (the suggested role was only to seed LDL with lipid peroxide). The studies by Sparrow and Olszewski [29] determined and correlated the inhibition of lipoxygenase reaction by ETYA and the oxidation of LDL in the medium in Ham's F-10 medium and radioiodinated lipoproteins. They observed that there was no correlation and concluded that lipoxygenases are not involved in the oxidation of LDL.

The several areas of oversight in their studies can be summarized as follows:

(i) The level of 15-LO activity was assumed to correlate with the degree of LDL oxidation. In the proposed role, 15-LO was only suggested to be involved in the initiation of lipid peroxidation by seeding LDL with a few molecules of LOOH. Regardless of the level of enzyme activity, if the amount of product released in the medium is comparable with or without substantial inhibition of the LO activity by ETYA, the results could be misinterpreted. The simplistic assumption that there should be a mathematical relationship between the oxidation of LDL and the lipoxygenase activity may be true only if 100% of lipoxygenase activity correlates with 100% oxidation. The latter is determined by the net availability of LDL (availability of fatty acid), by propagation reactions, and above all by the maximum oxidizability of the LDL preparation. Oxidized LDL preparations that gave identical degrees of degradation by macrophages very often vastly differed in their TBARS. If only 20% lipoxygenase activity was needed to generate enough lipid peroxides, an 80% inhibition would have no effect on the oxidation of LDL in the medium. In fact, recent *in vitro* studies by O'Leary et al. [30] have shown that addition of 1–12 nano moles of LOOH to LDL increased the LDL associated peroxides 4-fold. Despite a 400% increase in LOOH levels only a decrease of 5% in lag time was observed. Disproportionate increases were also noted in the electrophoretic mobility and TBARS levels. These results clearly imply that there may not be any correlation between the availability of LOOH from a specific pathway and the degree of oxidation. More importantly, the presence of peroxides in the radioiodinated LDL and the use of Ham's F-10 medium would have introduced considerable disproportionate availability of peroxides regardless of the extent of inhibition of lipoxygenase activity.

(ii) Inhibition observed by 0.5 μM ETYA in the presence of 10 μM substrate in 30-min incubations may bear no relation to the amounts of products formed in a 24-h incubation period, such as that carried out in the presence of LDL. Furthermore, the concentration of ETYA added in the presence of LDL may not represent the concentration that was available for cellular inhibition observed in the absence of LDL.

Studies based on in vitro *non-cell systems.* LDL can be readily oxidized by treatment with lipoxygenases. Studies by Cathcart et al. [31] have shown that the soybean lipoxygenase readily reacted with LDL and generated cytotoxic products. Superoxide radicals were generated during the reaction which eventually was converted to H_2O_2. The formation of H_2O_2 appeared to be necessary for the generation of the cytoxic component. Although it is very unlikely that lipoxygenase in the cell may come into contact with intact lipoprotein particle, experiments do show that enzyme readily reacts with esterified lipids in the lipoprotein. Several studies have now demonstrated that both soybean lipoxygenase and lipoxygenases derived from animal cells modify LDL to a form recognized by macrophages [32, 33]. The modification of the lipoprotein by an affinity-purified soybean lipoxygenase required pretreatment of the lipoprotein with phospholipase A_2. However, recent studies have shown that other lipoxygenases may oxidize the lipoprotein directly [33]. These studies also provided evidence that a single enzyme-catalyzed oxidation of the lipoprotein, in the absence of added metals, can lead to propagation and protein modifications. No claim of direct

oxidation of LDL by lipoxygenase has ever been made. In the likelihood
of cell damage it is, however, likely that the cytosolic enzyme may be
available for direct interaction with the lipoprotein.

In vivo *studies*. Several studies have reported that lipoxygenase products and activity are increased in atherosclerotic lesions [34]. The
increase in products could have resulted from nonspecific oxidation of
lipids. This increase could also, at least in part, be due to an inrease in
the activity of 15-lipoxygenase. Recent studies by Kuhn et al. [35, 36]
have identified specific lipoxygenase derived stereospecific products in
the athersoclerotic rabbit artery. Their results showed formation of
these specific products in early lesions, whereas in late lesions there were
more mixed isomers, suggesting non-specific lipid peroxidation. Ylä-
Herttuala et al. looked at the expression of 15-lipoxygenase in WHHL
rabbits using *in situ* hybridization and immunochemical techniques.
They used sense and anti-sense riboprobes for 5- and 15-lipoxygenase
enzyme messenger RNAs and antibodies directed towards the enzyme,
rabbit macrophages, and epitopes of oxidized LDL. Their results
showed positive evidence for the expression of 15-lipoxygenase gene in
the macrophage-rich lesions of the lesion. Antibodies against MDA-
LDL (which indicate that an epitope is present in oxidized LDL) were
also positive in the same areas represented by macrophages. Sense probe
for 15-lipoxygenase and the anti-sense probe for 5-lipoxygenase were
negative. Interestingly, weak hybridization was also noted in some areas
of the endothelium with the 15-lipoxygenase probe [38, 39]. The presence of the enzyme protein was established in these experiments by
Western blot analysis.

Monocytes, the progenitors of macrophages, do not express high
levels of 15-LO activity and generally produce products of 5-LO. Tissue
macrophages, particularly resident peritoneal macrophages, do express
substantial levels of 15-lipoxygenase activity. Anticipating that the
15-LO gene expression may be induced in these cells by appropriate
stimuli, Sigal and Conrad screened a number of cytokines and showed
that IL-4 (a T-lymphocyte product) was profoundly effective as a
stimulant [39]. Of the cytokines screened, only γ-interferon appeared to
counteract this effect. T-cells are present in the atherosclerotic lesion
and are temporally recruited at the same time as the monocytes. Thus,
the arterial macrophages have the potential to express 15-LO activity.

More recently, Ylä-Herttuala et al. reported the *in vivo* 15-lipoxyge-
nase gene transfection in rabbits. Their preliminary results indicated the
presence of increased amounts of immuno reactivity to antibodies
directed towards antigenic epitopes present in oxidized LDL at target
sites as compared to control areas [40].

Of various mechanisms proposed to explain the oxidation of LDL,
15-LO mediated initiation of oxidation has been the target of extensive
criticism. But, considering the lack of availability of direct inhibitors,

this proposal has more theoretical and experimental support from the physiological view point than other proposed mechanisms. More importantly, this is the only mechanism that has support from *in vivo* studies.

Peroxidases and metal-independent oxidation of LDL

Most of the studies on oxidation have suggested that close association of the cell and the LDL is necessary and that perhaps some prooxidant molecule is transferred from the cell to the lipoprotein particle. Yet the requirement of metals such as copper in the medium has been of major concern. Metal-containing proteins such as hematin, hemin, and hemoglobin can oxidize LDL under appropriate conditions, and such a mechanism may be appropriate when damaged red blood cells are present in the lesion [41, 42]. However, there is no evidence of necrosis in early atherosclerosis or under cell culture conditions. We and others recently observed that peroxidases such as horseradish peroxidase and myeloperoxidase can readily modify LDL in the presence of H_2O_2 or LOOH in simple buffers in the absence of added metals [43–45]. Furthermore, media conditioned in the presence of cells contained high levels of H_2O_2 and were capable of oxidizing LDL in the presence, but not in the absence, of added peroxidase. As with simple heme-containing proteins, the requirement of an active enzyme did not appear to be necessary.

There is no horseradish peroxidase in the body. However, myeloperoxidase is found in abundance in neutrophils and monocytes and in somewhat less abundance in macrophages due probably to the decrease in activity during the differentiation of monocytes into macrophages. Several studies have shown that hypochlorous acid (a product of myeloperoxidase reaction) can oxidize lipoproteins, including LDL and HDL [46, 47]. In addition to lipid peroxidation, these systems also induce the oxidation of tyrosine residues and generates a different type of modified LDL.

Hiramatsu et al. [15] studied the ability of monocytes to oxidize LDL from a subject suffering from myeloperoxidase deficiency. Monocyte from such a subject was able to oxidize LDL just as readily as the cells from a control subject. This result was interpreted to suggest that myeloperoxidase is not involved in the oxidation of LDL. The same author has recently suggested that myeloperoxidase may be involved in the generation of dityrosine via the formation of tyrosyl radicals and has proposed the participation of this enzyme in lipoprotein oxidation. Hydrogen peroxide is a substrate for this enzyme and the presence of SOD should increase the availability of H_2O_2. If myeloperoxidase reaction is indeed involved, one would expect an increased oxidation of LDL when SOD is present. Regardless whether this enzyme is involved in the oxidation of LDL or not, studies that utilize a single patient can often lead to misleading conclusions.

One important caveat in the peroxidase-mediated oxidation of LDL is that the mechanism itself may include the generation of a phenoxy radical, either from the apoprotein itself, or from the antioxidant present in the LDL or from extra cellular phenols, including phenol-red or tyrosine. If so, such a mechanism will rapidly deplete antioxidants from the lipoprotein and render the lipoprotein more readily oxidizable. In fact, under such conditions, a pro-oxidant function may be ascribed to the antioxidants when they are present at low concentrations and insufficient to inhibit the propagation [45].

Passive participation of cells

Extensive studies by Bannai and coworkers [48, 49] have established that cultured cells take up cystine and release cysteine. They established that cysteine is rapidly oxidized in cell culture medium and proposed a recycling system that would provide intracellular cysteine. Based on Bannai's model of cystine transport, a scheme for the involvement of cystine in the oxidation of LDL by these cells was proposed by Heinecke et al. [50]. According to the proposed mechanism, thiols would react with metals and generate superoxide and other radicals as earlier suggested [51].

The oxidation of LDL by a thiol-dependent mechanism was also suggested for other cell types. Sparrow and Olszewski [52], using a rabbit endothelial cell line and mouse peritoneal macrophages, established that Ham's F-10 medium devoid of cysteine was incapable of oxidizing LDL. Addition of cysteine in the presence or absence of cells resulted in the oxidation of LDL. However, addition of cystine, the oxidized form, stimulated only the cell-mediated oxidation of LDL and had no effect on the no-cell oxidation of LDL, demonstrating cellular dependence on cystine for oxidation of LDL. In support of the cellular "production" of reduced thiol from cysteine, evidence was presented for the time-dependent appearance of cystine in the medium when endothelial cells were incubated with cystine. The thiol production was specific for cysteine, and other thiols did not participate in the production of thiols. Additional evidence was provided for the cellular production of cysteine from cystine; inclusion of glutamate, which competes with cystine for entry into cells, blocked the modification. However, in the presence of glutamate, enzymatic reduction of cystine in the medium regenerated the thiols and reestablished the modification. These impressive studies are marred by one important observation. In several experiments, TBARS amounts were substantially high, often exceeding 20 nmols/mg LDL protein, which further increased when thiol production was increased in cell incubations that did not have any cysteine or cystine. This suggests either oxidation in the absence of thiols or that the LDL had large amounts of peroxides.

A number of questions needs to be answered before one can conclude that all cell types oxidize LDL by thiol-dependent mechanisms. First, is the requirement of very high concentrations of glutamate to inhibit cystine transport suggestive of the involvement of non-cystine dependent processes? For example, in the study of Sparrow and Olszewski [53], when 3 mM glutamate was able to prevent the release of thiols by over 75%, the inhibition of TBARS formation was less than 45%. Cystine transport across cell membrane is affected not only by glutamate, but also by other amino acids. Interestingly, the study noted a 15% reduction in 15-lipoxygenase activity in glutamate treated-cells, a reduction that was considered insufficient to account for the 45% decrease in TBARS production. It is difficult to comprehend the rationale for such interpretations involving enzyme assays carried out in the presence of large amounts of added substrates and determinations of TBARS at the end of a 24-h incubation that involved propagation reactions. It is my belief that correlations should reflect the initial peroxide content of the LDL and its oxidizability, and that correlations reflecting the rate of an intracellular enzyme reaction and the extracellular oxidation of LDL are relatively useless. Most of the oxidation occurs in the medium due to propagation reactions and, unless detailed analysis of lipid peroxide released and association with the lipoprotein is established with the incubated lipoprotein, no conclusion can be drawn regarding the role of cells in the initiation of oxidation by cells. Second, if the only role of cells is to generate reduced cysteine, how is it possible that oxidation of LDL separated from the cell by a dialysis membrane is prevented? Thiols, cysteine and cystine readily permeate a dialysis membrane, and the metal content of the membrane is the same as that present outside the membrane. The dialysis membranes used today are treated so that they will not preferentially block anions.

In summary, results indicate that cysteine or cysteine-derived products may enhance the rate of oxidation LDL and that cells may generate cysteine from cystine taken up from the medium. Unfortunately, the analytical techniques used do not permit evaluation of the specific role of the cell in the initiation of oxidation of LDL. These results simply mean that when LDL contains sufficient peroxides, cysteine or metals may enhance the rate of peroxidation and generate more TBARS and a modified LDL capable of being recognized by macrophages. If cysteine is rapidly oxidized to cystine in the medium, the formation of other mixed disulfides also should occur. Furthermore, when LDL is being oxidized in the medium, the oxidation of thiols should proceed beyond the formation of disulfides, to products that cannot be reduced back to cysteine.

The specific mechanism by which thiols may promote oxidation was recently described by Heinecke et al. using a variety of inhibitors [53]. These studies, predominantly repetitive in nature, confirmed that the addition of thiols in the absence of cells can result in the modification of LDL. A number of antioxidants and SOD inhibited the modification of LDL by cysteine and copper. In this *in vitro* oxidation system, in contrast to the oxidation of LDL by cysteine, SOD failed to inhibit the oxidation of LDL by other thiols such as glutathione or homocysteine. These results were interpreted to suggest that both superoxide-dependent and superoxide-independent mechanisms were available by which thiols could oxidize LDL. No explanation was offered for the differences between these thiols, and I am not aware of differences in their ability to form radicals with metals. It is inconceivable that only cysteine can form superoxide radicals, while other thiols form different radicals.

Potential involvement of thiyl radicals has also been suggested. However, unpublished studies by the author (Parthasarathy and Kalyanaraman) have also shown that thiyl radicals generated by chemical reactions are incapable of oxidizing LDL and that a thiyl radical spin trap DMPO has little effect on the oxidation of LDL by macrophages.

In summary, the oxidation of LDL by cells is complex and a number of issues has to be resolved. The use of freshly isolated lipoproteins, better analytical determination that can measure extremely low levels of peroxides and "metal-free" media may shed more light on this process.

References

1. Kappus, H. and Diplock, A.T. (1992) Tolerance and safety of vitamin E: A toxicological position report. *Free Radic. Biol. Med.* 13: 55–74.
2. Parthasarathy, S. (1994) *Modified lipoproteins in the pathogenesis of atherosclerosis*. R.G. Landes Company, Austin, Texas, pp 1–125.
3. Steinbrecher, U.P., Witztum, J.L., Parthasarathy, S. and Steinberg, D. (1987) Decrease in reactive amino groups during oxidation or endothelial cell modification of LDL. Correlation with changes in receptor-mediated catabolism. *Arteriosclerosis* 7: 135–143.
4. van Berkel, T.J., de Rijke, Y.B. and Kruijt, J.K. (1991) Different fate *in vivo* of oxidatively modified low density lipoprotein and acetylated low density lipoprotein in rats. Recognition by various scavenger receptors on Kupffer and endothelial liver cells. *J. Biol. Chem.* 266: 2282–2289.
5. Heinecke, J.W., Rosen, H. and Chait, A. (1984) Iron and copper promote modification of low density lipoprotein by human arterial smooth muscle cells in culture. *J. Clin. Invest.* 74: 1890–1984.
6. Parathasarathy, S., Wieland, E. and Steinberg, D. (1989) A role for endothelial cell lipoxygenase in the oxidative modification of low density lipoprotein. *Proc. Natl. Acad. Sci. USA* 86: 1046–1050.
7. Parthasarathy, S., Barnett, J. and Fong, L.G. (1990) High-density lipoprotein inhibits the oxidative modification of low-density lipoprotein. *Biochim. Biophys. Acta* 1044: 275–283.
8. Parthasarathy, S., Quinn, M.T., Schwenke, D.C., Carew, T.E. and Steinberg, D. (1989) Oxidative modification of beta-very low density lipoprotein. Potential role in monocyte recruitment and foam cell formation. *Arteriosclerosis* 9: 398–404.
9. Steinberg, D., Parthasarathy, S., Carew, T.E., Khoo, J.C. and Witztum, J.L. (1989) Beyond cholesterol. Modifications of low-density lipoprotein that increase its atherogenicity [see comments]. *New Eng. J. Med.* 320: 915–924.

10. Steinbrecher, U.P., Parthasarathy, S., Leake, D.S., Witztum, J.L. and Steinberg, D. (1984) Modifiation of low density lipoprotein by endothelial cells involves lipid peroxidation and degradation of low density lipoprotein phospholipids. *Proc. Natl. Acad. Sci. USA* 81: 3883–3887.

11. Heinecke, J.W., Baker, L., Rosen, H. and Chait, A. (1986) Superoxide-mediated modification of low density lipoproten by arterial smooth muscle cells. *J. Clin. Invest.* 77: 757–761.

12. Bedwell, S., Dean, R.T. and Jessup, W. (1989) The action of defined oxygen-centered free radials on human low-density lipoprotein. *Biochem. J.* 262: 707–712.

13. Cathcart, M.K., Morel, D.W. and Chisolm, G.M. (1985) Monocytes and neutrophils oxidize low density lipoprotein making it cytotoxic. *J. Leuk. Biol.* 38: 341–350.

14. Cathcart, M.K., McNally, A.K., Morel, D.W. and Chisolm, G.M. (1989) Superoxide anion participation in human monocyte-mediated oxidation of low-density lipoprotein and conversion of low-density lipoprotein to a cytotoxin. *J. Immunol.* 142: 1963–1969.

15. Hiramatsu, K., Rosen, H., Heinecke, J.W., Wolfbauer, G. and Chait, A. (1987) Superoxide initiates oxidation of low density lipoprotein by human monocytes. *Arteriosclerosis* 7: 55–60.

16. van Hinsbergh, V.W., Scheffer, M., Havekes, L. and Kempen, H.J. (1986) Role of endothelial cells and their products in the modification of low-density lipoproteins. *Biochim. Biophys. Acta* 878: 49–64.

17. Steinbrecher, U.P. (1988) Role of superoxide in endothelial-cell modification of low-density lipoproteins. *Biochim. Biophys. Acta* 959: 20–30.

18. Fruebis, J., Parthasarathy, S. and Steinberg, D. (1992) Evidence for a concerted reaction between lipid hydroperoxides and polypeptides. *Proc. Natl. Acad. Sci. USA* 89: 10588–10592.

19. McNally, A.K., Chisolm, III, G.M., Morel, D.W. and Cathcart, M.K. (1990) Activated human monocytes oxidize low-density lipoprotein by a lipoxygenase-dependent pathway. *J. Immunol.* 145: 254–259.

20. Beckman, J.S., Beckman, T.W., Chen, J., Marshall, P.A. and Freeman, B.A. (1990) Apparent hydroxyl radical production by peroxynitrite: implications for endothelial injury from nitric oxide and superoxide. *Proc. Natl. Acad. Sci. USA* 87: 1620–1624.

21. Darley Usmar, V.M., Hogg, N., O'Leary, V.J., Wilson, M.T. and Moncada, S. (1992) The simultaneous generation of superoxide and nitric oxide can initiate lipid peroxidation in human low density lipoprotein. *Free Radic. Res. Commun.* 17: 9–20.

22. Graham, A., Hogg, N., Kalyanaraman, B., O'Leary, V., Darley Usmar, V. and Moncada, S. (1993) Peroxynitrite modification of low-density lipoprotein leads to recognition by the macrophage scavenger receptor. *FEBS Lett.* 330: 181–185.

23. Jessup, W., Mohr, D., Gieseg, S.P., Dean, R.T. and Stocker, R. (1992) The participation of nitric oxide in cell free- and its restriction of macrophage-mediated oxidation of low-density lipoprotein. *Biochim. Biophys. Acta.* 1180: 73–82.

24. Jessup, W. and Dean, R.T. (1993) Autoinhibition of murine macrophage-mediated oxidation of low-density lipoprotein by nitric oxide synthesis. *Atherosclerosis* 101: 145–155.

25. Hogg, N., Kalyanaraman, B., Joseph, J., Struck, A. and Parthasarathy, S. (1993) Inhibition of low-density lipoprotein oxidaiton by nitric oxide. Potential role in atherogenesis. *FEBS Lett.* 334: 170–174.

26. Malo-Ranta, U., Ylä-Herttuala, S., Metsa-Ketela, T., Jaakkola, O., Moilanen, E., Vuorinen, P. and Nikkari, T. (1994) Nitric oxide donor GEA 3162 inhibits endothelial cell-mediated oxidation of low density lipoprotein. *FEBS Lett.* 337: 179–183.

27. Rankin, S.M., Parthasarathy, S. and Steinberg, D. (1991) Evidence for a dominant role of lipoxygenase(s) in the oxidation of LDL by mouse peritoneal macrophages. *J. Lipid. Res.* 32: 449–456.

28. Jessup, W., Darley Usmar, V., O'Leary, V. and Bedwell, S. (1991) 5-Lipoxygenase is not essential in macrophage-mediated oxidation of low-density lipoprotein. *Biochem. J.* 278: 163–169.

29. Sparrow, C.P. and Olszewski, J. (1992) Cellular oxidative modifiation of low density lipoprotein does not require lipoxygenases. *Proc. Natl. Acad. Sci. USA* 89: 128–131.

30. O'Leary, V., Graham, A., Darley-Usmar, V. and Stone, D. (1993) The effect of lipid hydroperoxides on the copper dependent oxidation of low density lipoprotein. *Biochem. Soc. Trans.* 21: 89S.

31. Cathcart, M.K., McNally, A.K. and Chisolm, G.M. (1991) Lipoxygenase-mediated transformation of human low density lipoprotein to an oxidized and cytotoxic complex. *J. Lipid. Res.* 32: 63–70.

32. Sparrow, C.P., Parthasarathy, S. and Steinberg, D. (1988) Enzymatic modification of low density lipoprotein by purified lipoxygenase plus phospholipase A2 mimics cell-mediated oxidative modification. *J. Lipid Res.* 29: 745–753.

33. Belkner, J., Wiesner, R., Rathman, J., Barnett, J., Sigal, E. and Kuhn, H. (1993) Oxygenation of lipoproteins by mammalian lipoxygenases. *Eur. J. Biochem.* 213: 251–261.

34. Simon, T.C., Makheja, A.N. and Bailey, J.M. (1989) The induced lipoxygenase in atherosclerotic aorta converts linoleic acid to the platelet chemorepellent factor 13-HODE. *Thrombos. Res.* 55: 171–178.

35. Kuhn, H., Belkner, J., Wiesner, R., Schewe, T., Lankin, V.Z. and Tikhaze, A.K. (1992) Structure elucidaton of oxygenated lipids in human atherosclerotic lesions. *Eicosanoids* 5: 17–22.

36. Kuhn, H., Belkner, J., Zaiss, S., Fahrenklemper, T. and Wohlfeil, S. (1994) Involvement of 15-lipoxygenase in early stages of atherogenesis. *J. Exp. Med.* 179: 1903–1911.

37. Ylä-Herttuala, S., Rosenfeld, M.E., Parthasarathy, S, Glass, C.K., Sigal, E., Witztum, J. L. and Steinberg, D. (1990) Colocalization of 15-lipoxygenase mRNA and protein with epitopes of oxidized low density lipoprotein in macrophage-rich areas of atherosclerotic lesions. *Proc. Natl. Acad. Sci. USA* 87: 6959–6963.

38. Ylä-Herttuala, S., Rosenfeld, M.E., Parthasarathy, S. Sigal, E., Sarkioja, T., Witztum, J. L. and Steinberg, D. (1991) Gene expression in macrophage-rich human atherosclerotic lesions. 15-lipoxygenase and acetyl low density lipoprotein receptor messenger RNA colocalize with oxidation specific lipid-protein adducts. *J. Clin. Invest.* 87: 1146–1152.

39. Conrad, D.J., Kuhn, H. Mulkins, M., Highland, E. and Sigal, E. (1992) Specific inflammatory cytokines regulate the expression of human monocyte 15-lipoxygenase. *Proc. Natl. Acad. Sci. USA* 89: 217–221.

40. Ylä-Herttuala, S., Luoma, J., Myllyharju, H., Hiltunan, T., Sisto, T. and Nikkari, T. Transfer of 15-lipoxygenase gene into rabbit iliac arteries results in the appearance of oxidation-specific lipid-protein adducts characteristic of oxidized LDL. Abstract presented at the American Heart Association meeting, November 1994.

41. Balla, G., Jacob, H.S., Eaton, J.W., Belcher, J.D. and Vercellotti, G.M. (1991) Hemin: a possible physiological mediator of low density lipoprotein oxidation and endothelial injury. *Arterioscler. Thromb.* 11: 1700–1711.

42. Paganga, G., Rice Evans, C.A., Rule, R. and Leake, D.S. (1992) The interaction between ruptured erythrocytes and low-density lipoproteins. *FEBS Lett.* 303: 154–158.

43. Wieland, E., Parthasarathy, S. and Steinberg, D. (1993) Peroxidase-dependent metal-independent oxidation of low density lipoprotein *in vitro*: a model for *in vivo* oxidation? *Proc. Natl. Acad. Sci. USA* 90: 5929–5933.

44. Savenkova, M.L., Mueller, D.M. and Heinecke, J.W. (1994) Tyrosyl radical generated by myeloperoxidase is a physiological catalyst for the initiation of lipid peroxidation in low density lipoprotein. *J. Biol. Chem.* 269: 20394–20400.

45. Santanam, N. and Parthasarathy, S. (1995) Paradoxical role of antioxidants in the oxidation of low density lipoprotein by peroxidases. *J. Clin. Invest.* 94: 1990–1995.

46. Hazell, L.J. and Stocker, R. (1993) Oxidation of low-density lipoprotein with hypochlorite causes transformation of the lipoprotein into a high-uptake form for macrophages. *Biochem. J.* 290: 165–172.

47. Panasenko, O.M., Evgina, S.A., Aidyraliev, R.K., Sergienko, V.I. and Vladimirov, Y.A. (1994) Peroxidation of human blood lipoproteins induced by exogenous hypochlorite or hypochlorite generated in the system of "myeloperoxidase + H_2O_2 + Cl^-". *Free Radic. Biol. Med.* 16: 143–148.

48. Bannai, S. (1984) Transport of cystine and cysteine in mammalian cells. *Biochim. Biophys. Acta* 779: 289–306.

49. Watanabe, H. and Bannai, S. (1987) Induction of cystine transport activity in mouse peritoneal macrophages. *J. Exp. Med.* 1675: 628–640.

50. Heinecke, J.W., Rosen, H., Suzuki, L.A. and Chait, A. (1987) The role of sulfur-containing amino acids in superoxide production and modification of low density lipoprotein by arterial smooth muscle cells. *J. Biol. Chem.* 262: 10098–10103.

51. Parthasarathy, S. (1987) Oxidation of low-density lipoprotein by thiol compounds leads to its recognition by the acetyl LDL receptor. *Biochim. Biophys. Acta.* 917: 337–340.

52. Sparrow, C.P. and Olszewski, J. (1993) Cellular oxidation of low density lipoprotein is caused by thiol production in media containing transition metal ions. *J. Lipid. Res.* 34: 1219–1228.
53. Heinecke, J. W., Kawamura, M., Suzuki, L. and Chait, A. (1993) Oxidation of low density lipoprotein by thiols: superoxide-dependent and -independent mechanisms. *J. Lipid. Res.* 34: 2051–2061.

Analysis of Free Radicals in Biological Systems
Favier et al. (eds)
© 1995 Birkhäuser Verlag Basel/Switzerland

Free radicals and antioxidants in human diseases

J. Pincemail

Department of Cardiovascular Surgery, University of Liège, CHU B35, Sart Tilman, B-4000 Liège, Belgium

Summary. Increased free radicals formation derived from abnormal metabolism of oxygen has been implicated in over 100 different human diseases which may be classified into six categories: chemical and xenobiotic toxicity, radiation injury, hyperoxygenation syndromes, inflammatory conditions, postischemic reperfusion injury and degenerative conditions. As their *in vivo* evidence always remains a difficult challenge, there is, however, a great deal of confusion about their precise role in the development of human diseases. Whatever the precise mechanism, preventing free radical generation by appropriate antioxidant therapy should provide beneficial clinical consequences.

Introduction: The "oxygen paradox"

Oxygen, friend or foe? Oxygen, a poison required for life. Oxygen, a dangerous friend. Oxygen, from Charybdis to Scylla. Oxygen represented as Janus. All these clichés are regularly used in the scientific literature and meetings discussing that oxygen that is so indispensable to our life, is also toxic ("oxygen paradox"). This is due to its ability to generate reactive intermediates (superoxide anion, hydroxyl radical, hydrogen peroxide, singlet oxygen, perferryl ion, nitrogen oxides) (see [1]).

In the 19th century, this "oxygen paradox" was described for the first time by Louis Pasteur, when he demonstrated that anaerobic organisms rapidly died after exposure to air containing the 20% oxygen required for the survival of aerobic organisms. In the middle of 1950s, Gerschman et al. [2] and Harman [3] resurrected this concept and respectively postulated the hypothesis that death of animals exposed to X-irradiation in presence of high oxygen tensions and the aging process were partially due to abnormal metabolism of oxygen, leading to toxic free-radical reactions. In France, Laborit et al. [4] attracted everyone's attention to the possible toxicity of high-pressure oxygen in divers through the production of free radicals, and suggested the use of certain antioxidants in biology. As discussed in detail in this book, a free radical (superoxide anion, hydroxyl radical) is defined as any species capable of independent existence that contains one or more unpaired electrons. This confers high reactivity to the molecule and, therefore, allows it to damage multiple

biological substrates, including proteins, lipoproteins, deoxyribonucleic acid (DNA), carbohydrates, and polyunsaturated fatty acids.
Moreover, they can also act as second messengers in the induction of
molecular processes [5].

The "free radical story" in biology and medicine really started with
the discovery by McCord, Keele and Fridovich in 1969 [6] of superoxide dismutase, an enzyme capable of destroying the superoxide anion
radical resulting from the univalent reduction of oxygen. Since then,
there has been an exponential increase in scientific articles indicating
that excessive production of "reactive oxygen species" (ROS) or "free
radical" is implicated in the pathogenesis of diseases in humans. Currently, *Free Radicals in Biology and Medicine* and *Free Radicals Research Communications* are the two specific journals encompassing
medical approaches to oxygen free radical research.

When does free radical production become dangerous?

If reactive oxygen species are generally regarded as dangerous, it must
be kept in mind that our metabolism produces free radicals under
normal conditions. Thus, a small amount of oxygen (2%) is continuously reduced to ROS in the mitochondrial electron transport chain.
Moreover, free radical species also play a key role in many physiologic
reactions. Typical examples are the killing of bacteria by granulocytes
and macrophages [7], the oxidation of xenobiotics by cytochrome P-450
[8], the regulation of smooth muscle by "endothelium-derived relaxing
factor" (EDRF), now recognised as the nitric oxide ($NO^{\cdot}$) radical, and
even fertilization [9]. To regulate these free radical reactions, our
organism has developed antioxidant defenses including not only enzymes (superoxide dismutase, catalase, glutathione peroxidase) and
small molecules (vitamins A, C, and E, uric acid, glutathione, albumine
or bilirubin), but also repair systems which prevent the accumulation of
oxidatively damaged molecules.

However, in human diseases, increased free radical activity can occur
either as a primary (e.g., excess radiation exposure) or a secondary (e.g.,
tissue damage by trauma) event, mediated by several biochemical processes: extracellular release of ROS by granulocytes, xanthine oxidase
activation, iron release from sequestered sites, phospholipase activation,
alteration of electron transport in the mitochondrion, etc. Consequently, antioxidant defenses can be rapidly overwhelmed, leading to
increased tissue injury. These situations can be considered as "oxidative
stress" since, as defined in detail by Sies [10], there is a profound
disturbance of the prooxidant-antioxidant balance in favour of the
former, leading to lipid peroxidation, denaturation of proteins or enzymes or mutagenic damage to nucleic acids. The dual nature of free

radicals in biological systems is perfectly illustrated by activated neutrophils, which can be involved in both physiologic (bacteria killing with controlled intracellular production of ROS) and pathophysiologic (inflammation with uncontrolled extracellular production of ROS) events. Similarly, excessive production of $NO^{\cdot}$, initiated by activation of the glutamate receptor after cerebral ischemia, is thought to participate in tissue injury in the brain (via the generation of oxidizing peroxynitrite anion from the reaction between the superoxide anion and nitric oxide).

A critical point: Assays for measuring free radical formation *in vivo*

Increased ROS formation has been implicated in over 100 pathophysiologic conditions [11]. The transient nature of free radicals makes *in vivo* assessment of their production a difficult challenge; the role of ROS in such a variety of processes may reflect this difficulty.

Most currently used techniques are called "fingerprint assays," because they examine the chemical changes that free radicals caused by interacting with targets such as lipids, proteins and DNA as measured by end-products of lipid peroxidation, protein carbonyls and 8-hydroxyguanine. Until recently, the most popular method involved the measurement of malondialdehyde (MDA) and thiobarbituric acid-reactivity as markers of lipid peroxidation [12], but this assay lacks specificity and accuracy. This has led to some controversy about the importance of free radicals in human diseases. Detection of other molecules, such as 4-hydroxynonenal or hydrocarbons (ethane, pentane), has also been proposed to assess lipid peroxidation, but these too require great care in data interpretation [13].

A more realistic strategy for monitoring free radical reactions *in vivo* is the measurement of antioxidant defences present in human plasma [14], red blood cells, and even tissue. As a typical example, vitamin E or (α-tocopherol), acting as a chain-breaking antioxidant, can protect membrane lipids against oxidative damage by virtue of its potent scavenging. Its consumption can thus be considered to be a specific, although indirect, index of *in vivo* peroxidative processes [15]. Measuring the product of attack of ROS on uric acid has also been proposed as a potential marker of oxidative damage [16].

All of this evidence is, however, indirect and does not afford detailed information about the exact role played by free radicals in including *in vivo* tissue injury. Efforts have been made recently to directly demonstrate the formation of free radicals in biological samples as complex as blood or plasma, using the techniques of aromatic hydroxylation (e.g., salicylate, see [17]) and spin trapping associated with electron spin resonance (ESR) spectroscopy, as trapping assays.

The ESR methodology is currently considered to be the technique of choice for detecting free radicals *in vivo* since it allows to detect the free radical by measuring the absorption of energy due to interaction of the unpaired electron present in the free radical with an applied external magnetic field produced by a magnet. Direct ESR can detect the ascorbyl radical in plasma at room temperature or the NO˙ radical in frozen blood [18]. More often, however, spin trapping agents are required in ESR to increase the half life of free radicals generated *in vivo*. Such agents react with short-lived free radicals to form more stable species (spin adducts) that can be detected, identified and sometimes quantitated. Numerous spin trapping agents, such as 5,5-dimethyl-1-pyrroline 1-oxide (DMPO) or alpha-phenyl-N-tert-butyl nitrone (PBN) are available. These agents cannot be administered safely to human patients, but this problem can be avoided by using a non-invasive spin-trapping technique involving drawing blood samples into syringes containing the spin-trap agent (*ex-vivo* reaction). (For a general review about all described methodologies, see [19].)

Some examples of human diseases associated with free radicals

Clinical conditions in which free radicals are though to be involved are numerous. For a complete listing, we suggest reviews by Halliwell [20] and Gutteridge [11]. Diseases may be classified into two groups depending on the target (Fig. 1). Inflammatory-immune injury [21], ischemia-reflow states [22], drug toxicity [23], iron overload [24], nutritional deficiences [25], alcohol toxicity [26], radiation injury [27], aging [28], cancer [29] and amyloid diseases [30] belong to the first group. In the primary single organ group are found erythrocytes [31], blood vessels [32, 33], lung [34–36], the heart and cardiovascular system [37], kidney [38, 39], the gastrointestinal tract [40], joint abnormalities [41, 42], brain [43–45], eye [46] and skin [47].

In 1990, Sinclair et al. [48] classified diseases associated with increased free radical production as having an intracellular (hyperoxygenation, hypo-oxygenation, chemicals and drugs, alcoholic liver disease, haemolytic iron overload, Parkinson's disease, aging), an extracellular (inflammatory states, immunological and autoimmune diseases, diabetes mellitus, atherosclerosis, cataractogenesis), or both intra- and extracellular (radiation, chemical carcinogens, smoking, air polluants) origin.

More recently, Bulkley [49] classified the human diseases into groups based on pathophysiologic categories: chemical and xenobiotic toxicity, radiation injury, hyperoxygenation syndromes, inflammatory conditions, postischemic reperfusion injury and degenerative conditions.

Herein only a limited number of examples will be discussed: ischemia-reperfusion states in cardiac surgery and organ transplantation, septic

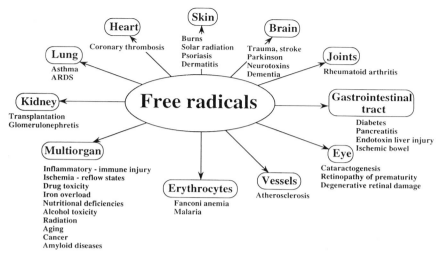

Fig. 1. Spectrum of human diseases where an excessive free radical production is thought to play a significant role in developing tissue injury. For each pathology, the reader will find references in text.

shock, rheumatoid arthritis and acquired immunodeficiency syndrome (AIDS).

Ischemia-reperfusion

Ischemia-reperfusion injury is a field that has been seized upon largely by clinical investigators. It is considered to play a major role in the organ dysfunction accompanying trauma, shock and sepsis, but also in cardiovascular diseases and organ transplantation.

Interruption of blood flow (ischemia) to an organ results in a cascade of biochemical events that predispose to production of increased reactive oxygen species on reperfusion. This has been unequivocally demonstrated by ESR spin trapping studies in both *in vitro* and in animal models [50–56a]. Sources of free radicals in the reperfused organ are: activation of xanthine oxidase and phospholipase, alteration in the electron transport chain, where superoxide anion is produced at two sites (ubiquinone and NADH dehydrogenase), increased metabolism of arachidonic acid by cyclooxygenase and lipooxygenase, accumulation and activation of granulocytes, hemoglobin oxidation and iron release from ferritin mediated by superoxide anion [56b].

Myocardial ischemia
Clinically, reperfusion of ischemic myocardium is recognized as beneficial because mortality is directly related to infarct size, which in turn is

related to the severity and duration of ischemia. However, restoration of normal blood flow to the heart using methods such as angioplasty, thrombolytic agents or aorta-coronary bypass grafting, can lead to specific lesions (arrhythmias, decreased in contractility, necrosis), the importance of which is dependent on the time of ischemia. Using ESR methodology with blood drawn into PBN, Coghlan et al. [57] showed an increased free radical production after reperfusion of infracting tissue in a patient undergoing delayed repair of a transected aorta. The same authors also demonstrated ESR signals in blood taken from the coronary sinus of patients undergoing percutaneous transluminal coronary angioplasty, as ideal model of myocardial ischemia-reperfusion [37]. Because only small ESR signals could be detected during ischemia, the authors concluded that reperfusion was a necessary condition for significant detection of radicals. In patients undergoing cardiopulmonary bypass (CPB) for cardiac surgery (repair of aortic aneurysms, coronary bypass grafting), an increased free radical activity in plasma has been shown to occur following aortic unclamping as shown by increased granulocyte activation [58], loss of antioxidant [59–61], iron overload [62], protein oxidation [63] and the appearance of ESR signals in blood drawn into spin-trapping agents [64, 65].

Organ transplantation
Organ transplantation is a typical example of ischemia-reperfusion, since between harvesting and reperfusion in the recipient, the procedure includes several steps during which severe damage to the organ can occur: cardioplegic arrest (only for heart transplantation) and cooling, storage for varying periods (cold ischemia), warm ischemia during the surgical procedure, and reperfusion. In humans, few studies have been conducted to investigate the generation of free radical under such condition [66–68]. During human kidney transplantation, we observed a decrease of antioxidant vitamin E early after the onset of reperfusion in blood samples specifically from the renal vein, when compared to the value observed in the renal artery just before reperfusion (Fig. 2). A similar decrease was shown in the systemic blood, indicating the occurrence of lipoperoxidation processes [39]. Using ESR methodology, we also directly demonstrated that a burst of free radical production occurred within the first minutes of reperfusion; this was related to increased granulocyte activation. Oxidative damage has also been shown to occur in human liver transplantation.

Septic shock

Severe shock or trauma are often associated with organ dysfunction occurring days after the initiating event. This is also seen after septic

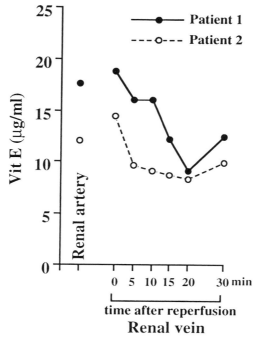

Fig. 2. Evidence of lipid peroxidation processes during human kidney transplantation (two patients) as assessed by the decrease of vitamin E in blood specifically taken from the renal vein since the beginning of graft reperfusion. Reference value was that found in the blood sample drawn from the renal artery just prior to inducing reperfusion.

shock, defined as sepsis with hypotension despite adequate fluid resuscitation, in the presence of perfusion abnormalities. Among the several syndromes which are related to septic shock in man are the adult respiratory distress syndrome (ARDS) or "shock lung syndrome" and multiple systems organ failure (MSOF).

Many papers provide evidence for increased free radical activity in such diseases, which often have a fatal outcome. Because of changes in microcirculatory perfusion, ischemia-reperfusion phenomena are present in sepsis and are considered to be the main source of radical production [69]. In ARDS and septic patients, neutrophils are also another important source; they have been shown to be in an activated state, as demonstrated by high levels of neutrophil proteins (myeloperoxidase, elastase) found in plasma [70] and in bronchoalveolar lavage (BAL) fluids [71] of such patients. α1-proteinase oxidized by excess production of hypochlorous acid (mediated by MPO activity) can be detected in BAL fluid of ARDS patients [71]. Other evidence for increased rates of lipid peroxidation and oxidative damage to proteins as well as depletion of plasma antioxidants has been described in critically ill patients, in ARDS patients, and during human septic shock [72–79]. Serum catalase was found to be increased in ARDS patients with sepsis [80]. Both hydrogen peroxide and pentane were found in the expired air of ARDS patients [81, 82]. Despite all this evidence, it is, however, not clear if oxidative stress is a significant mediator of organ injury in ARDS and/or MSOF. The clinical trials with antioxidant therapies are crucial to clarify this question.

Rheumatoid arthritis

The glycoaminoglycan hyaluronate rapidly loses its viscosity in the presence of free radical generating systems; this can be correlated with the decreased viscosity of synovial fluid from the joints of patients with rheumatoid arthritis. The presence of activated neutrophils [83] as well as the release of iron from lysed cells contribute to free radical activity [84]. Evidence consistent with oxidative stress in rheumatoid disease is abundant although indirect: increased exhalation of pentane [41], increased products of lipid peroxidation [85], loss of antioxidant in both serum and synovial fluid [85], the presence of oxidation products of uric acid [16], and evidence of aromatic hydroxylation [17].

Acquired immunodeficiency syndrome (AIDS)

Recently, a considerable number of papers has been devoted to oxygen free radicals in AIDS (see [86]). Caused by the human immunodeficiency virus (HIV), this disease is characterized by depletion of the T4 + T-cell population and cellular dysfunction that affects several cell types such as the T8 + T-cell subset.

Numerous studies have shown that the generation of activated oxygen species is impaired in mononuclear phagocytes from HIV-infected patients. In a very elegant study, Postaire et al. [87] have described increased release of pentane (a marker of lipid peroxidation) in the alveolar air of HIV patients. An increasing number of investigations has shown that HIV patients were severely depleted of antioxidants (for review, see [88]). GSH levels were profoundly depressed both in plasma and in peripheral blood mononuclear cells and lymphocytes [89, 90]. In comparison to healthy subjects, Favier et al. [91] reported, in stages II and IV of the disease, very low levels of all antioxidant micronutrients, particularly carotene. Retinol deficiency is also observed [92]. Moreover, several workers have clearly established a significant deficiency in trace element factors, especially of selenium [92]. This element is related to antioxidant activity via the enzyme glutathione peroxidase which is decreased significantly in plasma of patients in stages II and IV [91].

These data indicate the presence of an oxidative stress (a profound imbalance between prooxidant and antioxidant activities) during HIV infection. This is of primordial importance, since *in vitro* studies have revealed that oxidative stress can activate the HIV transactivation [93] by stimulating transcription of the nuclear factor κ-B (NF-κB) [5]. On the basis of all these observations, a place for an antioxidant therapy in HIV infection has been suggested.

Antioxidant therapy

The studies described above suggest that excessive free radical production occurs in many human diseases. Currently, the prevalent idea is that an initial insult induces a secondary increase in the rate of free radical production, and that oxidative damage exacerbates the primary tissue injury. Whatever the precise mechanism, preventing free radical generation should provide beneficial clinical consequences, and development of appropriate antioxidant therapy represents an important challenge for the future.

As previously reported, an antioxidant can be defined as "any substance that, when present at low concentrations compared to those of the oxidizable substrate, significantly delays, or inhibits, oxidation of that substrate" [1]. However, an antioxidant may act at a different stage of the production of free radicals, therefore making the design of appropriate antioxidant drugs for clinical studies difficult. Antioxidant substances can be divided into two large classes – those with enzymatic and those with non-enzymatic activities. In the first group are enzymes that remove ROS (superoxide dismutase, catalase, glutathione peroxidase), molecules blocking enzymatic activity (e.g., allopurinol, xanthine oxidase inhibitor) and molecules capable of trapping metal ions, which are potent catalysts of free radical reactions (desferrioxamine or lazaroid compounds). In the second group are molecules which interact mole by mole with the free radical and are, therefore, consumed during the reaction. Vitamin A (a quencher of singlet oxygen), vitamin C, glutathione, mannitol, albumin, probucol, N-Acetylcysteine are such free radical scavengers. Also in this group, vitamin E has a special status since, as a lipid soluble chain-breaking antioxidant, it is also consumed; it can, however, be regenerated via a catalytic cycle involving glutathione or ascorbic acid.

Meetings are increasingly organized throughout the world to critically consider the role of antioxidants in therapy. Several papers have recently reviewed in detail current studies using antioxidants in several human diseases [27, 33, 36, 49, 94–110].

Despite encouraging results [106, 109–113], most studies on the role of antioxidants in humans have been rather disappointing [114, 115].

In the concluding remarks of his talk at the meeting "Antioxidant Therapy—The Way Forward", which was held in Liverpool in 1993, Professor Gutteridge recognized "that, so far, antioxidants have made little or no impact on the treatment of serious diseases" (SFFR Newsletter nr. 14, 1994). The lack of efficacy could be attributed to difficulties in site-delivery of the drug (see [116] and [117] for SOD) at appropriate concentrations [56]. During ischemia-reperfusion studies on animals, the timing of antioxidant administration (before, during of after reperfusion, following short or long ischemia times) appears to be crucial to the outcome of the study. It is therefore crucial to define a therapeutic window for antioxidants during ischemia-reperfusion [107]. At least it should also be kept in mind that any antioxidant agent can be associated with a pro-oxidant action. An example is seen with ascorbic acid in the presence of iron [118], but also with N-acetylcysteine, which may generate thiol-derived free radical species [119]. This could therefore limit the beneficial effect of administration of thiol antioxidants. A pro-oxidant activity for desferrioxamine has also recently been described [120].

Conclusions

There is increasing evidence implicating free radical generation in clinical situations. However, there is also a great deal of confusion about the precise role of these species in the development of human diseases. Because of this, the use of reliable markers for detecting reactive oxygen species in clinical situations is absolutely necessary. Due to its direct nature, we feel that the use of electron spin resonance spectroscopy, especially the actual development of a new generation of devices, for *in vivo* studies, may become feasible in the near future. Indeed, because of its ability to provide an integrated measure of free radical production over a given interval of time (e.g., in ischemia-reperfusion states), this methodology can delineate the time-course of production of free radicals. Direct correlation between free radical production and clinical parameters should become possible; this would be helpful to clarify our knowledge about antioxidant therapy, which is now considered as controversial [121] or even a myth [122].

References

1. Gutteridge, J.M.C. (1994a) Biological origin of free radicals, and mechanisms of antioxidant protection. *Chemico-Biological Interactions* 91: 113–140.
2. Gerschman, R., Gilbert, D.L., Nye, S.W., Dwyer, P. and Fenn, W.O. (1954) Oxygen poisoning and x-irradiation: a mechanism in common. *Science* 119: 623–626.
3. Harman, D. (1956) Aging: a theory based on free radical and radiation chemistry. *J. Geront* 11: 298–300.

4. Laborit, H., Baron, C., Berthou, J., Drouet, J., Gerard, J., Jouany, J.M., Narvaes, C., Niaussat, P and Weber, B. (1960) Introduction à l'étude des antioxydants et des structures à électrons célibataires en biologie. 2ème partie: étude expérimentale et réanimation métabolique. *Agressologie* 1: 133–155.

5. Paeck, R., Rubin, P. and Bauerle, P.A. (1991) Reactive oxygen intermediates as apparently widely used messengers in the activation of the NF-κB transcription factor and HIV. *EMBO J.* 10: 2247–2258.

6. McCord, J.M., Keele, B.B., Jr. and Fridovich, J.J. (1969) Superoxide dismutase, an enzyme function for erythrocuprein. *J. Biol. Chem.* 244: 6049–6055.

7. Klebanoff, S.J. (1982) The iron-H2O2-iodide cytotoxic system. *J. Exp. Med.* 156: 1262–1267.

8. Bast, A. (1986) Is formation of reactive oxygen species by cytochrome P-450 perilous and predictable? *Biochem. Pharmacol.* 37: 569–571.

9. Shapiro, B.M. (1991) The control of oxidant stress at fertilization. *Science* 252: 533–536.

10. Sies, H. (1991) *In:* H. Sies (ed.): *Oxidative Stress. Oxidants and Antioxidants.* Academic Press, London, pp 15–22.

11. Gutteridge, J.M.C. (1993) Invited review free radicals in disease processes: a complication of cause and consequence. *Free Rad. Res. Comm.* 19: 141–158.

12. Janero, D.R. (1990) Malondialdehyde and thiobarbituric acid reactivity as diagnostic indices of lipid peroxidation and peroxidative tissue injury. *Free Rad. Biol. Med.* 9: 515–540.

13. Springfield, J.R. and Levitt, M.D. (1994) Pitfalls in the use of breath pentane measurements to assess lipid peroxidation. *J. Lip. Res.* 35: 1497–1504.

14. Stocker, R. and Frei, B. (1991) Endogenous antioxidant defences in human blood plasma. *In:* H. Sies (ed.): *Oxidative Stress. Oxidants and Antioxidants.* Academic Press, London, pp 213–243.

15. Burton, G.W. and Ingold, K.U. (1989) Vitamin E as an *in vitro* and *in vivo* antioxidant. *In:* A.T. Diplock, L.J. Machlin, L. Packer and W.A. Pryor (eds): *Vitamin E Biochemistry and Health Implications.* Vol. 570, Ann. New York Acad. Sci., pp 7–22.

16. Grootveld, M. and Halliwell, B. (1987) Measurement of allantoin and uric acid in human body fluids. A potential index of free radical reactions *in vivo? Biochem. J.* 243: 803–808.

17. Grootveld, M. and Halliwell, B. (1986) Aromatic hydroxylation as potential measure of hydroxyl radical formation *in vivo.* Identification of hydroxylated derivatives of salicylate in human body fluids. *Biochem. J.* 237: 499–504.

18. Cantilena, L.R., Smith, R.P., Frasur, S., Kruszyna, H., Kruszyna, R. and Wilcox, D.E. (1992) Nitric oxide hemoglobin in patients receiving nitroglycerin as detected by electron paramagnetic resonance spectroscopy. *J. Lab. Clin. Med.* 120: 902–907.

19. Pryor, W.A. and Godber, S.S. (1991) Noninvasive measures of oxidative stress status in humans. *Free Rad. Biol. Med.* 10: 177–184.

20. Halliwell, B. (1987) Oxygen radicals and metals ions: potential antioxidant intervention strategies *In:* C.E. Cross (ed.): *Oxygen Radicals and Human Disease.* Ann. Intern. Med. 107: 526–545.

21. Lunec, J. (1991) Free radicals and the immune response. *Molec. Aspects Med.* 12: 85–174.

22. Powell, S.R. and Tortolani, A.J. (1992) Recent advances in the role of reactive oxygen intermediates in ischemic injury. *J. Surg. Res.* 53: 417–429.

23. Hecht, S.M. (1986) DNA strand scission by activated bleomycin group antibiotics. *Fed. Proc.* 45: 2784–2791.

24. Young, I.S., Trouton, T.G., Torney, J.J., McMaster, D., Callender, M.E. and Trimble, E.R. (1994) Antioxidant status and lipid peroxidation in hereditary haemochromatosis. *Free Rad. Biol. Med.* 16: 393–397.

25. Chiu, D. and Lubin, B. (1980) Abnormal vitamin E and glutathione peroxidase levels in sickle cell anemia: evidence for increased susceptibility to lipid peroxidation *in vivo. J. Lab. Clin. Med.* 94: 542–548.

26. Peters, T.J., O'Connell, M.J., Venkatesan, S. and Ward, R.J. (1986) Evidence for free radical-mediated damage in experimental and human alcoholic disease. *In:* C. Rice-Evans (ed.): *Free Radicals, Cell Damage and Disease.* Richelieu Press, London, pp 99–110.

27. Korkina, L.G., Afanas'ef, I.B. and Diplock, A.T. (1993) Antioxidant therapy in children affected by irradiation from the Chernobyl nuclear accident. *Biochem. Soc. Trans.* 21(3): 314S.
28. Mecocci, P., MacGarvey, U., Kaufman, A. Koontz, D., Shoffner, J.M., Wallace, D.C. and Beal, M.F. (1993) Oxidative damage to mitochondrial DNA shows marked age-dependent increases in human brain. *Ann. Neurol.* 34: 609–616.
29. Frey, K.F., Brubacher, G.B. and Stähelin, H.B. (1987) Plasma levels of antioxidant vitamins in relation to ischemic heart disease and cancer. *Am. J. Clin. Nutr.* 45: 1368–1377.
30. Harman, D. (1984) Free radical theory of aging: the "free-radical diseases". *Age* 7: 111–131.
31. Clark, I.A. and Hunt, N.H. (1983) Evidence for reactive oxygen intermediates causing hemolysis and parasite death in malaria. *Infect. Immun.* 39: 1–6.
32. Janero, D.R. (1991) Therapeutic potential of vitamin E in the pathogensis of spontaneous atherosclerosis. *Free Rad. Biol. Med.* 11: 129–144.
33. Illingworth, D.R. (1993) The potential role of antioxidants in the prevention of atherosclerosis. *J. Nut. Sc. Vitam.* 39: S43–S47.
34. Pincemail, J., Bertrand, Y., Hanique, G., Denis, B., Leenaerts, L., Vankeerberghen, L. and Deby, C. (1989a) Evolution of vitamin E deficiency in patients with adult respiratory distress syndrome. *Ann. NY Acad. Sci.* 570: 498–499.
35. Malvy, J.-M., D., Lebranchu, Y., Richard, M.-J., Arnaud, J. and Favier, A. (1993) Oxidative metabolism and severe asthma in children. *Clin. Chim. Acta.* 218: 117–120.
36. Leuenberger, P. (1994) Respiratory diseases and oxidants. *Schweiz. Medizin. Wochenschrift.* 124: 129–135.
37. Coghlan, J.G., Flitter, W.D., Holle, A.E., Norell, M., Mitchell, A.G., Ilsley, C.D. and Slater, T.F. (1991a) Detection of free radicals and cholesterol hydroperoxides in blood taken from the coronary sinus of man during percutaneous transluminal coronary angioplasty. *Free Rad. Res. Comm.* 14: 409–417.
38. Johnson, K.J., Rehan, A. and Ward, P.A. (1987) Upjohn Symposium. The role of oxygen radicals in kidney disease. *Kidney Disease*, pp 115–121.
39. Pincemail, J., Defraigne, J.O., Franssen, C., Bonnet, P., Deby-Dupont, G., Pirenne, J., Deby, C., Lamy, M., Limet, M. and Meurise, M. (1993) Evidence for free radical formation during human kidney transplantation. *Free Rad. Biol. Med.* 15: 343–348.
40. Scott, P., Bruce, C., Schofield, D., Shiel, N., Braganza, J.M. and McCloy, R.F. (1993) Vitamin C status in patients with acute pancreatitis. *Br. J. Surg.* 80: 750–754.
41. Humad, S., Zarling, E., Clapper, M. and Skosey, J.L. (1988) Breath pentane excretion as a marker of disease activity in rheumatoid arthritis. *Free Rad. Res. Comm.* 5: 101–106.
42. Meier, B., Radeke, H.H., Selle, S., Raspe, H.H., Sies, H., Resch, K. and Habermehl, G. (1990) Human fibroblasts release reactive oxygen species in response to treatment with synovial fluids from patients suffering from arthritis. *Free Rad. Res. Comm.* 8: 149–160.
43. Hall, E.D. and Braughler, J.M. (1989) Central nervous system trauma and stroke. II. Physiological and pharmacological evidence for involvement of oxygen radicals and lipid peroxidation. *Free Rad. Biol. Med.* 6: 303–313.
44. Adams, J.D. and Odunze, I.N. (1991) Oxygen free radicals and Parkinson's disease. *Free Rad. Biol. Med.* 10: 161–169.
45. Girotti, M.J., Khan, N. and McLellan, B.A. (1991) Early measurement of systemic lipid peroxidation products in the plasma of major blunt trauma patients. *J. of Trauma* 31: 32–35.
46. Spector, A. (1991) The lens and oxidative stress. *In:* H. Sies (ed.): *Oxidative Stress. Oxidants and Antioxidants.* Academic Press, London, pp 529–558.
47. Fuchs, J. and Packer, L. (1991) Photooxidative stress in the skin. *In:* H. Sies (ed): *Oxidative Stress, Oxidants and Antioxidants.* Academic Press, London, pp 559–583.
48. Sinclair, A.J., Barnett, A.H. and Lunec, J. (1990) Free radicals and antioxidant systems in health and disease. *Brit. J. Hosp. Med.* 43: 334–344.
49. Bulkley, G.B. (1993) Free radicals and other reactive oxygen metabolites: clinical relevance and the therapeutic efficacy of antioxidant therapy. *Surgery* 113: 479–483.
50. Garlick, P.B., Davies, M.J., Hearse, D.J. and Slater, T.F. (1987) Direct detection of free radicals in the reperfused rat heart using electron spin resonance spectroscopy. *Cir. Res.* 61: 757–760.

51. Pietri, S., Culcasi, M. and Cozzone, P.J. (1989) Real-time continuous-flow spin trapping of hydroxyl radical in the ischemic and post-ischemic myocardium. *Eur. J. Biochem.* 186: 163–173.

52. Nilsson, U.A., Lundgren, O., Haglind, E. and Bylund-Fellenius, A.C. (1989) Radical production during *in vivo* intestinal reperfusion in the cat. *Am. J. Physiol.* 257: G409–G414.

53. Pincemail, J., Defraigne, J.O., Franssen, C., Defechereux, T., Canivet, J.-J., Philippart, C. and Meurisse, M. (1990) Evidence of *in vivo* free radical generation by spin trapping with alpha-phenyl N-tert-butyl nitrone during ischemia-reperfusion in rabbit kidneys. *Free Rad. Res. Comm.* 9: 3–6.

54. Tosaki, A., Bagchi, D., Pali, T., Cordis, G.A. and Das, D.P. (1993) Comparisons of esr and HPLC methods for the detection of OH˙ radicals in ischemic-reperfused hearts. *Biochem. Pharmacol.* 45: 961–969.

55. Li, X.Y., McCay, P.B., Zughaib, M., Jeroudi, M.O., Triana, J.F. and Bolli, R. (1993) Demonstration of free radical generation in the "stunned" myocardium in the conscious dog and identification of major differences between conscious and open-chest dogs. *J. Clin. Invest.* 92: 1025-1041.

56a Connor, H.D., Gao, W., Mason, R.P. and Thurman, R.G. (1993) New reactive oxidizing species causes formation of carbon-centered radical adducts in organic extracts of blood following liver transplantation. *Free Rad. Biol. Med.* 16: 871–875.

56b Omar, B., McCord, J. and Downey, J. (1991) Ischemia-reperfusion. *In:* H. Sies (ed.): *Oxidative Stress. Oxidants and Antioxidants.* Academic Press, London, pp 494–527.

57. Coghlan, J.G., Flitter, W.D., Isley, C.D., Rees, A. and Slater, T.F. (1991b) Reperfusion of infarcted tissue and free radicals. *Lancet* 338: 1145.

58. Faymonville, M.E., Pincemail, J., Duchateau, M.D., Paulus, J.M., Adam, A., Deby-Dupont, G., Deby, C., Albert, A., Larbuisson, R., Limet, R. and Lamy, M. (1991) Myeloperoxidase and elastase as markers of leukocyte activation during cardiopulmonary bypass in humans. *J. Thorac. Cardiovasc. Surg.* 102: 309–317.

59. Cavarocchi, N.C., England, M.D., O'Brien, J.F., Solis, E., Russo, P., Schaff, H.V., Orszulak, T.A., Pluth, J.R. and Kaye, M.P. (1986) Superoxide generation during cardiopulmonary bypass: is there a role for vitamin E. *J. Surg. Res.* 40: 519–527.

60. Pincemail, J., Faymonville, M., Deby-Dupont, G., Thirion, A., Deby, C. and Lamy, M. (1989b) Neutrophils activation evidenced by plasmatic myeloperoxidase release during cardiopulmonary bypass. Consequence on vitamin E status. *Ann. NY Acad. Sci.* 570: 501–502.

61. Murphy, M.E., Kolvenback, R., Aleksis, M., Hansen, R. and Sies, H. (1992) Antioxidant depletion in aortic crossclamping ischemia: increase of the plasma alpha tocopheryl quinone/alpha-tocopherol ratio. *Free Rad. Biol. Med.* 13: 95–100.

62. Pepper, J.R., Mumby, S. and Gutteridge, J.M.C. (1994) Transient iron-overload with bleomycin-detectable iron present during cardiopulmonary bypass surgery. *Free Rad. Res. Comm.* 21: 53–58.

63. Ward, A., McBurney, A. and Lunec, J. (1994) Evidence for the involvement of oxygen-derived free radicals in ischemia-reperfusion injury. *Free Rad. Res. Comm.* 20: 21–28.

64. Culcasi, M., Pietri, S., Carrière, I., d'Arbigny, P. and Drieu, K. (1993) Electron-spin-resonance study of the protective effects of Ginkgo biloba (EGb 761) on reperfusion-induced free-radical generation associated with plasma ascorbate consumption during open-heart surgery in man. *In:* C. Ferradini, M.T. Droy-Lefaix and Y. Christen (eds): *Advances in Gingko biloba Extract Research,* Vol 2. *Ginkgo Biloba (EGb 761) as Free Radical Scavenger* pp 153–162.

65. Hartstein, G., Pincemail, J., Deby-Dupont, G., Deby, C., Larbuisson, R. and Lamy, M. (1993) Evidence for free radical formation during cardiopulmonary bypass in man. *Anesthesio.* 79: no. 3A A140.

66. Heim, K.F., Makila, U.-M., Leveson, R., Ledley, G.S., Thomas, G., Rackley, C. and Ramwell, P.W. (1987) Detection of pentane as a measurement of lipid peroxidation in humans using gas chromatography with a photoionization detector. *In:* M. Paubert-Braqut (ed.): *Lipid Mediators in the Immunology of Shock.* Nato ASI Series vol., 139. Plenum Press, New York, pp 103–108.

67. Serino, F., Citterio, F., Lippa, S., Oradei, A., Agnes, S., Nanni, A., Pozzetto, A., Littaru, G. and Castagneto, M. (1990) Coenzyme Q, alpha tocopherol and delayed function in human kidney transplantation. *Transpl. Proc.* 22: 2224–2225.

68. Rabl, H., Khoschorur, G., Colombo, T., Tatzber, F. and Esterbauer, H. (1992) Human plasma lipid peroxide levels show a strong transient increase after successful revascularization operations. *Free Rad. Biol. Med.* 13: 281–288.
69. Zimmerman, J.J. (1992) Oxyradical pathogenesis in sepsis. *In:* M. Lamy and L.G. Thijs (eds): *Update in Intensive Care and Emergency Medicine. Mediators of Sepsis.* Vol. 16, pp 136–151.
70. Pincemail, J., Deby-Dupont, G., Deby, C., Thirion, A., Torpier, G., Faymonville, M., Damas, P., Tomassini, M., Lamy, M. and Franchimont, P. (1991) Fast double antibody radioimmunoassay of human granulocytes myeloperoxidase and its application to plasma. *J. Immunol. Meth.* 137: 181–191.
71. Cochrane, C.G., Spragg, R., Revak, S.D., Cohen, A.B. and McGuire, W.W. (1983) The presence of neutrophil elastase and evidence of oxidant activity in bronchoalveolar lavage fluid. *Am. Rev. Resp. Dis.* 127: S25–S27.
72. Takeda, K., Shimada, Y., Amano, M., Sakai, T., Okada, T. and Yoshiya, I. (1984) Plasma lipid peroxides and alpha-tocopherol in critically ill patients. *Crit. Car. Med.* 12: 957–959.
73. Bertrand, Y., Pincemail, J., Hannique, G., Denis, B., Leenaerts, L., Vankeerberghen, L. and Deby, C. (1986) Differences in tocopherol-lipid ratio in ARDS and no ARDS patients. *Int. Car. Med.* 15: 557–559.
74. Keen, R.R., Stella, L., Flanigan, P. and Lands, W.E.M. (1991) Differential detection of plasma hydroperoxides in sepsis. *Crit. Care Med.* 19: 1114–1119.
75. Ogilvie, A.C., Groeneveld, A.B.J., Straub, J.P. and Thijs, L.G. (1991) Plasma lipid peroxides and antioxidants in human septic shock. *Int. Care Med.* 17: 40–44.
76. Toonen, T.R., Lewandoski, J.R. and Zimmerman, J.J. (1991) Longitudinal analysis of plasma tocopherol levels in patients with septic shock. *Clin. Res.* 39: 688.
77. Krsek-Staples, J.A., Kew, R.R. and Wester, R.O. (1992) Ceruloplasmin and transferrin levels are altered in serum and bronchoalveolar lavage fluid of patients with the adult respiratory distress syndrome. *Am. Rev. Respir. Dis.* 145: 1009–1015.
78. Quinlain, G.J., Evans, T.W. and Gutteridge, J.M.C. (1994a) Oxidative damage to plasma proteins in adult respiratory distress syndrome. *Free Rad. Res. Comm.* 20: 289–298.
79. Quinlain, J.G., Evans, T.W. and Gutteridge, J.M.C. (1994b) Linoleic acid and protein thiol changes suggestive of oxidative damage in the plasma of patients with adult respiratory distress syndrome. *Free Rad. Res. Comm.* 20: 299–306.
80. Leff, J.A., Parsons, P.F., Day, C.E., Taniguchi, N., Jochum, M., Fritz, H., Moore, F.A., Moore, E., McCord, J.M. and Repine, J.E. (1993) Serum antioxidants as predictors of adult respiratory distress syndrome in patients with sepsis. *Lancet* 341: 777–780.
81. Baldwin, S.R., Simon, R.H., Grum, C.M., Ketai, L.H., Boder, L.A. and Devall, L.J. (1986) Oxidant activity in expired breath of patients with adult respiratory distress syndrome. *Lancet* 1: 11–14.
82. Pincemail, J., Deby, C., Dethier, A., Bertrand, Y., Lismonde, M. and Lamy, M. (1986) Pentane measurement in man as an index of lipid peroxidation. *Bioelectro. and Bioenerg.* 18: 117–126.
83. Nurcombe, H.L., Bucknall, R.C. and Edwards, S.W. (1991) Neutrophils isolated from the synovial fluid of patients with rheumatoid arthritis: priming and activation *in vivo*. *Ann. Rheum. Dis.* 50: 196–200.
84. Gutteridge, J.M.C. (1987) Bleomycin-detectable iron in the knee-joint synovial fluid from arthritic patients and its relationship to the extracellular antioxidant activities of caeruloplasmin, transferrin and lactoferrin. *Lancet* 2: 415–421.
85. Merry, P., Grootveld, M., Lunec, J. and Blake, D.R. (1991) Oxidative damage to lipids within inflamed human joints provides evidence of radical-mediated hypoxic-reperfusion injury. *Am. J. Clin. Nutr.* 56: 362S–369S.
86. The place of oxygen free radicals in HIV infections (1994), presented at a meeting organized in France by A.E. Favier, published in *Chemico-Biological Interactions* (Guest Editor A.E. Favier), Vol. 91, pp 77–232.
87. Postaire, E., Massias, L., Lopez, O., Mollerau, M. and Hazebroucq, G. (1994) Alcanes measurements in human immunodeficiency virus infection. *In:* C. Pasquier, R.Y. Olivier, C. Auclair and L. Packer (eds): *Oxidative Stress, Cell Activation and Viral Infection.* Birkhäuser Verlag, Basel, pp 333–340.

88. Müller, F. (1992) Reactive oxygen intermediates and human immunodeficiency virus (HIV) infection. *Free Rad. Biol. Med.* 13: 651–657.
89. Staal, F.J.T., Roederer, M., Israelski, D.M., Buop, J., Mole, L.A., McShane, D., Deresinski, S.C., Ross, W., Sussman, H., Rago, P.A., Anderson, M.T., Moore, W., Elsa, W. and Hersenberg, L.A. (1992) Intracellular glutathione levels in T cells subset decrease in HIV infected patients. *AIDS Res. Human Retroviruses* 2: 311.
90. Dröge, W., Eck, H.-P., Mihm, S. and Galter, D. (1994) Abnormal redox regulation in HIV infection and other immunodeficiency diseases. *In:* C. Pasquier, R.Y. Olivier, C. Auclair and L. Packer (eds): *Oxidative Stress, Cell Activation and Viral Infection.* Birkhäuser Verlag, Basel, pp 285–299.
91. Favier, A., Sappey, C., Leclerc, P., Faure, P. and Micoud, M. (1994) Antioxidant status and lipid peroxidation in patients infected with HIV. *Chemico-Biological Interactions* 91: 165–180.
92. Sergeant, C., Simonoff, M., Hamon, C., Peuchant, E., Dumon, M.F., Clerc, M., Thomas, M.J., Constans, J., Conri, C., Pelligrin, J.L. and Leng, B. (1994) Plasma antioxidant status (selenium, retinol and α-tocopherol) in HIV infection. *In:* C. Pasquier, R.Y. Olivier, C. Auclair and L. Packer (eds): *Oxidative Stress, Cell Activation and Viral Infection,* Birkhäuser Verlag, Basel, pp 341–351.
93. Legrand-Poels, S., Vaira, D., Pincemail, J., Van de Vorst, A. and Piette, J. (1990) Activation of human immunodeficiency virus type 1 by oxidative stress. *AIDS Res.* 6: 1389–1397.
94. Uden, S., Bilton, D., Nathan, L., Hunt, L.P., Main, C., Braganza, J.M. (1990) Antioxidant therapy for reccurent pancreatitis: placebo-controlled trial. *Alim. Pharmacol. Therapeu.* 4: 357–371.
95. Clemens, M.R. (1990) Antioxidant therapy in hematological disorders. *Ad. Exp. Med. Biol.* 264: 423–433.
96. Anonymous (1990) Antioxidants in therapy and preventive medicine. *Ad. Exp. Med. Biol.* 264: 1–576.
97. Müller, D.P. (1990) Antioxidant therapy in neurological disorders. *Adv. Exp. Med. Biol.* 264: 475–484.
98. Hearse, D.J. (1991) Prospects for antioxidant therapy in cardiovascular medicine. *Am. J. Med.* 91: 118S–121S.
99. Youn, Y.-K., LaLonde, C. and Demling, R. (1991) Use of antioxidant therapy in shock and trauma. *Circ. Shock* 35: 245–249.
100. Chow, C.K. (1991) Vitamin E and oxidative stress. *Free Rad. Biol. Med.* 11: 215–232.
101. Deucher, G.P. (1992) Antioxidant therapy in the aging process. *In:* I. Emerit and B. Chance (eds): *Free Radicals and Aging,* Birkhäuser Verlag, Basel, pp 428–437.
102. Manson, J.E., Gaziono, J.M., Jonas, M.A. and Hennkens, C.H. (1993) Antioxidants and cardiovascular disease. *J. Am. Coll. Nut.* 12: 426–432.
103. Goode, H.F. and Webster, N.R. (1993) Free radicals and antioxidants in sepsis. *Crit. Care Med.* 21: 1770–1776.
104. Packer, L. (1993) The role of anti-oxidative treatment in diabetes mellitus. *Diabetol.* 36: 1212–1213.
105. Levander, O.A. and Ager, A.L. (1993) Malarial parasites and antioxidant nutrients. *Parasitology* 107: S95–S106.
106. Sies, H. (1993) *Efficacy of vitamin E in the Human.* Veris, Lagrange, Illinois, USA.
107. Schiller, H.J., Reilly, P.M. and Bulkley, G.B. (1993) Antioxidant therapy. *Crit. Care. Med.* 21: S92–S102.
108. Yoshikawa, T., Naito, Y. and Kondon, M. (1993) Antioxidant therapy in digestive diseases. *J. Nutr. Sc. Vitam.* 39: S35–S41.
109. Rice-Evans, C.A. and Diplock, A.T. (1993) Current status of antioxidant therapy. *Free Rad. Biol. Med.* 15: 77–96.
110. Baruchel, S., Bounous, G. and Gold, P. (1994) Place for an antioxidant therapy in human immunodeficiency virus (HIV) infection. *In:* C. Pasquier, R.Y. Olivier, C. Auclair and L. Packer (eds): *Oxidative Stress, Cell Activation and Viral Infection.* Birkhäuser Verlag, Basel, pp 311–321.
111. Parthasarathy, S. and Rankin, S.M. (1992) Role of oxidised low density lipoproteins in atherogenesis. *Prog. Lipid Res.* 31: 127–143.
112. Dröge, W. (1993) Cysteine and gluthathione deficiency in AIDS patients: a rationale for the treatment with N-acetyl-cysteine. *Pharmacology* 46: 61–65.

113. Schneeberger, H., Schleibner, S., Schilling, M., Illner, W.D., Abendroth, D., Hancke, E., Janicke, U. and Land, W. (1990) Prevention of acute renal failure after kidney transplantation with rh-SOD; interim analysis of a double-bind placebo-controlled trial. *Transp. Proceed.* 22: 2224–2225.

114. Ganguly, P.K. (1991) Antioxidant therapy in congestive heart failure: is there any advantage? *J. Inter. Med.* 229: 205–208.

115. Hennekens, C.H., Buring, J.E. and Peto, R. (1994) Antioxidant vitamins—benefits not yet proved. *New Engl. J. Med.* 330: 1080–1081.

116. Flohé, L. (1988) Superoxide dismutase for therapeutic use: clinical experience, dead ends and hopes. *Mol. Cell. Biochem.* 84: 123–131.

117. Przyklenk, K. and Kloner, R.A. (1989) "Reperfusion injury" by oxygen-derived free radicals? *Cir. Res.* 64: 86–96.

118. Herbert, V. (1993) Dangers of iron and vitamin C supplements. *J. Am. Diet. Assoc.* 93: 526–527.

119. Turner, J.J.O., Rice-Evans, C., Davies, M.J. and Newma, E.S.R. (1991) The formation of free radicals by cardiac monocytes under oxidative stress and the effects of electron-donating drugs. *Biochem. J.* 277: 833–837.

120. Gutteridge, J.M.C., Quinlan, G.J., Swain, J. and Cox, J. (1994b) Ferrous ion formation by ferrioxamine prepared from aged desferrioxamine: a potential prooxidant property. *Free Rad. Biol. Med.* 16: 733–739.

121. Barinaga, M. (1991) Vitamin C gets a little respect. *Science* 254: 374–376.

122. Herbert, V. (1994) The antioxidant supplement myth. *Am. J. Clin. Nutr.* 60: 157–158.

Analysis of Free Radicals in Biological Systems
Favier et al. (eds)
© 1995 Birkhäuser Verlag Basel/Switzerland

How to demonstrate the occurrence of an oxidative stress in human?

A.E. Favier

GREPO (Research Group on Oxidative Diseases), Université de Grenoble, Faculté de Pharmacie, F-38700 La Tronche, France

Summary. There is an increasing need for accurate indicators of oxidative stress in human. Direct determination of free radical can be obtained by physical methods such as electron spin resonance (ESR) or chemiluminescence. But their utilization is limited to animal experiments, and only limited data have been obtained in humans, by ESR, for ascorbyl or peroxyl radicals.

Generally, evaluation of oxidative stress is done by determining damaged biological products. Various lipid derivatives can be determined: conjugated dienes, hydroperoxides, aldehydes such as malonaldehyde (MDA) or 4-hydroxynonenal, hydrocarbons such as ethane or pentane in exhaled air. Other oxidized biological compounds can be measured: derivatives of DNA such as 8HO-guanosine or 5HO-methyl uracil in urine, or derivatives of proteins in plasma such as carbonyl or thiol groups, or derivatives of free aminoacids such as O-tyrosine. The oxidized form of scavengers can be evaluated in biological fluid to prove the existence of oxidative stress: the ratio between oxidized glutathione and total glutathione, tocopheryl quinone, oxidized ascorbate. Many other indicators exist: HSP, haeme oxygenase, conjugates MDA-protein, or MDA-guanine, antibodies against MDA-proteins (anti AIP), and oxidized cholesterol. However, the choice of the diagnostic indicator depends mainly on the type of oxidative stress; acute stress such as ischemia needs fast reacting indicators whereas chronic stress such as rheumatism needs an index of median levels of peroxidation for a long period such as in the determination of lipid peroxidation or chemical trapping. Generally, MDA – although controversial – still remains the most commonly used indicator of oxidative stress.

The implication of free radicals in various pathological processes has been involved in an increasing number of human diseases. To demonstrate such an implication free radicals are chemically generated in the cell culture or the animal or specifically suppressed by addition of antioxidants in the model. A stimulating challenge is to demonstrate the free radical mechanism in humans. In the same way the efficiency of various new antioxidant agents needs to be demonstrated in patients. However, a major difficulty is to quantify the oxidative stress in humans. At present, no absolute and definitive index of oxidative stress has been discovered. Nonetheless, there exist numerous parameters proposed as indicators of oxidative stress which have been already described [1–4].

Analytical problems in measuring free radical stress in human

Problems specific to oxygen radical chemistry

The main difficulty is the short half-life of free radicals, which is often shorter than 10^{-4} s. Also, the production of free radical is not a regular and constant mechanism, but is often a fast and pulsed phenomenon. The term "oxidative burst" to describe their production by polymorphonuclear leukocytes is characteristic of the brevity of such a phenomenon. With these conditions, and because they are produced mainly in some tissues inside the body and less in plasma, it is difficult to assess the primary free radicals from a patient blood sample. Fortunately, primary radicals generated by activation of oxygen will form secondary and more stable radicals when encountering an organic compound. Others products are formed as demonstrated in Figure 1. A number of oxidized compounds is identified from the reaction with oxygen radicals. Many oxidized biological products can be produced by various reactions involving different radicals and are not typical of a particular stress. For instance, it is very difficult to know if singlet oxygen has been produced *in vivo* during an oxidative process. During *in vitro* tests, it will be necessary to use specific detoxifying enzymes (SOD, catalase) or selective scavenger to improve the specificity.

Oxidized compounds are often reactive, forming adducts with amino groups of proteins or with DNA bases. Such a reactive compound is denoted as secondary messenger of oxidative stress. Furthermore, the adducts themselves are biologically reactive even if they are not chemi-

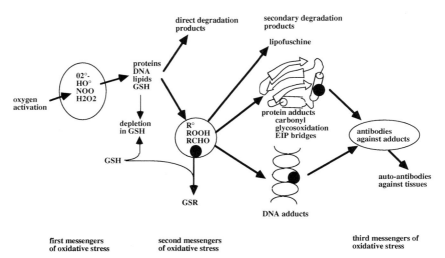

Fig. 1. Formation of radicals and toxic messengers from the initial activation of oxygen into a radical species.

cally reactive. As an unusual endogenous product they are recognized by lymphocytes as antigen. Therefore, we can characterize them as third messengers of oxidative stress. The resulting antibodies are often non-specific auto antibodies resulting in tissue damages.

General problems common to all clinical tests

From the point of view of the clinical chemist, the indicators of oxidative stress have to respond to different criteria of quality. First, we need good precision and good accuracy; in general, they are usually not achieved when determining peroxides or enzymes, i.e., some published techniques give a poor recovery of standard addition or precision. Few methods have been standardized or optimized by the standardization committees of scientific societies. The clinical results cannot be easily compared from one laboratory to another. The more important defect is the lack of reference material for quality control. We do not have any standardized biological material for MDA or GPx or SOD. Fortunately, some collaborative studies are in progress such as those done by the European FLAIR 10 group on total blood certified for GPx.

What parameters reflect oxidative stress

It will, of course, be interesting to demonstrate directly inside some tissue the production of free radicals and to identify them. Because of

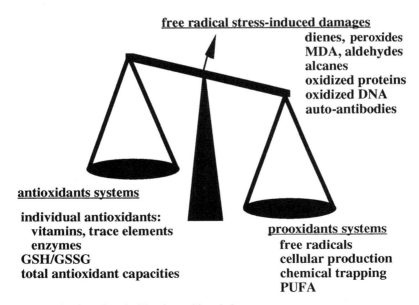

free radical stress-induced damages
dienes, peroxides
MDA, aldehydes
alcanes
oxidized proteins
oxidized DNA
auto-antibodies

antioxidants systems
individual antioxidants:
   vitamins, trace elements
   enzymes
GSH/GSSG
total antioxidant capacities

prooxidants systems
free radicals
cellular production
chemical trapping
PUFA

Fig. 2. Investigation of antioxidant/prooxidant balance.

the fast half-life of such species, we can rarely do so, but can more easily isolate them from blood cells producing free radicals. Another alternative consists of stabilizing primary or secondary radicals by a trapping agent when measuring ESR signal, but also to trap them *in vivo* by a safe chemical agent.

As free radical stress results from an imbalance between free radical production and antioxidant systems (Fig. 2), we measure the antioxidant levels separately or as a total global antioxidant capacity of biological tissue.

Generally, stress-generated oxidized products are rather stable and easier to determine. However, some are reactive and rapidly metabolized. Even after sampling, the oxidation can continue and the derivatives react. Thus, the determination of adducts and more antibodies against oxidized derivative will be a method of choice. Many parameters that can be investigated are presented in Table 1.

Table 1. List of indicators of oxidative stress in human or animal

---

Measurement of radicals:
    spin-trapping of peroxyl radicals in blood and ESR
    ascorbyl radical in blood by ESR
    nitrosohaemoglobin in tissue by ESR
    stimulated production of radicals by: total blood, PMN, monocytes

Measurement of oxidized biological compounds
    Lipid peroxidation: – decrease in polyunsaturated fatty acids in plasma
                                – conjugated dienes in plasma
                                – hydroperoxides in plasma
                                – aldehydes in plasma:
                                      TBA reactant substances
                                      malonaldehyde
                                      4 hydroxynonenal
                                – hexane and pentane in expired air
                                – oxide of cholesterol in plasma
                                – oxidized lipoprotein (LDL) in plasma
    Protein oxidation: – reduced thiol group of plasma proteins
                                  – protein carbonyls in plasma
                                – Formyl kynurenine in IgG
                                – ortho tyrosine in plasma proteins
                                – autoantibodies against iminopropene bridges
    DNA oxidation: – 8-hydroxy deoxy guanosine
                                – 5-hydroxymethyl uracil in urine
                                – MDA guanine adduct in urine
                                – autoantibodies against oxidized DNA

Dynamic loading test of oxidation:
                                – oxidation of salicylate into 2,3 dihydroxybenzoate

Oxidation of scavengers: – reduced/total glutathione
                                – tocopherylquinone/tocopherol
                                – decrease in ascorbate, tocopherol or carotene in plasma

Total antioxidant capacity of plasma or red blood cells

Direct determination of free radicals

Colorimetry or fluorimetry

Colorimetry permits to measure superoxide anion using the absorption spectrum of reduced cytochrome c or MTT, but also hydrogen peroxide by reaction with orange xylenol [5]. Nitric oxide can be measured by the absorption at 578 nm of its reaction product with haemoglobin, or as nitrite by a colorimetric reaction with sulfanilic acid (i.e., Gries's reaction).

Fluorometric techniques use dichlorofluorescein diacetate to measure hydrogen peroxide [6], or the enzymatic reaction with p-hydroxyphenylacetate catalysed by peroxidase [7].

Besides having problems with specificity or sensitivity, these techniques have little use in humans. They can, however, be used to test the production of radicals by cells from blood or in culture.

Electron spin resonance

Electron spin resonance (ESR or EPR) detects free radicals from the magnetic field induced by the non-paired electron characteristic of these species. It is the most frequently used technique by chemists and physicists because it permits them to identify the chemical nature of the radical and measure the kinetics of formation or decomposition of radicals. Many published works describe the fundamental basis and practical utilization of this technique [8]. In humans this technique cannot be used without using a spin trap agent to stabilize the secondary radicals. At present, spin-trap agents cannot be injected into humans; although they have an antioxidant effect [9], they can also be toxic in animals [10]. By rapid collection of blood in a tube containing a spin trap agent, abnormal ESR signals are detected after reperfusion during coronary angioplastia [11], and alkoxyl and carbon-centred free radicals are detected in coronary sinus blood from patients undergoing elective cardioplegia [12]. Ascorbate radical may also be a marker of oxidative stress. The one-electron oxidation of ascorbate produces the ascorbate free radical that has a longer lifetime than hydroxyl or peroxyl radicals and presents a characteristic doublet in ESR (aH4 = 1.8 G) [13]. It is found in human plasma at values around 0.1 to 0.03 mM. This radical is formed not only during radical reaction, but in all oxidative process concerning ascorbate (enzymatic, catalytic, photo oxidative) [14], and a decrease in ascorbyl radical is found after aortic valve replacement [15].

Chemiluminescence

Chemiluminescence is a very attractive way to evaluate an oxidative stress *in vivo*.

A direct ultra-weak emission of light by singlet oxygen can be measured using a germanium photo detector cooled by liquid nitrogen [16]. The emission of the bimol can be measured directly in cells or animals using a special apparatus with a red-sensitive photomultiplier [17]. The emission of light has been observed during stress realized in hepatic microsomes, isolated hepatocytes, tissues [18], but also in whole liver or heart after various stress [19] and after postischemic reoxygenation [20]. We used this technique to demonstrate the production of free radicals by human cells when stimulated by TNF [21].

In humans direct luminescence has been used to detect in expired air free radicals produced by tobacco smoke. Direct luminescence of centrifuged fresh urine is related to the amount of peroxides in urine and is suppressed by diethyldithiocarbamage but not by the usual scavengers, this light can come from the decomposition of peroxides [22].

The use of a chemilumigenic probe producing light when reacting with an oxygen radical increases the sensitivity of these techniques. *Luminol* (5-amino-2,3-dihydro-1,4-phtalazinedione) emits a light at 480 nm and is sensitive to various oxygen derivatives produced by PMN, H_2O_2, $^1O_2$, $HO°$, $HOCl$. Its emission is prone to many interferences [23]. *Lucigenine* (10,10'-dimethyl-9,9'-biacridinium dinitrate) is more specific to superoxide anions. According to some authors it is unable to enter the cell, but others claim it specifically measures the production of superoxide by the mitochondria and not the NADPH oxidase on membrane of macrophages [24].

Luminol has been widely used to determine free radical production by isolated PMN after stimulation [25]. But the separation step is time consuming and can change the native production, or the sensitivity of the cells to stimuli. Some authors obtained more physiological results when measuring the production of total blood after stimulation by PMA [26]. That technique gives interesting results in the clinical situation, demonstrating a priming effect of coronary angioplastia [27], an overproduction in AIDS [28], or in haemodialyzed patients after dialysis [29]. Chemiluminescence is very sensitive and thus can be applied to measure free radical production in human biopsies, such as with the catalase-inhibitable chemiluminescence produced by biopsy specimens obtained during colonoscopy from patients with ulcerative colitis [30]. But chemiluminescence suffers from some major difficulties, i.e., there is no reliable standard to compare the emission from day to day, and its high sensitivity to quenching agents.

Chemiluminescence has also been used as a detecting system to measure phospholipid hydroperoxides, mainly after HPLC separation because it reaches a high sensitivity [31]. A special chemiluminometer for gas permits the measurement of NO after extraction by a stream of inert gas and a reaction with ozone [32].

Determination of biological damage

When reacting with biological compounds of the body free radicals produce oxidized derivatives whose level can be used as a characteristic index of oxidative stress.

Oxidation of endogenous compounds

Lipid peroxidation
Because lipid oxidation is easier to measure, derivatives of lipid peroxidation continue to be the most popular diagnostic markers of oxidative stress [33].

Lipid peroxidation is a chain reaction initiated by the hydrogen abstraction or addition by an oxygen radical, resulting in the oxidative deterioration of polyunsaturated fatty acids [34]. Since polyunsaturated fatty acids are more sensitive than saturated ones, it is obvious that the activated methylene bridge represents a critical target site. In sequence of their appearance, alkyl, peroxyl and alkoxyl radicals are involved [35]. The resulting fatty acid radical is stabilized by rearrangement into a conjugated diene that retains the more stable products including hydroperoxides, alcohols, aldehydes, and alkanes. Many of these products can be found in biological fluids, as well as addition-derivatives of these very reactive end-products.

Conjugated dienes
Molecules expressing the diene configuration are characterized by an intense light absorption at 215–250 nm. They are easy to measure in body fluids after solvent extraction; unfortunately, the absorption maximum is not so intense and is superimposed on the end absorption of lipid extract [36]. Nevertheless, they can demonstrate increased peroxidation in blood and synovial fluids of patients with rheumatoid disease [37].

Hydroperoxides
Hydroperoxides are rather stable even if they become labile in the presence of metal ions. Different techniques exist to measure lipid hydroperoxides in serum. Iodometry needs careful precaution to prevent auto-oxidation by air of the excess unreacted iodide [38]. Hydroperoxides are measured by an enzymatic assay using glutathione peroxidase and glutathione reductase and monitoring the disappearance of NADPH [39]. This method is easy to run but not very sensitive and needs to inactivate endogenous enzymes present in biological fluids [40].

Aldehydes

As a result of lipid peroxidation a great variety of aldehydes can be produced, including hexanal, malonaldialdehyde, 5 hydroxynonenal, etc.

Malonaldehyde (MDA) is the widely used indicator of oxidative stress. His assay can be run in a specific determination or in a non-specific test with thiobarbituric acid, also measuring other aldehydes such as TBA-reactant substances (TBARs).

Malonaldehyde and TAB-reactant substances. MDA can act as a nucleophilic or electrophilic compound. Its reactivity is such that it non-specifically binds to various biological molecules present in the samples (proteins, nucleic acids, aminophospholipids, etc.) [41]. It reacts with neighbouring biological molecules once it is formed and is found in the plasma protein fraction and not in the lipoprotein fraction. MDA molecules also undergo self-condensation reactions, yielding polymers of variable molecular weight and polarity.

Free MDA. Many authors failed to measure a detectable amount of free MDA in biological fluids [42]. These results seem logical as the low level of MDA formed can react rapidly with amine and thiol groups [43], which are also metabolized in tissues by aldehyde dehydrogenase to form acetyl-CoA [44]. MDA can also be easily excreted in urine [45].

Liberated MDA. Conjugated or polymerized forms of MDA can be hydrolysed in acidic medium and are heat-labile. It is first necessary to hydrolyse directly in the presence of a reagent such as thiobarbituric acid in the TBA tests. The test can be performed after a previous precipitation of protein as in the technique of Yagi [46], or directly in total plasma as in the technique of Dousset [47]. The TBA-MDA conjugate can be detected by its colour, but more specifically by its fluorimetry [48]. It will be important to limit phenomena of *in vitro* peroxidation during the heating phase of the test, using reagents and materials which do not contain measureable quantities of iron, or chelators such as deferrioxamine [49], but the use of antioxidants as BHT is preferable.

A new patented derivative (R1) has been developed by R and D System Company (Abingdon, UK) that reacts specifically with MDA and 4-hydroxyalkenals at 40°C, yielding a stable chromophore absorbing at 586 nm, which prevents decomposition of peroxides by high temperature.

A more specific assay of malonaldialdehyde is performed by high performance liquid chromatography that satisfies the criteria of accuracy, specificity and sensitivity and is a method of choice for evaluating the oxidative stress status [50].

The determination of MDA gave excellent results in a great variety of clinical situations with oxidative stress: lipofushinosis [51], cystic fibrosis, Duchenne's myopathy [52], breast cancer [53], AIDS [54], rheumatoid arthritis [37], diabetes [55], coronary arteriosclerosis [56], leu-kaemia [57]. Determination of MDA permits to demonstrate the efficiency of antioxidants when treating patients, for example, for phenyketonuria [58] or cystic fibrosis [59], or those patients undergoing haemodialysis [60] treated with selenium.

Normal values of MDA depend on the technique, ranging from more than 4 μmol/l for TBA-reactants by colorimetric tests, to 2.5 μmol/l by fluorometry and to 0.60 to 1 μmol/l by HPLC. But normal values are different for men and women and increase with age [61, 62], and increase also during pregnancy [63]. All these variations may not be physiological and result in fact from a defect in antioxidant and from an increase in oxidative stress observed with age and pregnancy.

Other aldehydes. Several dozen aldehydes are formed during lipid peroxidation: n-alkanals (propanal, butanal, pentanal, hexanal, etc.), 2-alkenals (acrolein, pentenal, hexenal, etc.), 2−4 alkadienals (heptadi-enal, octadienal, decadienal), 4-hydroxy-2,5-undecadienal, 5-hydroxy octanal, 4-hydroxy-2-alkenals (4-hydroxyhexenal, 4-hydroxyoctenal, 4-hydroxynonenal), malonaldialdehyde. The major compounds found in biological samples are hexanal, MDA and 4 HO-nonenal, depending on the diet [64].

4-hydroxynonenal (HNE) is another interesting index of oxidative stress. It accumulates more than hexanal after oxidative stress, because it is not a substrate for aldehyde dehydrogenase [65]. It metabolizes as fast as MDA in tissue, mainly after reaction with glutathione [66]. HNE by itself is very cytotoxic and genotoxic [67, 68]. Generally HNE is determined by HPLC directly by detecting its strong absorbency at 220 nm, after formation of a coloured derivative with DNPH [69], or a fluorometric derivative with cyclohexanedione [70]. Normal values are around 0.68 ± 0.42 micromole/L in plasma [71].

Alkanes and alkenes
More than 371 volatile derivatives can be detected in the expired air from patients. Hydrocarbons such as ethane and pentane represent 0.1 to 10% of lipid peroxidation in the body. Usually respiratory gases are trapped on a column that is further desorbed inside a gas chro-matograph. This technique seems perfect as it does not need any reagent and is sensitive to changes in oxidative status and insensitive to fasting or meals [72]. But it is difficult to use because of the necessity of an uncomfortable apparatus to collect expired air that changes respiration by stressing the patients. The technique needs to use synthetically purified air for patients to breathe in order to decrease the important background of hydrocarbons in room air. Pentane and ethane are quickly metabo-lized in the liver [73]. The production of alkane is also strictly dependent on oxygen pressure when peroxide is not [74]. Normal values in adults range from 0.3 to 1.6 nmol ethane/L and from 0.2 to 4 nmol pentane/L

without lung washout, and from 24 pmol ethane/L and 4 to 11 pmol pentane/L using washout, or 3 pmol/Kg/mm ethane or pentane after taking into account the respiratory flow.

This technique permits to demonstrate the effect of changing the content of perfused lipids or tocopherol on peroxidation [75]. Patients with chronic heart failure present an increase in pentane inside expired air that decreases by using captopril treatment [76]. The pentane excreted by rheumatoid patients is proportional to the severity of the disease [77]. An increase in either pentane or hexane has been described in a great variety of oxidative diseases reviewed by Kneepkens.

Oxidized lipoproteins
The oxidation of lipoprotein is not only an important toxic mechanism involved in atheroma, but it can also be used as an index of oxidative stress. Different possibilities exist to measure the content in MDA or peroxides in low density lipoprotein (LDL) after separation by density centrifugation, or to measure the susceptibility to oxidation of plasma lipoprotein [78]. An interesting new way of evaluating oxidative stress is certainly the immunological determination of oxidized LDL using monoclonal antibodies [79, 80]. When commercially available, these techniques will permit rapid screening of patients.

Protein oxidation
Oxidative stress has a wide range of effects on proteins: it oxidizes the thiol group, creating a disulphide bridge [81], leads to fragmentation or aggregation changing the electrophoretic migration, destroys tryptophan residues, changes sensitivity to proteolitic enzymes, and creates bityrosine bridges. The disappearance of tryptophan or the formation of bityrosine [82] can be detected by change in the fluorometric spectrum of proteins. Fluorescence of N-formyl kynurenine in immunoglobuline isolated from plasma of patients undergoing vascular reconstructive surgery demonstrated production of free radicals. The oxidation of thiol group can be measured using Ellman's reagent (5,5'-dithiobis (2-nitrobenzoic acid) [83]. After separation on a column the content of plasma IgG converting into formylkinurenine is detected by its fluorescence (ex 360 nm, em 454 nm) and reflects its oxidation by oxidative stress during vascular surgery [84].

Ortho-tyrosine is an interesting derivative of protein oxidation because it can only be formed by a non enzymatic free radical oxidation of phenylalanine [85]. It cannot be detected in normal conditions by GC-MS or fluorometric HPLC, but its concentration in tissues increased after oxidative stress [86, 87].

The oxidative stress also indirectly contributes to modification of proteins. Different aldehydes such as HNE or MDA react with the amino groups of proteins containing lysyl residues, or can bind in a

Michael-type reaction to thiol groups. The number of protein-bound aldehyde molecules is measured via determination of protein carbonyl groups [88] using DNPH and can be clinically used on a plasma sample [89].

DNA oxidation

Bases of DNA are particularly sensitive to free radicals leading to a great number of derivatives by addition, oxidation or fragmentation: adenine N1 oxide, 5-hydroxymethyl desoxy uracile, 8-hydroxydesoxyguanine, thymine glycol, etc. [90, 91]. These modifications can be measured directly in DNA, but oxidized derivatives released by glycosylase can be directly identified in plasma and urine using post labelling, CG-MS, HPLC, HPLC-MS [92]. We have observed by post labelling the formation of adenine N1-oxide in fibroblasts subjected to a stress with H_2O_2 [93]. But this technique is cumbersome to run under clinical conditions where it is easier to determine 8 oxy deoxyguanosine in urine by HPLC with a coulometric detector [94]. By the same technique, its concentration was found to be increased in sperm DNA after an experimental vitamin C deficiency in human [95]. 8-hydroxyguanosine is an important biological compound [96], and its quantity found in tissues is proportional to the intensity of the oxidative process evaluated by MDA determination [97]. We have developed a precise method to measure 5 hydroxy methyluracile in urine by GC-MS using a stable-isotopic internal standard that gives reproducible results [98].

The adduct between MDA and deoxyguanosine can be measured by HPLC using a fluorometric detector. Its concentration in rat liver is seven times higher than that of 8-hydroxyguanosine after ischemia [99]. Oxidation of DNA *in vivo* results also in the formation of auto antibodies against oxidized DNA. An increase in anti-hydroxy-methyl-deoxyuridine antibodies of IgM type is detected in serum of patients with systemic lupus erythematosus and other inflammatory auto immune diseases [100].

Oxidation of natural scavengers

Oxidation of an endogenous antioxidant reflects an oxidative stress that is evaluated by measuring the decrease in the total level of the antioxidant or the increase in the oxidative form. The only way not to be influenced by nutritional status is to measure the ratio between oxidized and reduced antioxidants present in blood.

Vitamins. The levels of tocopherol and ascorbate in plasma are decreased after an acute oxidative stress such as after aorta clamping. During the same time tocopherylquinone (as well as the ratio between tocopherylquinone and tocopherol) measured by HPLC with electrochemical detection, increases in plasma [101]. In chronic stress during HIV infection, we observed that the more easily depletable antioxidant is beta-carotene, probably because this vitamin cannot be recycled after scavenging [102].

Glutathione. Many techniques can measure glutathione using nitro-prussiate [103], coulometry [104], spectrophotometry [105], ion exchange chromatography [106], enzymology [107], and HPLC [108, 109, 110].

The evaluation of oxidative stress is better done by measuring the ratio between reduced and oxidized glutathione. Unfortunately, reduced glutathione is very unstable, so the sample blood has to be stabilized rapidly by protein elimination and freezing at very low temperature, or directly by adding the thiol-reagent in the sampling tube.

Reduced glutathione disappears very quickly during oxidative stress, and returns as rapidly to basal level by recycling. This makes reduced glutathione a good index of a fast, transient stress, but during a long period we cannot be certain that a punctual determination is representative of the mean level.

In patients the content of leukocyte in reduced glutathione can be approximated by flux cytometry using fluorometric probes (dichloro-fluorescein diacetate, or dibromobimate) [111]. By this technique a defect in reduced glutathione can be demonstrated in patients with AIDS [112].

Other endogenous oxidized compounds

Different derivatives result from catecholamine oxidation: adreno-chrome increases in synovial fluid of patients with rheumatoid arthritis [113], adrenolutin is quantified in plasma by HPLC [114].

Oxidized derivatives of cholesterol are measured by thin-layer chromatography followed by gas chromatography, and they showed an increase in diabetic patients [115].

Oxidation of exogenous compounds

Some compounds present specific metabolites in body fluids after attack by a particular radical. For instance, DMSO (dimethyl sulfoxide) when oxidized by hydroxyl radical forms methane sulfinic acid, which can be measured by colorimetry [116]. Such compounds can be used by addition in cell culture, or by ingestion by the experimental animal to demonstrate the existence of oxidative stress. But their utilization in humans requires safe products with no risk of toxicity or reaction by the body. Fortunately, salicylate or acetyl salicylate can be used in such a way. Salicylate lead to three different derivatives by reaction with HO°: catechol (11%), 2,3-dihydrobenzoate (49%) and 2,5-dihydrobenzoate (40%). 2,3-salicylate is more specific for HO° production that 2,5-salicy-late, which can also be produced enzymatically by cytochrome P450

[117]. By giving *per os* 1 g aspirin and collecting blood 3 h later, 2,3-salicylate was found to be abnormally increased in the blood of diabetic patients [118].

Choice of an indicator

Among the numerous compounds appearing during oxidative stress, only a few can really be determined in clinical situation.

The choice of an indicator depends on the type of oxidative stress as presented in Figure 3. Determination of reduced glutathione is a good index of a fast and transient oxidation, as found after ischaemia-re-oxygenation, because it can decrease rapidly during the first minute after stress and recover to a normal level within a few minutes. Another interesting technique to monitor fast events is to collect blood with a

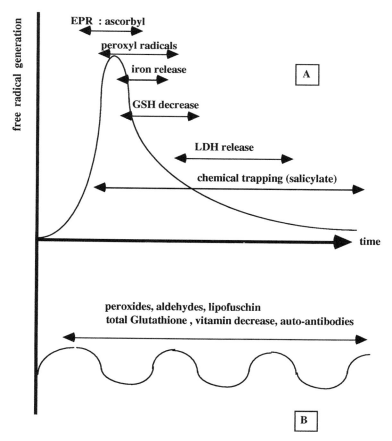

Fig. 3. Indicators of oxidative stress according to time in an acute (A) or chronic (B) stress.

spin-trapping agent and to measure ESR signals. To monitor a chronic stress such as in ageing or atheroma, MDA remains the more popular indicator despite its deficiencies and controversy. Clearly, some methods to measure MDA are not very specific, but mainly a majority of the compounds measured resulted from oxidation of lipids, aminoacids or ribose by the oxidative stress. More important is the fact that these compounds are toxic by themselves. Only for that last reason are they very important to monitor.

Some indicators as DNA damage are preferentially interesting to analyse when the oxidative stress is related to nucleus (irradiation) or when the disease involves a DNA modification (cancer). Lipid peroxidation is a better choice for an oxidative stress occurring close to the membrane during activation of macrophages or PMN. Protein oxidation is supposed to better reflect a stress in cytosol or extra cellular fluid.

Generally, a simultaneous determination of a few parameters is certainly recommended. When using different parameters (respiratory hydrocarbons, lipid peroxides, glutathione) to monitor acute hepatotoxicity in animals, conflicting results have been obtained [119]. Technical reasons are also to be considered. When making diagnosis in an individual patient we need a precise and discriminative method. Such a technique is not so necessary in epidemiological studies where a great number of patients and a statistical evaluation compensate for the lack of precision. In this latter case, we need easy-to-run and very stable parameters because a great number of samples is collected far from the laboratory.

In conclusion, we have to distinguish between clinical investigators who need an easy-to-run, inexpensive, useful technique, and basic scientists probing the mechanism at the origin of oxidative stress using various selective and sensitive techniques. We have to be pragmatic and realize that sophisticated and expensive techniques do not always offer more information than easy-to-run techniques. In many cases the determination of MDA still remains the indicator of choice, but in the near future immunological determination of oxidative damage will certainly become the more precise way to investigate oxidative stress in humans.

References

1. Halliwell, B. and Gutteridge, J.M.C. (1989) *Free Radicals in Biology and Medicine*. Clarendon Press, Oxford.
2. Greenwald, R.A. (1985) *Handbook of Methods for Oxygen Radical Research*. CRC Press, Florida, USA.
3. Rice-Evans, C., Diplock, A. and Symons, M.C. (1991) *Techniques in Free Radical Research*. Elsevier, Amsterdam.
4. Pryor, W.A. and Godber, S.S. (1991) Non invasive measures of oxidative stress status in humans. *Free Rad. Biol. Med.* 10: 177–184.
5. Jiang, Z., Woollard, A. and Wolff, S. (1990) Hydrogen peroxide production during experimental protein glycation. *FEBS Lett.* 268: 69–71.

6. Paraidathathu, T., De Groot, H. and Kehrer, J. (1992) Production of reactive oxygen by mitochondria from normoxic and hypoxic rat heart tissue. *Free Rad. Biol. Med.* 13: 289–297.
7. Sohal, R. (1993) Aging, cytochrome oxidase activity, and hydrogen peroxide release by mitochondria. *Free Rad. Biol. Med.* 14: 583–588.
8. Wertz, J.E. and Bolton, J.R. (1972) *Electron Spin Resonance*. MacGraw-Hill, New York.
9. Clough-Helfman, C. and Phillis, J. (1992) The free radical trapping agent N-tert-butyl-phenylnitrone(Pbn) attenuates cerebral ischaemic injury in gerbils. *Free Rad. Res. Comm.* 15: 177–186.
10. Li, X., Sun, J., Bradamante, S., Piccinini, F., Bollir, J. (1993) Effects of the spin-trap phenyl tert-butyl nitrone on myocardial function and flow: a dose response study in the open-chest dog and in the isolated rat heart. *Free Rad. Biol. Med.* 14: 277–285.
11. Coghlan, J.G., Flitter, W., Holley, A.E., Norell, M., Mitchell, A.G., Ilsley, C.D. and Slater, T.F. (1991) Detection of free radicals and cholesterol hydroperoxides in blood taken from the coronary sinus of man during percutaneous transluminal coronary angioplasty. *Free Rad. Res. Comm.* 14: 409–417.
12. Tortolani, A., Powell, S., Misik, V., Weglicki, W., Pogo, G. and Kramer, J. (1993) Detection of alkoxyl and carbon-centered free radicals in coronary sinus blood from patients undergoing elective cardioplegia. *Free Rad. Biol. Med.* 14: 421–426.
13. Buettner, G. and Jurkiewicz, B. (1993) Ascorbate free radical as a marker of oxidative stress: an EPR study. *Free Rad. Biol. Med.* 14: 49–55.
14. Roginsky, V. and Stegmann, H. (1994) Ascorbyl radical as natural indicator of oxidative stress: quantitative regularities. *Free Rad. Biol. Med.* 17: 93–103.
15. Pietri, S., Seguin, J., D'arbigny, P. and Ulcasi, M. (1994) Ascorbyl free radical: a non-invasive marker of oxidative stress in human open-heart surgery. *Free Rad. Biol. Med.* 16: 523–528.
16. Di Mascio, P., Kaiser, S. and Sies, H. (1989) Lycopene as the most efficient biological carotenoid singlet oxygen quencher. *Arch. Biochem. Biophys.* 274: 532–538.
17. Cadenas, E. and Sies, H. (1985) Detecting singlet oxygen by low level chemiluminescence. *In*: R. Greenwald (ed.): *CRC Handbook of Methods for Oxygen Radical Research*. CRC Press, Boca Raton, pp 191–195.
18. Gonzalez Flecha, B., Llesuy, S. and Boveris, A. (1991) Hydroperoxide-initiated chemiluminescence: an assay for oxidative stress in biopsies of heart, liver and muscle. *Free Rad. Biol. Med.* 10: 93–100.
19. Slawinski, J., Ezzahir, A., Godlewski, M., Kwiecinska, T., Rajfur, Z., Sitko, D. and Wierzuchowa, D. (1992) Stress-induced photon emission from perturbed organism. *Experientia* 48: 1041–1058.
20. Barsacchi, R., Caoassini, M., Maiorino, M., Pelosi, G., Simonelli, C. and Ursini, F. (1989) Increased ultra weak chemiluminescence emission from rat heart at post ischemic reoxygenation: protective role of vitamin E. *Free Rad. Biol. Med.* 6: 573–579.
21. Ferlat, S. and Favier, A. (1993) Tumor necrosis factor et radicaux libres: conséquences potentielles pour l'immunite cellulaire. *C.R. Soc. Biol.* 187: 296–307.
22. Lissi, E., Salim-Hanna, M., Sir, T. and Videla, L. (1991) Is spontaneous urinary visible chemiluminescence a reflection of *in vivo* oxidative stress. *Free Rad. Biol. Med.* 12: 317–322.
23. Vilim, V. and Wilheim, J. (1989) What do we measure by a luminol-dependent chemiluminescence of phagocytes. *Free Rad. Biol. Med.* 6: 623–629.
24. Rembish, S. and Trush, M. (1994) Further evidence that lucigenin-derived chemiluminescence monitor mitochondrial superoxide generation in rat alveolar macrophages. *Free Rad. Biol. Med.* 17: 117–126.
25. Trush, M. A., Wilson, M. E. and Van Dyke, K. (1978) The generation of chemiluminescence by phagocytic cells. *Methods Enzymol.* 57: 462.
26. Descamps-Latcha, V., Nguyen, A.T., Golub, R.M., Feuillet-Fieux, M.N. (1982) Chemiluminescence in microamounts of whole blood for investigation of the human phagocyte oxidative metabolism function. *Ann. Immunol.* 133: 349–364.
27. Steg, P., Pasquier, C., Huu, P.T., Chollet-Martin, S., Juliard, J.M., Himbert, D., Pocidalo, M.A., Gourgon, R. and Hakim, J. (1993) Evidence for priming and activation of nutrophils early after coronary angioplasty. *Eur. J. Med.* 2: 6–10.
28. Guillard, O., Sonnet, J. and Lauwerys, R. (1989) Reactive oxygen species production in whole blood from AIDS patients. *Clin. Chim. Acta* 185: 113–114.

29. Thu Nguyen, A., Lethias, C., Zingraff, J., Herbelin, A., Naret, C. and Descamps-Latscha, B. (1985) Hemodialysis membrane-induced activation of phagocyte oxidative metabolism detected *in vivo* and *in vitro* within microamounts of the whole blood. *Kidney Int.* 28: 158–167.

30. Keshavarzian, A., Sedghi, S., Kanofsky, J., List, T., Robinson, C., Ibrahim, C. and Winship, D. (1992) Excessive production of reactive oxygen metabolites by inflamed colon: analysis by chemiluminescence probe. *Gastroent.* 103: 177–185.

31. Frei, B., Yamamoto, Y., Niclas, D. and Ames, B. (1988) Evaluation of an isoluminol chemiluminescence assay for the detection of hydroperoxides in human blood plasma. *Anal. Biochem.* 175: 120–130.

32. Palmer, R.M., Ferridge, A.G. and Moncada, S. (1987) Nitric oxide release accounts for the biological activity of endothelium-derived relaxing factor. *Nature* 327: 524.

33. Slater, T.F. (1984) Overview of methods used for detecting lipid peroxidation. *Methods in Enzymology* 105: 283–305.

34. Porter, N. (1984) Chemistry of lipid peroxidation. *Methods Enzymol.* 105: 273–283.

35. Bors, W., Erben-Russ, M., Michel, C. and Saran, M. (1990) Radical mechanisms in fatty acid and lipid peroxidation. *In:* A. Crastes De Paulet, L. Douste-Blazy, R. Paoletti (eds): *Free Radicals, Lipoproteins and Membrane Lipids.* Plenum Press, New York.

36. Recknagel, R.O. and Glende, E.A. (1984) Spectrophotometric detection of lipid conjugated dienes. *Methods Enzymol.* 105: 331–337.

37. Lunec, J., Halloran, S.P., White, A.G. and Dormandy, T.L. (1981) Free radical oxidation (peroxidation) products in serum and synovial fluid. *J. Rheum.* 8: 233–245.

38. Pryor, W.A. and Castle, L. (1984) Chemical methods for the detection of lipid hydroperoxides. *Methods Enzymol.* 105: 293–299.

39. Heath, R.L. and Tapel, L. (1976) A new sensitive assay for the measurement of hydroperoxides. *Anal. Biochem.* 76: 184–191.

40. Seeeger, W., Roka, L. and Moser, U. (1984) Detection of organic hydroperoxides in rabbit lung lavage fluid, but not in lung tissue, using Gsh peroxidase and Gsh reductase. *J. Clin. Chem. Clin. Biochem.* 22: 711–715.

41. Janero, D. (1990) Malondialdehyde and thiobarbituric acid-reactivity as diagnostic indices of lipid peroxidation and peroxidative tissue injury. *Free Rad. Biol. Med.* 9: 515–540.

42. Larguilliere, C. and Melancon, S. (1988) Free malonaldehyde determination in human plasma by high performance liquid chromatography. *Anal. Biochem.* 170: 123–126.

43. Nair, V., Cooper, C.S., Vieitti, D.E. and Turner, G.A. (1986) The chemistry of lipid peroxydation metabolites: crosslink reactions of malonaldehydes. *Lipids* 21: 6–10.

44. Siu, G.M. and Draper, M.H. (1982) Metabolism of malonaldehyde *in vivo* and *vitro*. *Lipids* 17: 349–355.

45. Hadley, M. and Draper, M.H. (1988) Identification of N (2 propenal) serine as a urinary metabolite of malondialdehyde. *Faseb J.* 2: 138–140.

46. Yagi, K. (1976) A simple fluorimetric assay for lipoperoxide in blood plasma. *Biochem. Res.* 15: 212–216.

47. Dousset, J.C., Trouilh, M. and Foglietti, M.J. (1983) Plasma malonaldehyde levels during myocardial infarction. *Clin. Chim. Acta* 129: 319–322.

48. Richard, M.J., Portal, B., Meo, J., Coudray, C., Hadjian, A. and Favier, A. (1992) Malondialdehyde kit evaluated for determining plasma and lipoprotein fractions that react with thiobarbituric acid. *Clin. Chem.* 38: 704–709.

49. Halliwell, B. and Gutteridge, J.M.C. (1986) Oxygen free radicals and iron in relation to biology and medicine: some problems and concepts. *Arch. Biochem. Biophys.* 2: 501–14.

50. Janero, D.R. (1990) Malondialdehyde and thiobarbituric acid-reactivity as diagnostic indices of lipid peroxidation and peroxidative tissue injury. *Free Rad. Biol. Med.* 9: 515.

51. Favier, A., Wilke, B., Arnaud, J., Richard, M.J., Ducros, V. and Vidailhet, M. (1991) Antioxidizing then paradoxal prooxidizing effect of selenite supplementation in phenylketonuric (Pku) children. *In:* B. Momcilovic (ed.): *Trace Elements in Man and Animals 7*, Imi, Zagreb, pp 7–15/7–17.

52. Hunter, M.I. and Mohamed, J.B. (1986) Plasma antioxidants and lipid peroxidation products in Duchenne muscular dystrophy. *Clin. Chim. Acta* 155: 123–132.

53. Boyd, N.F. and McGuire, V. (1991) The possible role of lipid peroxidation in breast cancer risk. *Free Rad. Biol. Med.* 10: 185–190.

54. Revillard, J.P., Vincent, C.A., Favier, A., Richard, M.J., Zittoun, M. and Kazatchine, M. (1992) Lipid peroxidation in human immunodeficiency virus infection. *J. Acquired Immune Deficiency Syndromes* 5, 6: 637–638.
55. Faure, P., Corticelli, P., Richard, M.J., Arnaud, J., Coudray, C., Halimi, S. and Favier, A. (1993) Lipid peroxidation and trace element status in diabetic ketotic patients: influence on insulin therapy. *Clin. Chem.* 39/5: 789–793.
56. De Lorgeril, M., Richard, M.J., Arnaud, J., Boissonat, P., Guidollet, J., Dureau, G., Renaud, G., Renaud, S. and Favier, A. (1993) Lipid peroxides and antioxidant defenses in the accelerated transplantation-associated coronary arteriosclerosis. *Am. Heart J.* 125: 974–980.
57. Hammouda, A., Soliman, S., Tolba, K., El-Kabbany, Z., Makhlouf, M. (1992) Plasma concentrations of lipid peroxidation products in children with acute lymphoblastic leukemia. *Clin. Chem.* 38: 594–595.
58. Wilke, B.C., Vidailhet, M., Favier, A., Guillemin, C., Ducros, V., Arnaud, J. and Richard, M.J. (1992) Selenium, glutathione peroxidase (Gsh-Px) and lipid peroxidation products before and after selenium supplementation. *Clin. Chim. Acta* 207: 1/2, 137–142.
59. Portal, B., Richard, M.J., Ducros, V., Aguilaniu, B., Brunel, F., Faure, H., Gout, J.P., Bost, M. and Favier, A. (1993) Effect of a double-blind cross-over selenium supplementation on biological indices of selenium status in cystic fibrosis children. *Clin. Chem.* 39(9): 1023–1028.
60. Richard, M.J., Ducros, V., Foret, M., Arnaud, J., Coudray, C., Fusselier, M. and Favier, A. (1993) Reversal of selenium and zinc deficiencies in chronic hemodialysis patients by intravenous sodium selenite and zinc gluconate supplementation: time-course of glutathione peroxidase repletion and lipid peroxidation decrease. *Biol. Trace. Elem. Res.* 39: 149–159.
61. Knight, J.A., Smith, S.E., Kinder, V.E. and Austall, H.B. (1987) Reference intervals for plasma lipoperoxides = age, sex, and specimen related variations. *Clin. Chem.* 33: 2289–2291.
62. Poubelle, P., Chaintreuil, J., Bensadouri, J., Blotman, F., Simon, L. and Crastes De Paulet, A. (1982) Plasma lipoperoxides and age. *Biomed. Pharmaco.* 36: 164–166.
63. Ishihara, M. (1978) Studies on lipoperoxides of normal pregnant women and of patients with toxemia of pregnancy. *Clin. Chim. Acta* 84: 1–9.
64. Esterbauer, H., Zollner, H. and Schaur, R.J. (1990) Aldehydes formed by lipid peroxidation: mechanisms of formation, occurrence and determination. *In:* Vigo-Pelfrey, C. (ed.): *Membrane Lipid Oxidation.* CRC Press, Boca Raton, Florida, pp 239–283.
65. Yoshino, K., Sano, M., Fujita, M. and Tomita, I. (1986) Formation of aldehydes in rat plasma and liver due to vitamin E deficiency. *Chem. Pharm. Bull.* 34: 5184–5187.
66. Spitz, D., Sullivan, S., Malcom, R. and Roberts, R. (1991) Glutathione dependent metabolism and detoxification of 4-hydroxy-2-nonenal. *Free Rad. Biol. Med.* 11: 415–423.
67. Emerit, I., Khan, S.H. and Esterbauer, H. (1991) Hydroxynonenal, a component of clastogenic factors? *Free Rad. Biol. Med.* 10: 371–377.
68. Zollner, H., Schaur, R.J. and Esterbauer, H. (1991) Biological activities of 4-hydroxyalkenals. *In:* H. Sies (ed.): *Oxidative Stress: Oxidants and Antioxidants.* Academic Press, New York, pp 337–369.
69. Esterbauer, H. and Zollner, H. (1989) Methods for determination of aldehydic lipid peroxidation products. *Free Rad. Biol. Med.* 7: 197–203.
70. Yoshino, K., Matsuura, T., Sano, M., Satto, S. and Tomita, I. (1986) Fluorometric liquid chromatographic determination of aliphatic aldehydes arising from lipid peroxidation. *Chem. Pharm. Bull.* 34: 1694–1699.
71. Selley, M.L., Bartlett, M.R., McGuiness, J.A., Hapel, A.J., Ardlie, N.G. and Lacey, M.J. (1989) Determination of the lipid peroxidation product trans-4-hydroxy-2-nonenal in biological samples by high-performance liquid chromatography and combined capillary gas-chromatography-negative ion chemical ionization mass spectrometry. *J. Chromat.* 488: 329–340.
72. Zarling, E., Mobarhan, S., Bowen, P. and Sugerman, S. (1992) Oral diet not alter pulmonary pentane or ethane excretion in healthy subjects. *J. Am. Coll. Nutr.* 11: 349–352.
73. Kneepers, C.M., Lepage, G. and Roy, C. (1994) The potential of the hydrocarbon breath test as a measure of lipid peroxidation. *Free Rad. Biol. Med.* 17: 127–160.

74. Reiter, R. and Burk, R. (1987) Effect of oxygen tension on the generation of alkanes and malonaldialdehyde by peroxidizing rat liver microsomes. *Biochem. Pharmacol.* 36: 925–929.

75. Jeejeebhoy, K.N. (1991) *In vivo* breath alkane as an index of lipid peroxidation. *Free Rad. Biol. Med.* 10: 191–193.

76. Sobotka, P., Brottman, M., Weitz, Z., Birnbaum, A., Skosey, J. and Zarling, E. (1993) Elevated breath pentane in heart failure reduced by free radical scavenger. *Free Rad. Biol. Med.* 14: 643–647.

77. Humad, S., Zarling, E., Clapper, M. and Skosey, J.L. (1988) Breath pentane excretion as marker of disease activity in rheumatoid arthritis. *Free Rad. Res. Comm.* 5: 101–106.

78. Esterbauer, H., Rotheneder, M., Waeg, G., Striedl, G. and Juergens, G. (1990) Biochemical, structural, and functional properties of oxidized low-density lipoproteins. *Chem. Res. Toxicol.* 3: 77–92.

79. Palinski, W., Rosenfeld, M.E., Yla-Herttuala, S., Gurtner, G.C., Socher, S.A., Butler, S.W., Parthasarathy, S., Carew, T.E., Steinberg, D. and Witztum, J.L. (1989) Low density lipoprotein undergoes oxidative modifications *in vivo*. *Proc. Natl. Acad. Sci. USA* 86: 1372–1376.

80. Juergens, G., Ashy, A. and Esterbauer, H. (1990) Detection of new epitopes formed upon oxidation of low-density lipoprotein, lipoprotein (A) and very-low-density lipoprotein. Use of an antiserum against 4-hydroxynonenal-modified low-density lipoprotein. *Biochem. J.* 265: 605–608.

81. Di Simplicio, P., Cheeseman, K.H. and Slater, T.F. (1991) The reactivity of the Sh group of bovine serum albumin with free radicals. *Free Rad. Res. Comm.* 14: 253–262.

82. Prutz, W.A., Butier, J. and Land, E.J. (1983) Phenol coupling initiated by one electron oxidation of tyrosine units in peptides and histone. *Int. J. Radiat. Biol. Relat. Stud. Phy. Chem. Med.* 44: 183–196.

83. Susuki, Y., Lyall, V., Biber, T. and Ford, G. (1990) A modified technique for the measurement of sulfhydryl groups oxidized by reactive oxygen intermediates. *Free Rad. Biol. Med.* 9: 479–484.

84. Ward, A., Mc Burney, A. and Lunec, J. (1994) Evidence for the involvement of oxygen derived free radicals in ischaemia-reperfusion injury. *Free Rad. Res.* 20: 21–28.

85. Hoskins, J.A. and Davis, L.J. (1988) Analysis of the isomeric tyrosines in mammalian and avian systems using high performance chromatography with fluorescence detection. *J. Chromat.* 426: 155–161.

86. Ishimitsu, S., Fujimito, S. and Ohara, A. (1984) Hydroxylation of phenylalanine by the hypoxanthine-xanthine oxidase system. *Chem. Pharm. Bull.* 32: 4645–4649.

87. Meier, W., Burgin, R. and Frolich, D. (1988) Analysis of O-tyrosine as a method for the identification of irradiated chicken meat. *Beta Gamma* 1: 34–36.

88. Benedetti, A., Esterbauer, H., Ferrali, M., Fulceri, R., Comporti, M. (1982) Evidence for aldehydes bound to liver microsomal protein following Ccl4 or Brccl3 poisoning. *Biochim. Biophys. Acta* 711: 345–356.

89. Lenz, A.G., Costable, V., Swaltel, S. and Levine, R.L. (1989) Determination of carbonyl group in oxidatively modified proteins by reduction with tritied sodium borohydride. *Anal. Biochem.* 177: 419–425.

90. Demple, B. and Levin, J. (1991) Repair systems for radical-damaged DNA. *In:* H. Sies (ed.): *Oxidative Stress*, Academic Press, London, pp 119–154.

91. Cadet, J., Berger, M. and Decarroz, C. (1986) Photosensitized reactions of nucleic acids. *Biochemie* 68: 813–814.

92. Cadet, J., Incardona, M.F., Odin, F., Molko, D., Mouret, J.F., Polverrelli, M., Faure, H., Ducros, V., Tripier, M. and Favier, A. (1993) Measurement of oxidative base damage to DNA by using HPLC 32-P postlabeling and Gc-Ms-Sim Assays. *In:* D.H. Phillips, M. Castegno and H. Bartsh (eds): *Post-Labelling Methods for Detection of DNA Adducts.* Iarc, Lyon, pp 271–276.

93. Cadet, J., Odin, F., Mouret, J.F., Polverelli, M., Audic, A., Giacomoni, P., Favier, A. and Richard, M.J. (1992) Chemical and biochemical postlabelling methods for singling out specific oxidative DNA lesions. *Mut. Res.* 275: 343–354.

94. Shinega, M. and Ames, B. (1991) Assays for 8-hydroxy-2'-deoxyguanosine: A biomarker of *in vivo* oxidative DNA damage. *Free Rad. Biol. Med.* 10: 211–216.

95. Jacob, R., Kelley, D., Pianalto, F., Swendseid, M., Henning, S., Zhang, J., Fraga, C. and

Peters, J. (1991) Immunocompetence and oxidant defence during ascorbate depletion of healthy men. *Am. J. Clin. Nutr.* 54: 1302S–1309S.

96. Kasai, H. and Nishimura, S. (1991) Formation of 8-hydroxydeoxyguanosine in DNA by oxygen radicals and its biological significance. *In:* H. Sies (ed.): *Oxidative Stress.* Academic Press, London, pp 99–116.

97. Park, J. and Floyd, R. (1992) Lipid peroxidation products mediate the formation of 8-hydroxydeoxyguanosine in DNA. *Free Rad. Biol. Med.* 12: 245–250.

98. Faure, H., Cadet, J., Boujet, C. and Favier, A. (1993) Gas chromatographic-mass spectrometric determination of 5-hydroxymethyluracil in human urine by isotope-dilution. *J. Chromat.* 616: 1–7.

99. Agarwal, S. and Draper, H. (1992) Isolation of a malonaldialdehyde-deoxyguanosine adduct from rat liver DNA. *Free Rad. Biol. Med.* 13: 695–699.

100. Frenkel, K., Karkoszka, J., Kim, E. and Taioli, E. (1993) Recognition of oxidized DNA bases by sera of patients with inflammatory diseases. *Free Rad. Biol. Med.* 14: 483–494.

101. Murphy, M., Kolvenbach, R., Aleksis, M., Hansen, R. and Sies, H. (1992) Antioxidant depletion in aortic crossclamping ischemia: increase of the plasma A tocopheryl quinone/ A tocopherol ratio. *Free Rad. Biol. Med.* 13: 95–100.

102. Favier, A., Sappey, C., Leclerc, P., Faure, P. and Micoud, M. (1994) Antioxidant status and lipid peroxidation in patients infected with HIV. *Chemico. Biol. Interact.* 91: 165–180.

103. Grunert, R. and Philips, P.H. (1951) A modification of the nitroprussiate method of analysis for glutathione. *Arch. Biochem.* 1730: 217.

104. Ladenson, S.H. and Purdy, W.C. (1971) Microcoulometric argentimetric titrations. *Anal. Chim. Acta* 57: 465.

105. Mergel, D., Anderrnann, G. and Anderrnann, C. (1979) Simultaneous spectrophotometric determination of oxidized and reduced glutathione in human and rabbit red cells. *Methods Find. Exp. Clin. Pharmacol.* 1: 277.

106. Tabor, C.W. and Tabor, H. (1977) An automated ion exchange assay for glutathione. *Anal. Biochem.* 78: 543.

107. Tietze, F. (1969) Enzymatic methods for quantitative determination of nanogramme amount of total oxidized glutathione: application to mammalian blood and other tissues. *Anal. Biochem. Biophys.* 27: 502–522.

108. Nishiyama, J. (1984) Assay of biological thiols by a combination of HPLC and post column reaction with 6-6'-dithiodinicotinic acid. *Anal. Biochem.* 138: 95.

109. Reed, D.J., Babson, J.R., Jr., Beatty, P.W. (1980) HPLC analysis of nanomole levels of glutathione, glutathione disulfide and related thiols and disulfides. *Anal. Biochem.* 106: 55.

110. Reeve, J. and Kuhlenkamp, J. (1980) Estimation of glutathione in rat liver by reversed phase HPLC: separation from cystein and glutamyl cystein. *J. Chrom.* 194: 424.

111. Poot, M. (1991) Flow cytometric analysis of cell cycle dependent changes in cell thiol level by combining a new laser dye with Hoeschst 3342. *Cytometry* 12: 184.

112. Roderer, M., Anderson, M., Rabin, R., Herzenberg, L. and Herzenberg, L. (1993) Redox plays a central role in T-cell signaling regulation of HIV and possibly the progression of AIDS. Conference on Oxidative Stress in HIV (Bethesda).

113. Matthews, S., Hallett, M., Henderson, A. and Campbell, A. (1985) The adrenochrome pathway: a potential catabolic route for adrenaline metabolism in inflammatory disease. *Adv. Myocardiol.* 6: 367–381.

114. Dhalla, K., Ganguly, P., Rupp, H., Beamish, R. and Dhalla, N. (1989) Measurement of adrenolutin as an oxidation product of catecholamines in plasma. *Mol. Cell. Biochem.* 87: 85–92.

115. Arshad, M., Bhadra, S., Cohen, R. and Subbiah, M. (1991) Plasma lipoprotein peroxidation potential: a test to evaluate individual susceptibility to peroxidation. *Clin. Chem.* 37: 1756–1758.

116. Steiner, M. and Babbs, C. (1990) Hydroxy radical generation by postischemic rat kidney slices *in vitro. Free Rad. Biol. Med.* 9: 67–77.

117. Halliwell, B., Kaur, H. and Ingelman-Sundberg, M. (1991) Hydroxylation of salicylate as an assay for hydroxyl radicals: a cautionary note. *Free Rad. Biol. Med.* 10: 439–441.

118. Ghiselli, A., Laurenti, O., De Mattia, G., Maiani, G. and Ferro-Luzzi, A. (1992) Salicylate hydroxylation as an early marker of *in vivo* oxidative stress in diabetic patients. *Free Rad. Biol. Med.* 13: 621–626.

119. Smith, C.V. (1991) Correlation and apparent contradictions in assessment of oxidant stress status *in vivo. Free Rad. Biol. Med.* 10: 217–224.

Analysis of Free Radicals in Biological Systems
Favier et al. (eds)
© 1995 Birkhäuser Verlag Basel/Switzerland

EPR evidence of generation of superoxide anion radicals ($O_2^{\cdot-}$) by irradiation of a PDT photosensitizer: Hypericin

C. Hadjur[1] and A. Jeunet[2]

[1]*Laboratoire de Photochimie and* [2]*LEDSS, Université Joseph Fourier, BP 53, F-38041 Grenoble Cédex 9, France*

Summary. Irradiation of hypericin in EYL liposomes with visible light under aerobic conditions at pH 7, produced superoxide anion radicals which were detected with the spin trap 5,5-dimethyl-1-pyrroline-1-oxide (DMPO).

Introduction

Photodynamic therapy (PDT) is a promising cancer treatment involving visible light irradiation of malignant cells following selective uptake of a photosensitizer by the tumor tissue [1–4]. There is a considerable emphasis to find new photosensitizers for PDT. In general, the photocytotoxicity of these molecules is singlet oxygen-dependent.

In a recent study, we have investigated the photodynamic effects of a new photosensitizer, hypericin (HYP), on a human fibroblast cell line MRC5 [5]. The cell survival was assessed by measuring cell proliferation capacity. In the dark, no cytotoxicity was observed in the range of $10^{-6}-10^{-9}$ M HYP. A concentration of 5×10^{-9} M HYP was found to kill 100% of the irradiated cells with visible light [5].

With regard to the phototherapeutic potential, it is essential to study the photosensitizing mechanisms to elucidate the photocytotoxic activity. Although a Type II reaction (singlet oxygen) has been demonstrated to play an important role in the activity of hypericin [6, 7], the exact mechanism of the photodynamic action (Type I or/and Type II) still remains to be determined [8]. In this study, we have applied EPR (electron paramagnetic resonance) spin trapping technique to measure free radical photosensitization (Type I reaction) from HYP embedded in egg yolk lecithin (EYL) liposomes.

In this technique, a nitrone or a nitroso compound (the spin trap) reacts with a short-lived free radical to produce a nitroxide (the spin adduct) whose lifetime is considerably greater than that of the parent

free radical and therefore, detectable by conventional EPR spectroscopy
[9]:

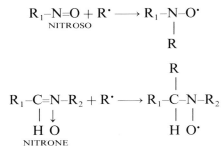

Because of increased stability of oxy radical adducts, nitrones are
better traps than nitroso spin traps [10]. Among the various spin traps,
5,5-dimethyl-1-pyrroline-1-oxide (DMPO) has received the most attention [11–13].

Reaction of this spin trap with either superoxide or hydroxyl radical
produces spin-trapped adducts with characteristic EPR spectra [14]. The
EPR spectra of the spin adducts exhibit a primary ^{14}N triplet which is
split into a secondary doublet due to the β proton of DMPO. The
magnitude of the hyperfine splitting constants, i.e., a^N and a_β^H, are
characteristic of the nature of the trapped radical.

Although this methodology is rather straightforward, this approach
has some limitations:

(1) In aqueous solution, at pH 7, the rate constant of DMPO with
$O_2^{\cdot-}$ is low: 10 $M^{-1}s^{-1}$ [9]. Nevertheless, under the same conditions, the
rate constant of the reaction of DMPO with $^\cdot$OH is approximately
$2 \times 10^9 M^{-1}s^{-1}$ [14].

(2) The half-life of the DMPO-OOH signal is approximately 80 s at
pH 6 and only about 35 s at pH 8. However, it is significantly more
persistent (i.e., half-life of 2 h) [9].

(3) Spin-trapping procedures give indirect information which make
data interpretation difficult [15, 16]. Reaction of DMPO with hydroxyl
radicals leads to the formation of DMPO-OH. But, previous works
have shown that the superoxide spin trap adduct DMPO-OOH is
unstable especially in the presence of transition metals and rapidly
decomposes into various species, including DMPO-OH (Fig. 1) [15, 16].
Thus, detection of DMPO-OH does not unequivocally prove the formation of hydroxyl radicals. To determine if the observed $^\cdot$OH adduct of
a spin trap is the result of the initial trapping of superoxide and of the
subsequent decomposition of the superoxide spin adduct, the SOD and
catalase enzymes could be regarded as valuable tools. If it is superoxide
dismutase and not catalase that inhibits the generation of DMPO-OOH
and DMPO-OH, then this indicates that DMPO-OH arose from
DMPO-OOH degradation and does not represent hydroxyl radical

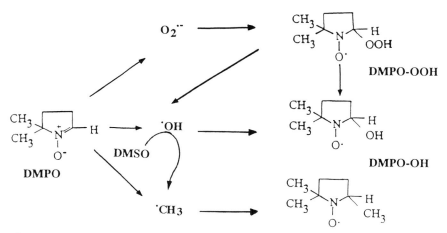

Fig. 1. General reaction scheme of generation of spin trap adducts as a consequence of the interaction between superoxide and hydroxyl radical with DMPO.

trapping. If a hydroxyl radical is generated, then one can use DMSO which would react with hydroxyl radical to produce methyl radicals [9], which can be spin trapped by DMPO, giving DMPO-CH$_3$ instead of DMPO-OH (Fig. 1). If hydroxyl radicals were not produced, then addition of DMSO would not significantly decrease the peak height of DMPO-OH. Ethanol can be substituted for DMSO, yielding the α-hydroxyethyl radical in the presence of hydroxyl radical [9].

Materials

Chemicals

The following chemicals were obtained and used without purification: L-α-lecithin (TYPE XIII E) or egg yolk lecithin, superoxide dismutase (SOD) and catalase from Sigma Co. 5,5-dimethyl-1-pyrroline-1-oxide (DMPO) from Sigma Co., was purified before use by means of the Buettner's procedure [10]. The high-purity ethanol solvent required was prepared by further purification of the commercial grade product [17] and no impurities were detected by absorption and fluorescence spectroscopies. Hypericin was synthesized and purified according to the method described by Brockman and Adams [18, 19].

Liposome preparation

EYL liposomes were prepared following the injection method previously described [20], in an aqueous solution at pH 7 containing 0.026 M K_2HPO_4 and 0.041 M Na_2HPO_4, and the concentration of EYL is 1.0×10^{-4} M.

Analytical techniques

Spectrometer settings
EPR spectra (microwave power: 20 mW; modulation amplitude: 1.6 G; time constant: 0.2 s; scan rate: 200 s; scan width: 100 G) were obtained with a Bruker ER 100 D spectrometer at room temperature ($22-24°C$) (X band, microwave frequency 9.4 GHz, modulation frequency 100 kHz).

Detection of DMPO-OOH adducts
Photo-induced EPR spectra of DMPO-OOH radicals were obtained from the samples (150 μL) injected into quartz capillaries designed especially for EPR analysis and illuminated directly inside the microwave cavity.

Detection of DMPO-OH adducts
In the DMPO spin-trapping experiments, sample irradiations were carried on outside the microwave cavity with the same illumination conditions. Following irradiation, 150 μL of the solution was introduced in the sample cell for EPR measurements.

Irradiation procedures

Irradiation was carried out using a 250 W tungsten halogen lamp, with a maximum emission centered at 600 nm (ORIEL, USA). Wavelengths shorter than 590 nm were cut off using a Shott filter (model 16588). Light intensity in the central area of the illuminator was about 10 mW/ cm^2 as determined by chemical actinometry according to Parker [21].

Notes

Optimisation of the analytical detection of DMPO-OOH signal

As already mentioned, a number of factors may affect the results in a superoxide-generating aqueous system. The optimal conditions were determined in the following experiments:

(1) The DMPO-OOH signal was recorded from the samples introduced into the aqueous flat quartz cell and illuminated directly inside the microwave cavity, because the half-life of the EPR signal is approximately 80 s at pH 6 and only about 35 s at pH 8 [10].

(2) Great effort was made to get rid of the redox-active metal ion impurities which are always present in buffers and commercial preparations. We thus used desferrioxamine, an iron chelator which stabilizes ferric ion and blocks its reduction by superoxide. Thus desferrioxamine strongly inhibited the production of the hydroxyl radical spin adduct. Moreover, the DMPO-OOH adduct is unstable, especially in the presence of transition metals, and will decompose to form the ˙OH spin adduct. Thus, no EPR signal characteristic of DMPO-OOH could be detected when desferrioxamine was omitted.

(3) The partition coefficient of DMPO between lipid and water is only 0.09 [22]. This suggests that in dispersion of EYL liposomes, DMPO is exclusively present in water and not encapsulated in liposomal

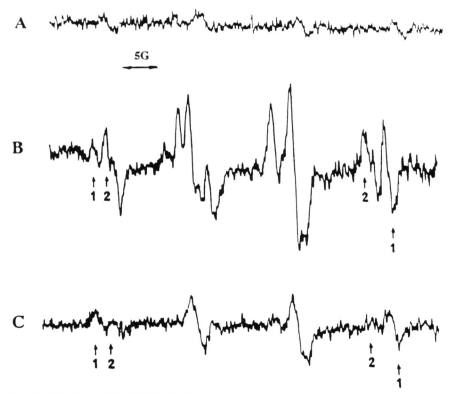

Fig. 2. Detection of the DMPO-OOH adducts formed during the aerobic illumination of a dispersion of EYL liposomes at pH 7 containing HYP (1.1×10^{-5} M) desferrioxamine (3×10^{-6} M) and DMPO (0.2 M).

membranes in which $O_2^{\cdot-}$ is formed by HYP photosensitization. Hence, it was necessary to use a high concentration of DMPO (0.2 M) to detect the superoxide radical.

Production of DMPO-OOH adducts

The illumination of an aerobic dispersion of EYL liposomes containing HYP (1.1×10^{-5} M) and desferrioxamine (3×10^{-6} M) in the presence of DMPO (0.2 M) leads to the formation of DMPO-OOH spin adducts (Fig. 2).

- The EPR spectrum is characterized by hyperfine coupling constant of $a^N = 14.2$ G, $a_\beta^H = 11.2$ G and $a_\gamma^H = 1.25$ G, demonstrating that the radical species is identical to DMPO-OOH, the DMPO-trapped superoxide radical (Fig. 2B). Scan B is a combination of two products: (1) DMPO-OH adduct, $a^N = 14.9$ G, $a_\beta^H = 14.9$ G; (2) DMPO-OOH adduct, $a^N = 14.2$ G, $a_\beta^H = 11.2$ G and $a_\gamma^H = 1.25$ G. Receiver gain: 5×10^5.
- No typical EPR spectra were detected in the absence of HYP, in the absence of oxygen or in darkness (Fig. 2A).
- An addition of SOD (30 μg/mL) caused the DMPO-OOH signal to entirely disappear and only the DMPO-OH spin adduct could be observed (Fig. 2C).

Production of of DMPO-OH adducts

The illumination of the aerobic dispersion of EYL liposomes containing HYP in the presence of DMPO leads to the formation of a large amount of DMPO-OH spin adducts (Fig. 3).

- The EPR spectra are a combination of two products (Fig. 3B): (1) DMPO-α-hydroxyethyl adduct, $a^N = 15.8$ G, $a^H = 22.8$ G (peak 2) and (2) DMPO-OH radical adduct, $a^N = 14.9$ G, $a_\beta^H = 14.9$ G (peak 1). Receiver gain: 5×10^5.
- Using the injection method to form EYL liposomes, ethanol was added (5%) (15). Under such conditions, and EPR signal was detected due to the trapping of hydroxyethyl radicals by DMPO (Fig. 3A, peak 1). These radicals may derive form the reaction of ·OH free radicals with ethanol. Ethanol is considered to be effective at inhibiting the formation of DMPO-OH in the Fenton system [16].
- No radical could be trapped in the absence of either HYP, light or DMPO (Fig. 3A).
- Neither SOD (30 μg/mL) nor catalase (10 μg/mL) could be totally reduce the intensity of the DMPO-OH signal. The DMPO-OH signal decreased by about 30%. When SOD was added along with catalase, a higher signal reduction was observed (data not shown).

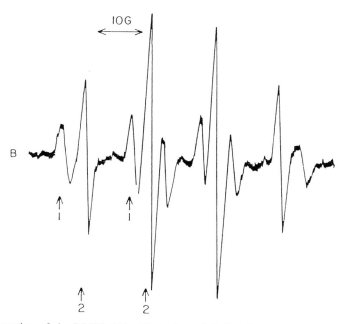

Fig. 3. Detection of the DMPO-OH adducts formed during the aerobic illumination of a dispersion of EYL liposomes at pH 7 containing HYP (1.1×10^{-5} M) and DMPO (0.2 M).

– Desferrioxamine (3×10^{-6} M) completely inhibited DMPO-OH formation, whereas the addition of hydrogen peroxide or ferric iron stimulated the reaction (data not shown).

Conclusions

(1) Irradiation of hypericin with visible light under aerobic conditions at pH 7, produced superoxide anion radicals which were detected with the spin trap 5,5-dimethyl-1-pyrroline-1-oxide (DMPO).

(2) The DMPO-OH adduct is probably formed by two main processes: (a) The Fenton reaction or (b) the decomposition of the DMPO-

OOH adduct. However, in a hetereogeneous system such as liposomes, one cannot unequivocally determine the mechanism of formation of DMPO-OH.

Acknowledgment
This research was supported by a grant from the National Cancer League of France.

References

1. Dougherty, T.J. (1993) Photodynamic therapy. *Photochem. Photobiol.* 58: 895–900.
2. Moan, J. and Berg, K. (1992) Photochemotherapy of cancer: Experimental research. *Photochem. Photobiol.* 55: 931–948.
3. Gomer, C.J. (1991) Preclinal examination of first and second generation photosensitizers used in photodynamic therapy. *Photochem. Photobiol.* 54: 1093–1107.
4. Henderson, B.W. and Dougherty, T.J. (1992) How does photodynamic therapy work? *Photochem. Photobiol.* 1: 145–157.
5. Hadjur, C., Richard, M.J., Parat, M.O., Favier, A. and Jardon, P. (1995) Photodynamic induced cytotoxicity of hypericin dye on human fibroblasts cell line MRC5. *J. Photochem. Photobiol. B. Biol.* 27: 139–146.
6. Racinet, H., Jardon, P. and Gautron, R. (1988) Formation d'oxygène singulet photosensibilisée par l'hypéricine. Etude cinétique en milieu micellaire non ionique. *J. Chim. Phys.* 85: 971–977.
7. Duran, N. and Song, P.-S. (1986) Hypericin and its photodynamic action. *Photochem. Photobiol.* 43: 677–680.
8. Hadjur, C., Jeunet, A. and Jardon, P. (1994) Photosensitization by hypericin: ESR evidence for singlet oxygen and superoxide anion radicals formation in an *in vitro* model. *J. Photochem. Photobiol. B. Biol.* 26: 67–74.
9. Pou, S., Hassett, D.J., Britigan, B.E., Cohen M.S. and Rosen, G.M. (1989) Problems associated with spin trapping oxygen-centered free radicals in biological systems. *Anal. Biochem.* 177: 1–6.
10. Buettner, G.R. and Oberley, L.W. (1978) Considerations in the spin-trapping of superoxide and hydroxyl radical in aqueous systems using DMPO. *Biochem. Biophys. Res. Commun.* 83: 69–74.
11. Janzen, E.G., Wang, Y.Y. and Shetty, R.V. (1978) Spin trapping with α-pyridyl 1-oxide N-ter-butyl nitrones in aqueous solutions: a unique electron spin resonance spectrum for the hydroxyl radical adduct. *J. Am. Chem. Soc.* 100: 2923–2925.
12. Harbour, J.R., Chow, V. and Bolton, J.R. (1974) An electron spin resonance study of the spin adducts of OH and HO$_2$ radicals with nitrones in the ultraviolet photolysis of aqueous hydrogen peroxide solutions. *Can. J. Chem.* 52: 3549–3553.
13. Finkelstein, E., Rosen, G.M., Rauckman, E.J. and Paxton, J. (1979) Spin trapping of superoxide. *Mol. Pharmacol.* 16: 676–685.
14. Finkelstein, E., Rosen, G.M. and Rauckman, E.J. (1980) Spin-trapping. Kinetics of the reaction of superoxide and hydroxyl radicals with nitrones. *J. Am. Chem. Soc.* 102: 4994–4999.
15. Finkelstein, E., Rosen, G.M. and Rauckman, E.J. (1981) Production of hydroxyl radical by decomposition of spin-trapped adducts. *Mol. Pharmacol.* 21: 262–265.
16. Finkelstein, E., Rosen, G.M. and Rauckman, E.J. (1980) Spin trapping of superoxide and hydroxyl radical: practical aspects. *Arch. Biochem. Biophys.* 200: 1–16.
17. Casey, M., Leonard, J. and Procter, G. (1990) *Advanced Practical Organic Chemistry.* Chapman and Hall, New York, pp 28–42.
18. Brockman, H. and Spitzner, D. (1975) Die Konstitution des pseusohypericins. *Chem. Ber.* 37–40.
19. Adams, R. and Graves, G.D. (1923) Trihydroxy-methyl-anthraquinones I. *J. Am. Chem. Soc.* 45: 2439.
20. Batzri, S. and Korn, E.D. (1980) Single bilayer liposomes prepared without sonication. *Biochim. Biophys. Acta.* 594: 53–84.
21. Parker, C.A. (1986) *Photoluminescence of Solutions.* Elsevier, Amsterdam, pp 208–216.
22. Rosen, G.M., Finkelstein, E. and Rauckman, E. J. (1982) A method for the detection of superoxide in biological systems. *Arch. Biochem. Biophys.* 215: 367–378.

Analysis of Free Radicals in Biological Systems
Favier et al. (eds)

Nitrone spin traps as reagents for the study of oxidative modification of low density lipoproteins: Implications for atherosclerosis

C.E. Thomas

Marion Merrell Dow Research Institute, Cerebrovascular Biology Department, 2110 E. Galbraith Rd., Cincinnati, OH 45215, USA

Introduction

Atherosclerosis and its complications, including ischemic heart disease, represent the leading cause of death in the industrialized world. Accordingly, considerable resources have been expended in an effort to delineate causality and to devise effective therapeutic modalities. Several trials have demonstrated that lowering of LDL-cholesterol or an elevation of HDL-cholesterol reduces the risk of adverse cardiovascular events. Nonetheless, it is recognized that greater than 50% of the survivors of myocardial infarction have normal lipid levels, thus implicating other or additional factors in the pathogenesis of atherosclerosis.

Over the years hypotheses have been put forward which address the etiology of atherosclerosis. These include the "response to injury hypothesis" as described by Ross et al. [1], the "lipid peroxidation hypothesis" [2], the "LDL oxidative modification hypothesis" of Steinberg and colleagues [3] and the "antioxidant hypothesis" as proposed by Gey [4]. The latter three are similar in that they center on the propensity of LDL to undergo lipid peroxidation. Oxidatively modified LDL have been demonstrated to impact on numerous biologic events which could contribute to atherogenesis. These include the enhancement of monocyte [5] and leukocyte [6] binding to the endothelium, upregulation of interleukin-1β release [7, 8] and foam cell formation [9]. Recently, it has become clear that IL-1β and other cytokines may play an active role in the etiology of atherosclerosis which, while generally considered to be a lipid disorder, also involves a number of inflammatory modulators. This aspect was highlighted by Ku and Jackson [10] who championed the "cytokine hypothesis" of atherosclerosis in which, interestingly, LDL oxidation also plays a pivotal role.

Oxidative modification of LDL involves peroxidation of its polyunsaturated fatty acids which exist both as esters or as free acids, as well as modification of apolipoprotein B. It appears that oxidation of both components can contribute to the aforementioned biologic activity of oxidized LDL. Lipid oxidation generates a number of aldehydic species, including malondialdehyde and 4-hydroxynonenal, and other products such as 9-hydroxyoctadecadienoic acid (9-HODE). Several LDL-associated aldehydes [8] and 9-HODE [7] have been shown to promote the release of IL-1β release from human peripheral blood monocytes. This is of significance as IL-1β can subsequently lead to the smooth cell proliferation which characterizes atherosclerotic plaques. With respect to the protein component of LDL, lipid-derived alkoxyl or peroxyl radicals may directly modify apolipoprotein B which can also be altered by derivatization of its lysine residues with lipid aldehydes. This leads to unregulated uptake of the LDL by macrophages resulting in intracellular lipid accumulation and foam cell formation, a hallmark of atherosclerosis.

The clear importance of the role of radical-mediated modification of LDL in atherosclerosis has led us and others to examine in detail the phenomenon of LDL oxidation. One approach is to utilize antioxidants which retard or inhibit oxidation as tools to probe the peroxidative process in a temporal fashion. Many *in vitro* and animal studies have been conducted with phenolic antioxidants such as probucol [11], α-tocopherol [12, 13] and BHT [14]. While the studies have generally suggested that minimization of LDL oxidation translates favorably in terms of decreasing plaque formation, they have done little to help understand the relative importance of lipid versus protein oxidation and to demonstrate whether site-specific oxidation of the LDL occurs in the face of various oxidative stimuli.

Recently, we have reported on the use of nitrone spin traps for the study of oxidative modification of LDL [15, 16]. In contrast to antioxidants such as probucol which function as H atom donors, spin traps are diamagnetic molecules containing either nitroso or nitrone functionalities. This paper will be limited to nitrone spin traps which react with a radical (R$^{\cdot}$) to generate a more stable nitroxide spin adduct as shown:

nitrone   nitroxide spin adduct

The adduct can be detected and quantitated by electron spin resonance (ESR) spectroscopy. The ESR spectrum of the adduct can yield valuable information on the nature and location of the trapped radical within the LDL matrix although the utility of spin traps in the study of LDL oxidation will depend greatly upon the concentration and location of the spin trap in the LDL, the type of radicals which the nitrone can trap, the efficiency of radical trapping and the stability of the resultant adduct. Herein will be described studies using the nitrone spin traps α-phenyl-N-*tert*-butyl nitrone (PBN), α-(4-pyridyl 1-oxide)-N-*tert*-butyl nitrone (POBN) as shown below, and a series of novel cyclic nitrones. We have used these agents to study LDL oxidation and to examine their potential to serve as anti-atherogenic agents, particularly with regards to their ability to modulate the release of IL-1β by oxidized LDL.

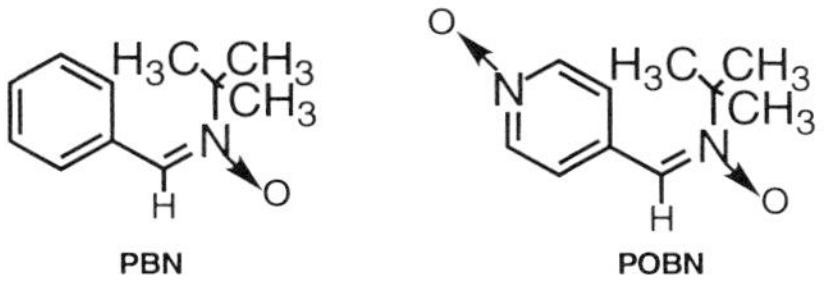

PBN   POBN

Materials and methods

LDL preparation

Human plasma was obtained from 18–20 normolipidemic, fasting donors. To each pool of plasma was added 1 mM EDTA, 0.5 mM phenylmethylsulfonyl fluoride, 50 units/ml aprotinin and 0.01% sodium azide. LDL were isolated by ultracentrifugal flotation in KBr (d 1.019–1.063 g/ml) as described by Cardin et al. [17]. After isolation the LDL were stored in 1 mM EDTA and dialyzed against deoxygenated phosphate buffered saline (PBS) (4 × 4 l over 24 h) at 4°C in the dark immediately before use. After dialysis, the LDL are used within 6 h as they are highly susceptible to transition metal-dependent oxidation once the EDTA is removed. Unfortunately, virtually all laboratory reagents and water sources are contaminated with trace amounts of metals, particularly iron. To minimize spurious oxidation solutions can be passed over a column containing Chelex 100 resin (BioRad, Richmond, CA) to remove metals and the solutions extensively purged with N_2 gas.

Assessment of LDL oxidation

For determination of the effects of the nitrone spin traps or phenolic antioxidants on oxidation, LDL (100 μg protein) were incubated at 37°C for 18 h with 5 μM Cu^{2+} or for 30 min with 20 mM 2,2'-azobis-2-amidinopropane hydrochloride (ABAP) in 1 ml of PBS. In some experiments the LDL were oxidized at 2 mg protein/ml. One h prior to the addition of the oxidant, the antioxidant to be tested was added to the LDL in PBS or 20 μl of ethanol and the tube capped under N_2 and kept on ice. Generally, the nitrones were at 100 mM for ESR experiments. Any insoluble material was removed by carefully decanting the LDL/nitrone mixture to a clean tube. It should be noted that for PBN and other hydrophobic cyclic nitrones, not all of the added compound can be incorporated using this method. For example, we determined that 50% of the PBN was incorporated which was sufficient to allow analysis by ESR (15). However, this method served to minimize disruption of the LDL which would result from the use of solvents such as DMSO. At appropriate times Cu^{2+}-dependent oxidation was terminated with 10 μM EDTA and ABAP-dependent oxidation by placing the reaction mixture on ice to stop thermal generation of peroxyl radicals.

Determination of LDL oxidation

Oxidation was quantitated using the thiobarbituric acid (TBA) test (18). In 13 $\times$ 100 mm borosilicate glass test tubes was added an aliquot of the reaction mixtures which contained 100 μg of LDL protein and, if necessary, sufficient PBS to achieve a volume of 1 ml. This was followed by 0.05 ml of 2% butylated hydroxytoluene and 2 ml of 0.67% TBA and 10% trichloroacetic acid (2:1) in 0. 25 N HCl. The tubes were covered with marbles and heated at 100°C for 20 min in a heating block. After cooling, the tubes were centrifuged at 3000 rpm for 10 min and the absorbance of the resultant supermatant read at 532 nm minus 580 nm to account for any turbidity using a Hewlett Packard 8452A diode array spectrophotometer. Quantitation of thiobarbituric acid reactive substances (TBARS) was determined by comparison to a standard curve of malondialdehyde equivalents generated by acid-catatyzed hydrolysis of 1,1,3,3-tetraethoxypropane (Aldrich). The IC_{50} values were determined from a 15 min timepoint and the calculations performed using Graph-Pad InPlot (GraphPad Software, Inc., San Diego, California).

ESR measurements

For ESR experiments, LDL (12 mg) were incubated with 500 μM Cu^{2+} for 18 h in a final volume of 2 ml PBS. The nitrones were added 1 h prior to initiation of oxidation as described above. Where indicated, 1 ml of the reaction mixtures was extracted twice with 2 ml of $CHCl_3$:CH_3OH (2:1). The lipid, protein and the aqueous fractions were separated by centrifugation (2000 rpm for 10 min). The organic fractions were combined and dried under N_2. The dried fraction was solubilized in 200 μl of N_2 purged ethanol and taken up in an aqueous flat cell. The protein pellet was removed from the interface of the organic and aqueous layers and washed twice with cold acetone, dried under nitrogen and placed inside a 4 mm cylindrical quartz tube. Spectra were recorded on a Varian E-109 spectrometer operating at 9.5 GHz and employing 100 KHz field modulation.

To aid in determination of whether spin trap-LDL-lipid adducts are located within core lipids, lipid-protein or lipid-aqueous interphase, the effect of chromium oxalate (CROX) on the spectral line-shape of the spin adduct can be used. Chromium oxalate is a charged, water soluble but membrane-impermeant, paramagnetic relaxing agent (19). Upon collision with a nitroxide, CROX will shorten the effective spin-lattice relaxation time and broaden the ESR spectrum. Since CROX cannot enter the LDL particle, collision will occur only with adducts which are free in solution or oriented towards the aqueous phase. Thus, the spectra of nitroxide adducts buried within the LDL matrix would not be affected by the addition of CROX. Typically, 25–75 mM CROX are added to the LDL oxidation mixture following termination of oxidation and immediately prior to ESR analysis.

Chromatographic separation of spin adducts

Spin adducts generated during oxidation of LDL and Cu^{2+} were extracted with 2 volumes of $CHCl_3:CH_3OH$ (2:1) mixture and the solvent removed under N_2 to yield a yellow residue. The residue were dissolved in hexane and chromatographed on silica gel 60 using hexane:ethyl acetate (4:1) as an eluent. The fraction eluting with a fast moving yellow band was scraped and examined by ESR.

The fraction isolated from the silica gel can also be subjected to HPLC separation and analysis. We have used a Waters 625 system consisting of a Model 625 pump/controller and a Model 996 photodiode array detector. Separation can be acheived using a C_{18} reverse phase column such as a Waters Novapak (5 micron, 15 cm) with a mobile phase consisting of acetonitrile/water. The exact composition of the mobile phase varies depending upon the nitrone under study, but typically involves a gradient going from 5%–50% acetonitrile over a 40 min run time with a flow rate of 1.5 ml/min. For PBN, the absorbance at 302.5 nm was used for quantitation based on comparison to areas obtained for a standard curve of PBN.

Isolation of human peripheral blood monocytes

Human peripheral blood monocytes were prepared in 10 mM sodium citrate from blood collected from healthy volunteers. In order to minimize non-experimental variation of IL-1β, donors were asked to refrain from consuming caffeine prior to blood draw and were not currently taking anti-inflammatory drugs. Erythrocytes and neutrophils were removed by low-speed centrifugation in Leucoprep tubes (Becton Dickinson, Oxnard, California) according to the protocol suggested by the manufacturer. The resultant mixture of platelets and mononuclear cells was incubated in tissue culture dishes (2×10^6 cells per well of 24-well plates, Corning, Corning New York) for 1 h at 37°C. Following the removal of non-adherent cells, fresh medium RPMl-1640 (Gibco, Grand Island, New York) was added to the plated cells.

Measurement of IL-1β release from monocytes

Oxidized LDL or 9-HODE (Cayman Chemical, Ann Arbor, Michigan) were utilized to induce IL-1β release from monocytes. Free Cu^{2+} is generally toxic to cells in culture, thus 15 μM EDTA was added to chelate the metal at the end of the oxidation period. When studying cytokine release from monocytes it is imperative to ensure that stimuli are free of endotoxin contamination as it is a potent inducer of cytokine release. To safeguard against possible endotoxin contamination, we often chromatographed oxidized LDL over a pre-packed endotoxin affinity column (Detoxigel, Pierce Chemical Co., Rockford, Illinois). As determined by the *Limulus polyphemus* amebocyte lysate test (Sigma, St. Louis, Missouri; sensitivity 1 ng of endotoxin), a column has the capacity of removing 2 mg of endotoxin. We also determined that the efficiency of the column in removing endotoxin was not affected by the LDL. The protein concentration of the LDL was determined after chromatography as there is unavoidably some dilution of the preparation.

To the culture dishes containing the adherent cells was added 350-400 μg of oxidized LDL for approximately 24 h. When 9-HODE was used as the stimulus it was added from a 1 mg/ml stock solution in ethanol at 33 μM. Our previous work demonstrated that oxidized LDL induces IL-1β release, but not TNFα, thus culture supernatants were routinely assayed for both cytokines to monitor for potential endotoxin contamination. If this is done, it is not necessary to use Detoxigel as the presence of endotoxin will lead to TNFα release. However, while we had little problem with endotoxin, this may not be universally true and experiments may be rendered useless if TNFα expression is enhanced, indicating endotoxin contamination. For this reason, the use of the endotoxin affinity column is recommended. The human IL-1β and TNFα ELISA kits were obtained from Cistron (Pinebrook, New Jersey) and used in accordance with the manufacturer's suggested protocol.

Table 1. Effect of PBN or POBN on Cu^{2+}-induced LDL oxidation

Treatment	TBARS (nmol/mg LDL protein)
$-Cu^{2+}$	0.3 ± 0.3
$+Cu^{2+}$	35.1 ± 0.6
PBN, 5 mM	19.7 ± 0.5
PBN, 10 mM	13.3 ± 0.5
PBN, 20 mM	12.1 ± 0.7
POBN, 5 mM	36.3 ± 0.6
POBN, 10 mM	36.6 ± 0.7
POBN, 20 mM	36.8 ± 0.3

LDL (100 μg) were oxidized for 18 h in the presence of 5 μM Cu^{2+}. Oxidation was terminated by the addition of EDTA and quantitated by determination of TBARS. Values represent the mean $\pm$ SE for four separate incubations. (Data taken from reference [15] with permission.)

Results and discussion

Work by Kalyanaraman and colleagues [15, 20–22] has demonstrated that PBN could prevent *in vitro* Cu^{2+}-dependent oxidation of LDL and recognition by the macrophage scavenger receptor. Likewise, the lipophilic nitroso spin trap 2-methyl-2-nitroso propane also trapped lipid-derived radicals and thereby prevented LDL oxidation. Conversely, the corresponding hydrophilic analogs POBN and 2-hydroxy-methyl-2-nitroso propane appeared to trap only secondary alkyl radicals and aldehydes, respectively.

The above data led us to examine in more detail the potential utility of spin traps to inhibit oxidative modification of LDL and its subsequent biologic activity, namely, stimulation of IL-1β release from monocyte/macrophages. In our initial studies, we confirmed that PBN, and not POBN, could prevent Cu^{2+}-mediated LDL oxidation as determined by TBARS (Tab. 1). The LDL subjected to Cu^{2+}, in the presence and absence of PBN or POBN, were also studied by fluorescence spectroscopy. As shown in Figure 1, incubation with Cu^{2+} led to a dramatic increase in fluorescence emission at 430 nm (excitation, 360 nm). The fluorescence changes were little affected by POBN but decreased by inclusion of PBN. It should be noted that, in control experiments, when PBN was added to oxidized LDL it had an autofluorescence while POBN quenched the fluorescence of the oxidized lipids. Thus, the difference in ability to prevent Cu^{2+}-induced modification is greater than is evident from the figure. These data indicate that the spin traps react with reactive species that could otherwise form adducts with apolipoprotein B. This contention is supported by the dose-dependent ability of PBN, but not POBN, to prevent oxidation enhanced electrophoretic mobility in agarose gels [15].

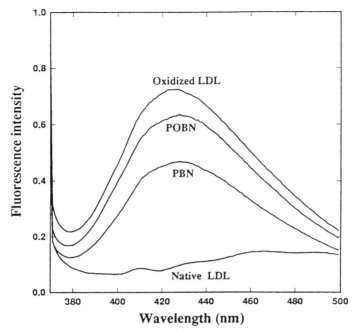

Fig. 1. Effects of PBN or POBN on Cu^{2+}-induced fluorescence changes in LDL. LDL (2 mg/ml) were incubated with 5 μM Cu^{2+} for 18 h in PBS. Where indicated, incubations also contained 5 mM PBN or POBN. (Reprinted from reference [15] with permission.)

The ESR spectra of adducts of PBN and POBN in LDL were also evaluated. As seen in Figure 2 (top left), the nitroxide adduct of PBN demonstrates rotational restriction, indicating its presence within the LDL matrix. Upon extraction and resolubilization in ethanol, a more well-resolved spectrum was obtained (middle left). The spectra obtained from LDL incubated with Cu^{2+} and POBN was distinct from that of PBN. A well-resolved spectrum was evident without extraction (top right) suggesting that POBN traps radicals which have diffused into the aqueous phase. An adduct was observed in the lipid extract also, indicating that some POBN-lipid adducts are soluble in chloroform (middle right). No adduct was detectable in the absence of Cu^{2+} with either nitrone (bottom spectra).

Further validation that the PBN-derived adducts represent trapping of lipid-derived radicals was obtained using CROX. Figures 3A–D show the effect of CROX on the ESR spectra of the PBN-LDL adduct. Addition of CROX broadened the fast tumbling component of the nitroxide with minor effect on the immobilized component. This differential sensitivity to CROX implies that a portion of the PBN-LDL adduct is in proximity to the aqueous phase (fast tumbling) with the

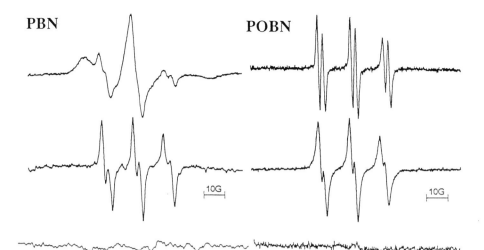

Fig. 2. The ESR spectra of PBN (left side) or POBN (right side) adducts in Cu^{2+}-oxidized LDL. In the upper scans LDL (6 mg/ml) were incubated with 0.5 mM Cu^{2+} and 100 mM of PBN or POBN for 18 h. Spectrometer conditions were: modulation amplitude, 2.5 G; microwave power, 20 mW; scan range, 100 G; time constant, 0.25 sec; scan time, 4 min. In the middle traces the incubation mixtures were extracted in chloroform:methanol 2:1, dried under N_2 and taken up in 0.2 ml of N_2-purged ethanol. Spectrometer conditions: modulation amplitude, 1 G; microwave power 5 mW; scan range, 100 G; time constant 0.128 s; scan time 4 min. In the lower spectra, the incubation mixture lacked Cu^{2+} and the spectrometer conditions were as for the middle trace. (Adapted from reference [15] with permission.)

majority of the spin adduct (immobilized component) either buried in the LDL core lipid or embedded near the lipid-protein interface. Either environment would prevent collision between the adduct and CROX.

The studies with PBN, POBN and the nitroso spin traps indicated that lipophilicity of the radical traps was a primary determinant of efficacy towards prevention of metal-dependent oxidation of LDL. Recently, we have prepared a series of cyclic variants of PBN which encompass a wide range of lipophilicity (Fig. 4). These novel nitrones were examined for their ability to inhibit both Cu^{2+}-dependent and peroxyl radical (ABAP)-dependent oxidation of LDL. As highlighted in Table 2, all the cyclic nitrones were much more potent than PBN, irrespective of the oxidant. In general, activity positively correlated with lipophilicity as determined by a computer generated log P (cLogP). However, it is apparent that cyclization of the nitrone leads to an inherent improvement in activity as MDL 101,002 and MDL 102,073 have a lower cLogP than does PBN, yet are much more active. In other studies, we have shown that, indeed, the cyclic nitrones are more efficient radical traps than PBN (unpublished observations).

The ESR spectra for many cyclic nitrone-LDL adducts have been evaluated. In general, they are similar to that of PBN (see Fig. 2), and

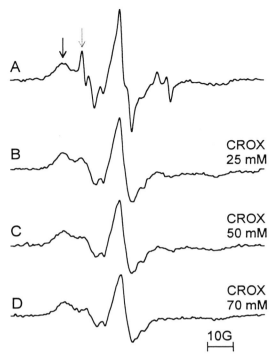

Fig. 3. Effect of chlomium oxalate on the ESR spectrum of the PBN-LDL-lipid adduct. (A) ESR spectrum obtained 18 h after the addition of Cu^{2+} (0.5 mM) to a mixture containing LDL (6 mg/ml) and PBN (100 mM) in PBS; (B) the same as above, but in the presence of 25 mM CROX; (C and D) in the presence of 50 and 70 mM CROX, respectively. Solutions of CROX were added to the incubation mixture in (A) after 18 h (B–D). Arrows denote the line-positions of the immobilized (left arrow) and the fast tumbling (right arrow) components of the PBN-LDL adduct. (Reproduced from reference [15] with permission.)

representative of a highly immobilized, rotationally restricted nitroxide. This is illustrated with MDL 105,185 in the upper spectrum in Figure 5 which, again, exhibits β-hydrogen coupling when extracted and resolubilized in ethanol. Following extraction with CHCl$_3$:MeOH, the upper aqueous phase consisting of MeOH and PBS was also studied by ESR and no adduct was detected. Somewhat surprisingly, when the protein fraction of the extracted LDL was isolated and subjected to analysis, a strong, immobilized adduct was detected ($2T_t = 68.0 \pm 0.5$ G). This is the first demonstration of trapping of an apolipoprotein-derived radical and suggests the intriguing possibility that the protein radicals may be critical in propagating radial chain reactions within the LDL. It is plausible that the protein-derived radical represents trapping of cysteinyl residues as the cyclic nitrones can trap thiyl radicals (data not shown). Studies to examine this possibility are currently ongoing. These findings also suggest that nitrone spin traps may be used to study the

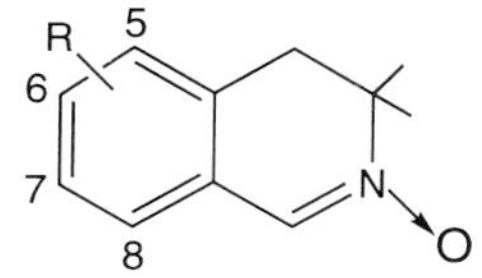

MDL 101,002; R=H

MDL 102,073; R=7-OCH₃

MDL 100,777; R=7-Cl

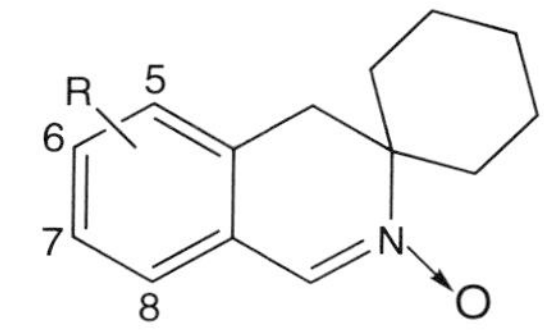

MDL 102,832; R=H

MDL 101,694; R=7-Cl

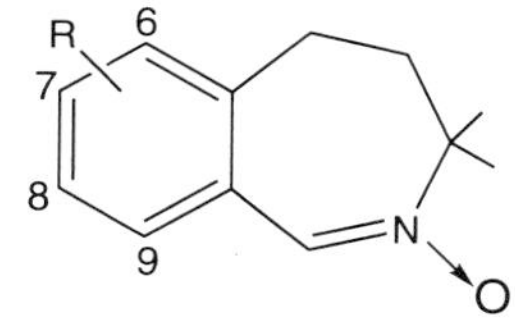

MDL 102,389; R=H

MDL 104,342; R=7,9-Cl₂

MDL 105,185; R=8-Cl

Fig. 4. Structures of cyclic nitrone spin traps. The structures of the novel cyclic nitrones are shown. Phenyl ring substitutions are listed below the unsubstituted parent molecule for each class of nitrone. (Modified from reference [16] with permission.)

role of protein oxidation in LDL modification. For example, a study of the time-dependent formation of lipid adducts versus protein-derived radicals could provide valuable insight into the process of Cu^{2+}-mediated oxidation which remains controversial in spite of intensive study.

Table 2. Inhibition of Cu^{2+} or ABAP-induced LDL oxidation by PBN and cyclic nitrones

| Compound | $ICc_{50}(\mu M)$ | | cLogP |
	Cu^{2+} oxidation	ABAP oxidation	
PBN	10.1×10^3	2.2×10^3	1.23
MDL 101,002	3.8×10^3	90	1.01
MDL 102,073	3.4×10^3	24	1.08
MDL 102,389	nd	81	1.57
MDL 102,832	128	2	1.97
MDL 105,185	125	nd	2.29
MDL 101,694	14	1	2.68
MDL 104,342	51	7	3.00

LDL (100 μg) were incubated with 5 μM Cu^{2+} for 18 h or ABAP (20 mM) for 30 min and oxidation determined by TBARS. The computer-determined octanol:water partition coefficients (cLogP) were determined using the MEDCHEM utilities in SYBYL. (Modified from reference [16] with permission.)
nd = not determined

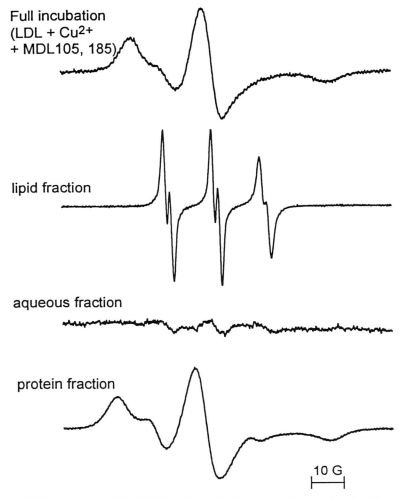

Fig. 5. ESR spectra of MDL 105,185 adducts obtained from Cu^{2+}-mediated oxidation of LDL. LDL (12 mg) were incubated with MDL 105,185 (100 mM) for 1 h and the contents decanted to a clean tube to obtain a clear solution. To this was added Cu^{2+} (0.5 mM) and the mixture incubated for 18 h. The ESR spectra were obtained from the lipid, aqueous, and protein fractions as described under "Materials and Methods." Spectrometer conditions: scan range, 100 G; modulation amplitude, 2 g; and microwave power, 2 mW. (From reference [16] with permission.)

Next, we examined the effect of nitrones on oxidized LDL-dependent induction of IL-1β release from monocyte/macrophages. Shown in Table 3 are the effects of Cu^{2+}-oxidized LDL, the lipid oxidation product 9-HODE, and lipopolysaccharide (LPS) on IL-1β release from human cells. 9-HODE, either free or esterified, is a major component in oxidized LDL [7] and likely contributes to their ability to induce IL-1β. We have recently

Table 3. Induction of IL-1β release from human peripheral blood monocyte-derived macrophages by Cu²⁺-oxidatized LDL or 9-HODE

	(pg IL-1β/culture)		
Donor	Vehicle	Oxidized-LDL	9-HODE
R. S.	33 ± 5	228 ± 14	117 ± 36
T. M	84 ± 2	228 ± 30	372 ± 53
M. T.	33 ± 3	486 ± 26	nd
M. K.	74 ± 9	355 ± 20	nd
D. M.	12 ± 2	nd	247 ± 17
C. B.	17 ± 3	nd	212 ± 13

Macrophages from various donors were incubated with Cu²⁺-oxidized LDL (350–400 μg) or 9-HODE (33 μM) for approximately 24 h. The amount of secreted IL-1β in the supernatant was determined by ELISA. Two donors (R. S. and T. M.) were tested against both stimuli in separate experiments. (Data compiled from reference [7] with permission.)
nd = not determined

determined that other lipid-derived aldehydes also promote the release of this cytokine [8]. When LDL were oxidized with Cu²⁺ in the presence of PBN, a dose-dependent decrease in the ability of the LDL to induce IL-1β was observed (Fig. 6). On the other hand, POBN was without effect up to 10 mM. These results mirror the effect of the nitrones on LDL oxidation. As the nitrones do not totally prevent LDL oxidation, we have determined that mediators of IL-1β release including 9-HODE are still found within

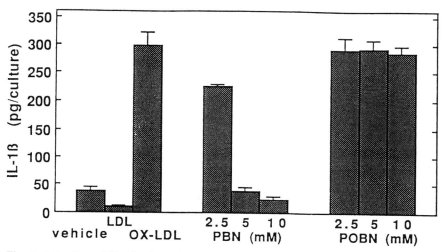

Fig. 6. Induction of IL-1β release from human monocyte-derived macrophages by LDL oxidized with Cu²⁺ in the presence or absence of PBN or POBN. LDL (2 mg/ml) were incubated with 5 μM Cu²⁺ for 18 h in the presence of 0–10 mM PBN or POBN. LDL (350 μg/ml) were added to the cells for 22 h and the amount of IL-1β in the supernatant determined by ELISA. The values represent triplicate determinations for each of two LDL samples. (Reproduced in part from reference [15] with permission.)

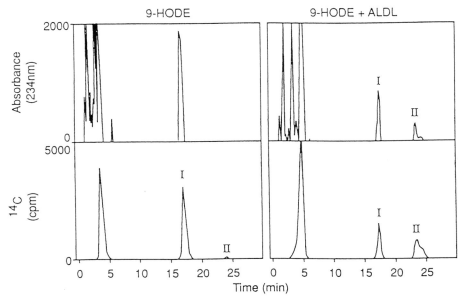

Fig. 7. Effect of acetylated LDL on distribution of ^{14}C in human monocyte-derived macrophages incubated with ^{14}C-9-HODE. Radiolabelled 9-HODE was prepared using potato 5-lipoxygenase and ^{14}C-linoleic acid with sodium borohydride reduction. Human peripheral blood monocyte-derived macrophages (28×10^6 cells/dish) were incubated with 33 μM of ^{14}C-9-HODE for 3 h in the presence or absence of acetylated LDL (900 μg). The cells were scraped into 0.1 M SDS and extracted with chloroform:methanol (2:1). The lipid extract was subjected to normal phase HPLC with detection at 234 nm and by on-line detection of [14]C. A ^{14}C-containing peak (peak II), which was more polar than 9-HODE (peak I), was detected and was markedly increased in the presence of acetylated LDL. The early eluting ^{14}C-containing peak represents 9-HODE which has been esterified.

the LDL. However, the degree of oxidation is reduced sufficiently to prevent recognition and uptake of the LDL by the scavenger receptor of the macrophage. Uptake of the oxidized lipids is required for activation of cytokine release as mixing of the 9-HODE with acetylated LDL (which is taken up by the scavenger receptor) greatly stimulated IL-1β release over that observed with equimolar 9-HODE only. It has yet to be determined whether this simply reflects greater cellular uptake of 9-HODE or alters subsequent processing of the lipid to a penultimate "signaling agent". Very preliminary experiments with ^{14}C-9-HODE reveals the presence of an additional intracellular ^{14}C-containing peak when acetylated LDL are included (Fig. 7). The identity of this peak has yet to be ascertained.

Several of the cyclic nitrones were examined for their ability to prevent IL-1β release. In Figure 8 it is shown that these compounds could significantly ameliorate the induction of IL-1β. It must be noted that the concentrations of the nitrones were above those reported to inhibit LDL oxidation in Table 2. Those values were determined with 100 μg of LDL while the

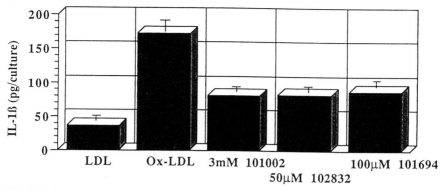

Fig. 8. Effect of cyclic nitrones on the ability of LDL incubated with Cu^{2+} to induce IL-1β release from human monocyte-derived macrophages. LDL (2 mg/ml) were incubated with 5 μM Cu^{2+} for 18 h in the presence of the cyclic nitrone spin traps. Human peripheral blood monocyte-derived macrophages (2 × 10^6 cells/well) were incubated with 400 μg of the various LDL preparations for 24 h prior to analysis of IL-1β in the culture medium. The LDL oxidized with Cu^{2+} had a TBARS value 58.6 nmol/mg protein. The TBARS values for the cyclic nitrones were 13.9, 18.5 and 16.0 nmol/mg protein for MDL 101,002; MDL 102,832, and MDL 101,694 respectively.

IL-1β experiments required oxidation with LDL at 2 mg/ml. As a result, oxidation was not totally prevented and a partial amelioration of IL-1β release ensued. Nonetheless, these concentrations are still much lower than the amount of PBN required to prevent LDL oxidation and IL-1β release.

Conclusions

The studies described herein demonstrate that nitrone spin traps can be useful tools in the study of *in vitro* LDL oxidation. It was of great interest that at least one of the cyclic nitrones (MDL 105,185) was also capable of trapping apolipoprotein B-derived radicals. Importantly, the lipid and protein-derived adducts of the cyclic nitrones appear to be quite stable. We have observed that there is no loss in signal intensity in the ESR spectra over a period of months, particularly if the samples are kept at − 20°C. Thus, the nitrones could be invaluable in helping to identify trapped radicals. With nitrone spin traps, little structural information can be gleaned from the ESR spectra which are typically dominated by a triplet of doublets with the trapped radical influencing primarily the β-hydrogen coupling. The apparent stability of the cyclic nitrone-LDL-adducts should render them amenable to other analyses, for example, mass spectral identification of trapped adducts. Our recent studies have focused on HPLC separation of the lipid fraction extracted from Cu^{2+} oxidized LDL which exhibit a strong ESR signal. If successful, this methodology could be applied to a kinetic study of the peroxidative process. For example, it would be of interest to evaluate the formation of lipid and/or protein-derived radicals in the outer monolayer of the LDL versus the formation of cholesteryl ester-derived radicals in the LDL core.

It has yet to be determined whether the spin traps could also serve to identify trapped radicals in *in vitro* studies. Isolation of LDL samples from animal models of atherogenesis such as the cholesterol-fed rabbit and subsequent ESR and mass spectral analysis could provide new insight into the oxidative modification of LDL *in vivo*; of which much has been hypothesized but little conclusively demonstrated. While pure speculation, the continued improvement of low frequency ESR to allow spectroscopy directly in living tissue may one day allow on-line monitoring of circulating LDL oxidation with concurrent development of appropriate ESR reagents. Finally, our ability to design spin traps which are as effective inhibitors of LDL oxidation as are probucol and α-tocopherol, and which also inhibit IL-1β release, opens new avenues for a unique pharmacologic class of anti-atherogenic drugs.

Acknowledgements
I would like to thank Dr. B. Kalyanaraman (Medical College of Wisconsin) for his many contributions to this work and Dr. C. Channa Reddy (Pennsylvania State University) for the gift of the potato 5-lipoxygenase.

References

1. Ross, R. (1986) The pathogenesis of atherosclerosis – an update. *N. Engl. J. Med.* 314: 488–500.
2. Yagi, K. (1984) Increase serum lipid peroxides initiate atherogenesis. *BioEssays* 1: 58–60.
3. Steinberg, D., Parthasarathy, S., Carew, T.E., Khoo, J.C. and Witztum, J.L. (1989) Beyond cholesterol. Modifications of low-density lipoprotein that increase its atherogenicity. *N. Engl. J. Med.* 320: 915–924.
4. Gey, K.F. (1986) On the antioxidant hypothesis with regard to arteriosclerosis. *Bibl. Nutr. Diets* 37: 53–91.
5. Liao, F., Berliner, J.A., Mehrabian, M., Navab, M., Demer, L.L., Lusis, A.J. and Fogelman, A.M. (1991) Minimally modified low density lipoprotein is biologically active *in vivo* in mice. *J. Clin. Invest.* 87: 2253–2257.
6. Lehr, H.A., Hubner, C., Finckh, B., Angermuller, S., Nolte, D., Beisegel, U., Kohlschutter, A. and Messmer, K. (1991) Role of leukotrienes in leukocyte adhesion following systemic administration of modified human low density lipoprotein in hamsters. *J. Clin. Invest.* 88: 9–14.
7. Ku, G., Thomas, C.E., Akeson, A.L. and Jackson, R.L. (1992) Induction of interleukin 1 beta expression from peripheral blood monocyte-derived macrophages by 9-hydroxyoctadecadienoic acid. *J. Biol. Chem.* 267: 14183–14188.
8. Thomas, C.E., Jackson, R.L., Ohlweiler, D.F. and Ku, G. (1994) Multiple lipid oxidation products in low density lipoproteins induce interleukin-1 beta release from human blood mononuclear cells. *J. Lipid Res.* 35: 417–427.
9. Rosenfeld, M.E., Khoo, J.C., Miller, E., Parthasarathy, S., Palinski, W. and Witztum, J.L. (1991) Macrophage-derived foam cells freshly isolated from rabbit atherosclerotic lesions degrade modified lipoproteins, promote oxidation of low-density lipoproteins and contain oxidation-specific lipid-protein adducts. *J. Clin. Invest.* 86: 90–99.

10. Ku, G. and Jackson, R.L. (1994) The pathogenesis of atherosclerosis. A cytokine hypothesis. *In:* D.L. Laskin (ed.): *Xenobiotics in Inflammation,* Academic Press, Inc., San Diego, pp 233–248.

11. Mao, S.J.T., Yates, M.T., Rechtin, A.E., Jackson, R.L. and Van Sickle, W.A. (1991) Antioxidant activity of probucol and its analogs in hypocholesterolemic Watanabe rabbits. *J. Med. Chem.* 34: 298–302.

12. Jialal, I. and Grundy, S.M. (1992) Effect of dietary supplementation with alpha-tocopherol on the oxidative modification of low density lipoprotein. *J. Lipid Res.* 33: 899–906.

13. Verlangieri, A.J. and Bush, M.J. (1992) Effect of d-α-tocopherol supplementation on experimentally induced primate atherosclerosis. *J. Amer. Coll. Nutr.* 11: 131–138.

14. Bjorkhem, I., Henriksson-Freyschuss, A., Breuer, O., Diczfalusy, U., Berglund, L. and Henriksson, P. (1991) The antioxidant butylated hydroxytoluene protects against atherosclerosis. *Arterioscler. Thromb.* 11: 15–22.

15. Thomas, C.E., Ku, G. and Kalyanaraman, B. (1994) Nitrone spin trap lipophilicity as a determinant for inhibition of low density lipoprotein oxidation and activation of interleukin-1β release from human monocytes. *J. Lipid Res.* 35: 610–619.

16. Thomas, C.E., Ohlweiler, D.F. and Kalyanaraman, B. (1994). Multiple mechanisms for inhibition of low density lipoprotein oxidation by novel cyclic nitrone spin traps. *J. Biol. Chem.* 269: 28055–28061.

17. Cardin, A.D., Holdsworth, G. and Jackson, R.L. (1984) Isolation and characterization of plasma lipoproteins and apolipoproteins. *Methods Pharmacol.* 5: 141–166.

18. Aust. S.D. (1985) Lipid peroxidation. *In:* R.A. Greenwald (ed.): *Handbook of Methods for Oxygen Radical Research,* CRC Press, Inc., Boca Raton, FL, pp 203–207.

19. Keana, J.F.W., Tamura, T., McMillen, D.A. and Jost, P.C. (1981) Synthesis and characterization of a novel cholesterol nitroxide spin label. Application to the molecular organisation of human high density lipoprotein. *J. Amer. Chem. Soc.* 103: 4904–4912.

20. Kalyanaraman, B., Joseph, J. and Parthasarathy, S. (1991) The spin trap, α-phenyl N-*tert*-butylnitrone, inhibits the oxidative modification of low density lipoprotein. *FEBS Lett.* 280: 17–20.

21. Kalyanaraman, B., Joseph, J. and Parthasarathy, S. (1993) Site-specific trapping of reactive species in low-density lipoprotein oxidation: biological implications. *Biochim. Biophys. Acta* 1168: 220–227.

22. Kalyanaraman, B., Joseph, J. and Parthasarathy, S. (1993) The use of spin traps to investigate site-specific formation of free radicals in low-density lipoprotein oxidation. *Biochem. Soc. Trans.* 21: 318–321.

Analysis of Free Radicals in Biological Systems
Favier et al. (eds)

Ascorbate radical: A valuable marker of oxidative stress

G.R. Buettner and B.A. Jurkiewicz

Free Radical Research Institute, EMRB 68, The University of Iowa, Iowa City, IA 52242-1101, USA

Summary. The ascorbate anion is an endogenous water-soluble antioxidant that is present in biological systems. The one-electron oxidation of ascorbate produces the ascorbate free radical that is easily detectable by electron paramagnetic resonance (EPR), even in room temperature aqueous solution. The ascorbate radical has a relatively long lifetime compared to other free radicals, such as hydroxyl, peroxyl, and carbon-centered lipid radicals. This longer lifetime in conjunction with its relatively narrow EPR linewidth makes it easily detectable by EPR. In this essay we describe the EPR detection of the ascorbate radical and its use as a marker of oxidative stress.

Ascorbate, the terminal small-molecule antioxidant

Introduction

Ascorbate (Asc H$^-$) is ubiquitous, yet there is still much to be learned about its chemistry, biochemistry, and biology. Ascorbate is an excellent reducing agent [1–5]. It readily undergoes two consecutive, yet reversible, one-electron oxidation processes to form the ascorbate radical (Asc$^{\bullet-}$) as an intermediate. Loss of a second electron yields dehydroascorbic (DHA) [1].

Because Asc$^{\bullet-}$ has its unpaired electron in a highly delocalized π-system, it is a relatively unreactive free radical. These properties make ascorbate a superior biological, donor antioxidant [3–17].

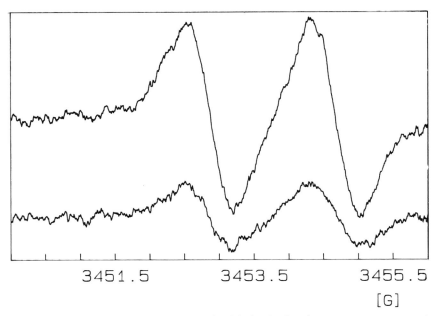

3451.5   3453.5   3455.5

[G]

Fig. 1. The ascorbate radical doublet EPR signal in bovine lens increases upon exposure to UV light. An approximate 1 cm² section of lens was placed into the well of an EPR tissue cell, then positioned in an EPR TM_{110} cavity, and subsequently exposed to UV light. Lower: ambient Asc· ⁻; Upper: Asc· ⁻ during exposure to UV light. See the section on lens below for details.

The ascorbate free radical is a strong acid having a pK_a of -0.86 [18]. Thus, it will exist as a monoanion, Asc· ⁻, over the entire biological pH range.[1]

When biological fluids or tissues are examined by electron paramagnetic resonance spectroscopy (EPR), Asc· ⁻ will most likely be observed. See Figure 1 below. This is consistent with ascorbate's role as the terminal small-molecule antioxidant [3, 4].

Ascorbate thermodynamics and kinetics

As can be seen in Table 1, ascorbate is thermodynamically at the bottom of the pecking order of oxidizing free radicals. That is, all oxidizing free radials with greater reduction potentials, which includes, HO·, RO·, LOO·, GS·, the urate radical, and even the tocopheroxyl

[1]A note on nomenclature: Asc· ⁻ is usually referred to in brief as the ascorbate free radical. The ending "ate" being used because it is a charged species. The short name ascorbyl radical would be used for AscH·, the neutral protonated form of Asc· ⁻. The ending "yl" being used for this neutral species.

Table 1. One-electron reduction potentials at pH 7.0 for selected radical couples

Redox couple	$E^{0'}/mV$
HO⋅, H^+/H_2O	+2310
RO⋅, H^+/ROH (aliphatic alkoxyl radical)	+1600
ROO⋅, $H^+/ROOH$ (alkyl peroxyl radical)	+1000
GS⋅/GS$^-$ (glutathione)	+920
PUFA⋅, $H^+/PUFA$-H (bis-allylic-H)	+600
HU$^{⋅-}$, H^+/UH_2^- (Urate)	+590
TO⋅, H^+/TOH (Tocopherol)	+480
H_2O_2, H^+/H_2O, HO⋅	+320
Ascorbate$^{⋅-}$, $H^+/$Ascorbate monoanion	+282
Fe(III)EDTA/Fe(II)EDTA	+120
$O_2/O_2^{⋅-}$	−330
Paraquat/Paraquat$^{⋅-}$	−448
Fe(III)DFO/FE(II)DFO (Desferal)	−450
RSSR/RSSR$^{⋅-}$ (GSH)	−1500
H_2O/e_{aq}^-	−2870

This table is adapted from references [3, 12, 19].

Table 2. Rate constants for the reaction of the equilibrium mixture of $AscH_2/AscH^-/Asc^{2-}$ at pH 7.4 unless noted otherwise

Radical	$k_{obs}/M^{-1} s^{-1}$	(pH 7.4)	Ref.[a]
HO⋅	1.1×10^{10}		[20]
RO⋅ (tert-butyl alkoxyl radical)	1.6×10^9		[21]
ROO⋅ (alkyl peroxyl radial, e.g., $CH_3OO⋅$)	$1-2 \times 10^6$		[22]
$Cl_3COO⋅$	1.8×10^8		[23]
GS⋅ (glutathiyl radical)	6×10^8	(5.6)	[24, 25]
PUFA⋅	—[b]		
UH$^{⋅-}$ (Urate radical)	1×10^6		[26]
TO⋅ (Tocopheroxyl radical)	2×10^{5} [c]		[3]
Asc$^{⋅-}$ (dismutation)	2×10^5 [d]		[27]
CPZ$^{⋅+}$ (Chlorpromazine radical cation)	1.4×10^9	(5.9)	[28]
Fe(III)EDTA/Fe(II)EDTA	$\approx 10^2$ [e]		
$O_2^{⋅-}/HO_2^⋅$	1×10^5 [d]		[29, 30]
	2.7×10^5		[31]
Fe(III)DFO/Fe(II)DFO	Very slow		[32, 33]

[a]A complete summary of free radical solution kinetics can be found in [34].
[b]We were unable to find data that addresses this reaction directly.
[c]Estimated k_{obs} for TO⋅ when in a biological membrane.
[d]k is pH dependent, thus this is k_{obs} at pH 7.4.
[e]Estimated from data in [35, 36, 63].

radical (TO⋅), can be repaired by ascorbate. Therefore, we have:

$$AscH^- + X⋅ \longrightarrow Asc^{⋅-} + XH,$$

where X⋅ can be any of these oxidizing free radicals. From Table 2, we see that the kinetics of these electron (hydrogen atom) transfer reactions are rapid. Thus, both thermodynamically and kinetically, ascorbate can be considered to be an excellent antioxidant.

Although ascorbate itself forms a radical in this reaction, a potentially very dangerous oxidizing radical ($X^•$) is replaced by the domesticated $Asc^{•-}$. $Asc^{•-}$ does not react by an addition reaction with O_2 to form dangerous peroxyl radicals. Ascorbate (probably Asc^{2-}, *vida infra*) and/or $Asc^{•-}$ appear to produce very low levels of superoxide [37, 38]. But by removing $O_2^{•-}$, superoxide dismutase provides protection from this possibility [39, 40]. Thus, the biological organism is protected from further free radical-mediated oxidations. In addition. $Asc^{•-}$ as well as dehydroascorbic can be reduced back to ascorbate by enzyme systems. Thus, it is recycled. Ascorbate's ubiquitous presence in biological systems in conjunction with its role as an antioxidant suggests that the ascorbate free radical would also be present.

Equilibrium

The ascorbate free radical will be present in solutions due to both the autoxidation and the metal catalyzed oxidation of ascorbate. Forester et al. observed that $Asc^{•-}$ can also arise from comproportionation of $AscH^-$ and DHA [41],

$$AscH^- + DHA \rightleftharpoons 2\ Asc^{•-} + H^+$$

$$K = \frac{[Asc^{•-}]^2}{[AscH^-][DHA]}$$

Using EPR, they determined the equilibrium constant for this process and noted that it was pH dependent. The equilibrium constant K was found to vary from 5.6×10^{-12} at pH 4.0 to 5.1×10^{-9} at pH 6.4. Later, after the acid-base properties of ascorbic acid and ascorbate free radical were understood, it was then possible to develop an expression for K at any pH value [27].

$$K = \frac{[Asc^{•-}]^2\,[H^+] + \{1 + [H^+]/10^{-pK_1}\}}{[DHA]\,[AscH_2]_{total,}} = 2.0 \times 10^{-15}\ M^{-2}$$

where pK_1 is the first ionization constant of ascorbic acid and $[AscH_2]_{total}$ is the analytical concentration of $AscH_2$, i.e., $[AscH_2]_{total} = [AscH_2] + [AscH^-] + [Asc^{2-}]$. [27, 41].

Using $Asc^{•-}$ as a marker of oxidative stress

Overview

The ascorbate free radical is naturally detectable by EPR at low steady-state levels in biological samples, such as leaves from crops [42], plasma [14, 43, 44], synovial fluid [45], skin [46, 47], and lens of the eye, *vida infra.* As oxidative stress increases in a system, the steady-state Asc$^{\cdot-}$ concentration increases [4]. These findings are consistent with ascorbate's role as the terminal small-molecule antioxidant (see Tab. 1). It is proposed that ascorbate, i.e., the ascorbate free radical, which is naturally present in biological systems, can be used as a noninvasive indicator of oxidative stress [4, 48].

The ascorbate radical as a marker of oxidative flux has been shown to be useful in the study of free radical oxidations in many biological systems including mouse skin [46, 47, 49], hepatocytes [50], and ischemia reperfusion of hearts [51–53][2]. Human sera and rat plasma intoxicated with paraquat and diquat, known superoxide generators, have increased ascorbate radical levels [56]. In animal experiments, sepsis has also been shown to increase Asc$^{\cdot-}$, indicating the involvement of oxidative stress with this health problem [57]. Sasaki et al. have investigated in human serum the use of Asc$^{\cdot-}$ signal intensity in combination with measurements of AscH$^-$ and DHA as an indicator of oxidative stress in human health problems that range from aging to xenobiotic metabolism [58–62]. Taken together, these studies demonstrate that the asorbate radical level in biological systems may be useful for monitoring free radical oxidations *in vivo*, particularly when free radical production is low and other methods are insensitive.

Absorption spectra

Pure ascorbic acid solutions are colorless as neither the diacid nor the monoanion have significant absorbances in the visible region of the spectrum. However, each has an absorbance in the ultraviolet region.

(1) Ascorbic acid: The diacid has an approximately symmetrical Gaussian absorption spectrum with $\varepsilon_{244} = 10\,800$ M$^{-1} \cdot$ cm^{-1} in aqueous solution [1].

(2) Ascorbate monoanion: Compared to the diacid, the peak of the absorption curve for the monoanion is red-shifted to 265 nm. A wide range of molar extinction coefficients have been reported, ranging from 7500–20 400 M$^{-1} \cdot$ cm^{-1} [1]. We find that $\varepsilon_{265} = 14\,500$ M$^{-1} \cdot$ cm^{-1} best reflects our experimental observations when doing experiments in near-neutral buffered aqueous solutions [63].

(3) Ascorbate radical: The ascorbate free radical has an approximately symmetrical Gaussian shaped absorption curve with $\varepsilon_{360} =$

---

[2]In a quite different approach Pietri et al. [54, 55] have used Asc$^{\cdot-}$ as a probe for plasma ascorbate concentrations. In their approach, a 1:1 mixture of plasma and dimethylsulfoxide is examined for Asc$^{\cdot-}$ by EPR. They claim that the Asc$^{\cdot-}$ is an index of the transient changes in plasma ascorbate status during ischemia/reperfusion. Whereas, in our studies the Asc$^{\cdot-}$ levels reflect the ongoing free radical flux in the system being examined [4, 47, 53].

3300 $M^{-1} \cdot cm^{-1}$ and a half-width at half-maximum of about 50 nm
[27]. With this small extinction coefficient, Asc$^{\cdot -}$ will not be observable by standard UV-VIS spectroscopy in steady-state experiments.
(4) Dehydroascorbic: Dehydroascorbic (acid) has a weak absorption at
300 nm, $\varepsilon_{300} = 720\ M^{-1} \cdot cm^{-1}$ [1].

EPR Detection of the ascorbate free radical

The ascorbate free radical is usually detected by EPR as a doublet signal
with $a^H = 1.8$ G, $\Delta H_{pp} \approx 0.6$ G and g = 2.0052, Figure 1. However, each
line of the ascorbate doublet is actually a triplet of doublets,
$a^{H4} = 1.76$ G, $a^{H6}(2) = 0.19$ G, and $a^{H5} = 0.07$ G [64].

In most biological experiments where the Asc$^{\cdot -}$ EPR signal will be
weak, a compromise is made in the choice of modulation amplitude.
The usual choice is to sacrifice resolution of the hyperfine structure for
improved sensitivity. We find that a modulation amplitude of ≈ 0.65 G
maximizes the ascorbate free radical double peak-to-peak signal amplitude [65].

The EPR power saturation curve of Asc$^{\cdot -}$ in room temperature
aqueous solutions shows that saturation effects begin at ≈ 16 mW and
maximum signal height is achieved at 40 mW nominal power when
using an aqueous flat cell and a TM cavity, see Figure 2. Thus, if
quantitation of the Asc$^{\cdot -}$ levels is desired, appropriate corrections for
saturation effects must be included in the calculations.

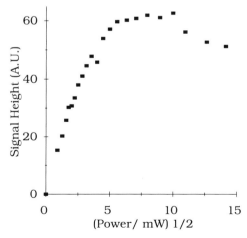

Fig. 2. EPR Power Saturation Curve for Asc$^{\cdot -}$. Signal heights are arbitrary units. Asc$^{\cdot -}$ was
observed in a demetalled 50 mM pH 7.8 phosphate buffer containing 10 mM ascorbate.
Bruker ESP-300 instrument settings were: 0.65 modulation amplitude; 10 G/167 s scan rate;
0.167 s time constant [65].

Applications

$Asc^{\cdot-}$ in solution

Stock ascorbate solutions

In our work with ascorbate in solution we have found that the quality of the stock solution determines the quality and reproductivity of the results. We prepare ascorbate stock solutions using only the diacid. It is prepared as a 0.100 M stock solution (10 mL) using high purity water. This solution is colorless, having a pH of ≈ 2. It is stored in a volumetric flask with a tight-fitting plastic stopper, thus oxygen is kept from the solution during long-term storage. As the solubility of oxygen in air-saturated water is ≈ 0.25 mM, the solution will become anaerobic with loss of $< 1\%$ of the original ascorbate. If the flask is indeed clean, we have found that the solution can be kept for several weeks without significant loss of ascorbate due to the low pH and lack of oxygen. The appearance of a yellow color is an indication of ascorbate deterioration. We avoid the use of sodium ascorbate as it invariably contains substantial quantities of oxidation products as evidenced by the yellow color of the solution. [63].

Autoxidation and metal catalyzed oxidations

Before beginning this discussion it must be understood that we use the term *autoxidation* to mean oxidation in the absence of metal catalysts [66]. The term oxidation is used more broadly and includes all oxidations, with or without catalysts.

Ascorbate is readily oxidized. However, the rate of this oxidation is dependent upon pH and the presence of catalytic metals [32, 33, 35, 36, 63, 67–70]. The diacid is very slow to oxidize. Consequently, at low pH, i.e., less than 2 or 3, ascorbic acid solutions are quite stable, assuming catalytic transition metal ions are not introduced into the solutions. However, as the pH is raised above pK_1 (4.2), $AscH^-$ becomes dominant and the stability of the ascorbate solution decreases. This loss of stability is usually the result of the presence of adventitious catalytic metals (on the order of 1 μM) in the buffers and salts that are typically employed in studies at near neutral pH [63]. For example, we have found that in room temperature aerated, aqueous solutions at pH 7.0 (50 mM phosphate buffer) 10–30% of 125 μM ascorbate is lost in just 15 min. This large variation is the result of different sources and grades of phosphate used in the buffer preparation. However, if care is taken to remove these trace levels of transition metals, this rate of loss can be lowered to as little as 0.05%/15 min [63], thus demonstrating the extreme importance of metals in controlling ascorbate stability. At pH 7.0 we have set an upper limit for the observed rate constant for the

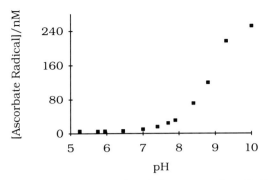

Fig. 3. Background [Asc$^{\cdot\,-}$] *vs*. pH: Each solution was made with 50 mM demetalled phosphate buffer that contained 50 μM desferoxamine mesylate, for at least 12 hours. To these solutions 500 μM ascorbate was added and the EPR spectra were collected. The points represent the Asc$^{\cdot\,-}$ concentration observed in the second of three EPR scans, where the values had a standard deviation of less than 1 nonomolar (adapted from [4]). These data demonstrate the importance of pH control. At pH values greater than ≈ 8 [Asc$^{\cdot\,-}$]$_{ss}$ is not a good indicator of oxidative stress, but at near neutral pH it is excellent.

oxidation of ascorbate to be 6×10^{-7} s^{-1} under our experimental conditions [63]. However, even in carefully demetalled solutions as the pH is varied the rate of oxidation increases, Figure 3 [4].

We attribute this increase in rate at higher pH values to the increasing concentrations of the ascorbate dianion. Williams and Yandell have made an estimate based on the Marcus theory of electron transfer that the ascorbate dianion would undergo true autoxidation at a significant rate [38].

$$k \approx 10^2 \text{ M}^{-1} \text{ s}^{-1}$$

$$\text{Asc}^{2-} + \text{O}_2 \longrightarrow \text{Asc}^{\cdot\,-} + \text{O}_2^{\cdot\,-}$$

Our experimental results are consistent with these estimates [4, 33, 63]. Marcus theory would predict that the rate of the true autoxidation of AscH$^-$ would be much slower.

Thus, at pH ≈ 7.4 the rate of autoxidation of an ascorbate solution is determined predominantly by Asc^{2-}.

Typical buffers employed in biochemial and biological research have on the order of 1 μM iron and <1 μM copper [63]. But because copper is ≈ 80 times more efficient as a catalyst for ascorbate oxidation than iron, it is the adventitious copper that is the biggest culprit in catalyzing ascorbate oxidation [63].

We have developed two assays that take advantage of this chemistry:

Iron analysis at the nM level
Fe-EDTA is an excellent catalyst of ascorbate oxidation, while Cu-EDTA is a very poor catalyst. We have found that with careful

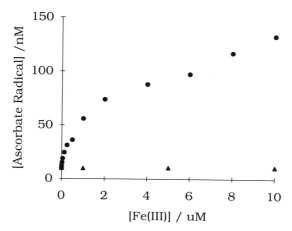

Fig. 4. These data were gathered using EPR spectroscopy to quantitate the steady state level of Asc$^{\cdot-}$. The curves were obtained in 50 mM demetalled phosphate buffer, pH 7.40 with 250 μM EDTA (●) or 50 μM Desferal (▲) with 125 μM ascorbate present (adapted from [33]).

attention to detail to ensure that all glassware, pipettes and pipette tips are scrupulously clean, we can estimate iron levels in phosphate buffer to a lower limit of ≈ 100 nM using UV-Vis spectroscopy [63]. However, using EPR spectroscopy this limit can be as low as ≈ 5 nM [33], Figures 4 and 5. For the EPR method of analysis we add EDTA to the solution to be assayed. This converts the iron to a "standard" catalytic form. We

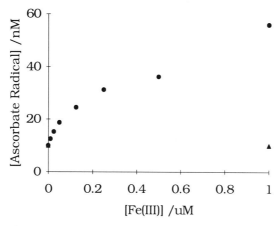

Fig. 5. This figure is an expansion of the 0–1 μM [Fe(III)] region of Figure 4. The experimental conditions are the same as in Figure 4. The curves were obtained in 50 mM demetalled phosphate buffer, pH 7.40 with 250 μM EDTA (●) or 50 μM Desferal (▲) with 125 μM ascorbate present (adapted from [33]).

then introduce ascorbate and determine by EPR the steady-state concentration (i.e., signal height) of Asc$^{•-}$. From a standard curve we can
then estimate the iron concentration from $\approx 5\,nM \rightarrow \approx 10\,\mu M$. To
achieve estimates at the lowest end of this range, extreme care must be
taken with each step and the EDTA must be pure; recrystallized at least
three times using methods that will produce the best result. The standard curve must be obtained using the same buffer/salt system and exact
pH. This buffer/salt must be demetalled using a chelating resin such as
Chelex 100 [63]. This method is useful if there is interference from
standard colormetric assays of iron, or if only "loosely bound" iron is
to be estimated [71].

For the UV visible method, the experiment is similar except the rate
of loss of ascorbate is followed at 265 nm. This rate is plotted vs.
Fe(III)EDTA concentration for the standard curve, from which unknown concentrations of iron are estimated.

Removal of Trace Metals
We have also found that ascorbate is an excellent tool to ascertain the
effectiveness of adventitious catalytic metal removal from near-neutral
buffer systems. In this method we follow the loss of ascorbate due to
oxidation by monitoring its absorbance at 265 nm. In our standard test
we add $\approx 3.5\,\mu L$ of 0.100 M ascorbic acid solution to 3.00 mL of the
buffer in a standard 1 cm quartz cuvette. This results in an initial
absorbance of 1.8. The loss of ascorbate is followed for 15 min. A loss
of more than $\approx 0.5\%$ in this time indicates significant metal contamination. (If using a diode array spectrometer, interrogate the solution only
a few times as the UV radiation near 200 nm will itself initiate ascorbate
photooxidation.) [63, 72].

Plasma
The free radical initiator AAPH (2,2'-azo-bis(2-amidinopropane) dihydrochloride) undergoes thermal decomposition at a constant rate (at a
fixed temperature) producing carbon-centered sigma radicals that react
with O_2 at nearly diffusion-controlled rates yielding peroxyl radicals
[73]. Thus, AAPH, in an oxygen-containing system, produces a constant
flux of oxidizing free radicals that can oxidize ascorbate or produce spin
adducts with the spin trap DMPO (5,5-dimethylpyrroline-1-oxide).
When using AAPH as a source of oxidizing radicals in plasma a linear
increase in $[Asc^{•-}]_{ss}$ is seen with increasing concentrations of AAPH
(Fig. 6). This plasma sample contained 58 μM ascorbate, a value typical
of physiological conditions. Thus, in plasma $[Asc^{•-}]_{ss}$ is indeed an
excellent indicator of oxidative stress. [4].

Cells
Iron and ascorbate are well-known as a prooxidant combination that
will initiate lipid peroxidation [9, 74–79]. Lipid-derived radicals from

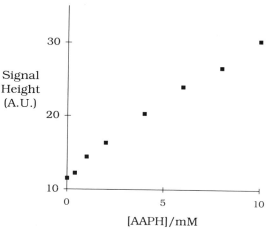

Fig. 6. Azo Initiator-Produced Asc•⁻ Radicals in Plasma: Asc•⁻ EPR signal height (arbitrary units) *versus* AAPH concentration. The plasma contained 58 μM ascorbate and varying amounts of AAPH [4].

cells have been detected using EPR spin trapping techniques when cells are exposed to iron and ascorbate [78, 79]. The introduction of edelfosine, an ether lipid drug being investigated for use in cancer treatment, to an L1210 murine leukemia cell suspension with 20 μM iron and 100 μM ascorbate present, results in a burst of Asc•⁻ production within 1–2 min after the addition. This burst of Asc•⁻ production corresponds with an increase in the rate of cellular lipid peroxidation as observed by EPR spin trapping, consistent with [Asc•⁻]$_{ss}$ being a real time reflection of the oxidation flux in the system [78].

Asc•⁻ in tissues

To examine tissues by EPR, e.g., skin, lens or samples whose viscosity precludes the use of an aqueous EPR flat sample cell, tissue cells (sometimes called cavity cells) such as produced by Wilmad Glass Co. (Buena, New Jersey) are available. These cells generally have a 0.5 mm depth sample cavity well and two supporting rods. In our experience, those cells with two stems are prone to breakage. Thus, we use a one stem tissue cell, i.e., no lower positioning rod. This reduces the incidence of breakage and facilitates tuning of the sample in the EPR cavity. A cover slip fits over the sample cavity well. Phosphor-bronze clips are provided to hold the cover over the well. However, we find that to prevent potential scraping of the inside of the EPR cavity with these clips that Parafilm ties can be used to provide an even more secure fit.

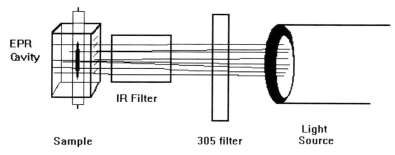

Fig. 7. EPR and light source setup for EPR tissue cell experiments. The 305 filter cuts off those wavelengths below ≈ 300 nm. The IR filter is a water-filled large diameter (50 mm) cylindrical, quartz UV-visible cell having a 50 mm path length. Our light source is an Oriel Photomax system with a 150 W Xe lamp. This system requires no special air handling due to ozone production. For our skin and lens experiments the lamp was operating at 3 mW/cm².

We find Parafilm to be an excellent tool in EPR experiments as it yields no significant EPR signals. These ties are made by cutting ≈ 2 mm wide strips of Parafilm and then wrapping them tightly around the cell and cover plate at the indentions that are provided for the clips. A diagram of the experimental setting is given in Figure 7.

Skin
Whole skin harvested from SKH-1 hairless male mice (Charles River Laboratories, Portage, Michigan) is cut into EPR usable pieces ($\cong 1.0$ cm², epidermis and dermis), placed in a Wilmad Glass Co. (Buena, New Jersey) one stem tissue cell, and positioned in the EPR cavity. EPR spectra are obtained at room temperature. The EPR spectrometer settings for the ascorbate radical experiments are: microwave power, 40 milliwatts; modulation amplitude, 0.66 G; time constant 0.3 s; scan rate 8 G/41.9 s; receiver gain, 2×10^6. The epidermal surface of the skin is exposed to UV light while in the EPR cavity. The light source is a Photomax 150 W xenon arc lamp (Oriel Corporation, Stratford, Connecticut) operating at 32 W; wavelengths below 300 nm are filtered out using a Schott WG 305 filter (Duryea, Pennsylvania). Infrared radiation from the light is removed by a 5 cm water filter. The filtered light fluence rate, including the visible wavelengths, as measured using a Yellow Springs Instrument (Yellow Springs, Ohio) model 65A radiometer with a 6551 probe, was 3 mW/cm², assuming the cavity grid transmits 75% of the incident light.

Lens
There is considerable evidence that UV-induced epithelial damage can be related to lens opacity and subsequent cataract formation [80]. The involvement of free radicals in cataract formation has been suggested [81, 82]. Ascorbic acid is clearly of importance as an antioxidant in the

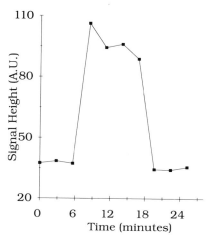

Fig. 8. Asc·⁻ in bovine lens subjected to UV light-induced oxidative stress. The Asc·⁻ signal height is shown in arbitrary units. The lens was exposed to the UV light after collection of the third data point. The light was turned off after collection of the seventh data point. The light source and EPR experimental setup is depicted in Figure 7.

lens of the eye, present at steady-state concentrations of 1–2 mM in the human lens and adjacent aqueous and vitreous humors. To examine Asc·⁻ levels in bovine lens tissue, lens tissue is placed in an EPR tissue cell and irradiated as described above in the skin experiments. A low steady-state level of the ascorbate free radical is detectable by EPR in the lens of the bovine eye. During UV photooxidative stress the levels of ascorbate free radical significantly increase (Fig. 8). When the light is turned off, the ascorbate free radical signal returns to baseline levels.

Whole bovine lens was cut into EPR usable pieces (≈ 1.0 cm²), placed in a Wilmad Glass Co. (Buena, New Jersey) one stem tissue cell, and positioned in a TM_{110} EPR cavity. EPR spectra were obtained at room temperature using a Bruker ESP 300 spectrometer operating at 9.74 GHz with 100 kHz modulation frequency. The EPR spectrometer settings for the ascorbate radical experiments were: microwave power, 40 milliwatts; modulation amplitude, 0.63 G; time constant, 1.3 s; scan rate, 6 G/167.7 s; receiver gain, 2×10^6. While in the EPR cavity, the lens was exposed to UV light after the third consecutive scan, and turned off after the seventh scan.

Asc·⁻ in vivo

Rat in vivo/ex vivo Asc·⁻
Mori et al. [83, 84] have observed Asc·⁻ in the circulatory blood of living rats with EPR. In these experiments a 1 mm tube was used to

**EPR
SPECTROMETER**

**OPEN-CHEST
ANIMAL PREP**

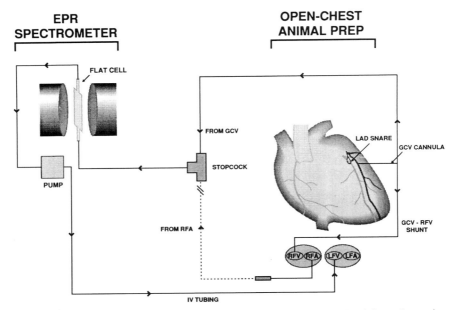

Fig. 9. Schematic of EPR setup for monitoring of Asc$^{\cdot -}$ in whole blood from the canine myocardium *ex vivo*. Note that the blood is returned to the animal. The lower end of the flat cell (Wilmad WG-813) is connected to the coronary venous cannula using thin (0.5 mm outer diameter) Teflon tubing of ≈ 1 meter in length. The end of the Teflon tubing is placed completely into the lower stem of the flat cell so the blood emerges directly into the bottom of the flat portion of the cell. The pump is an IV infusion pump; we draw from the top of the flat cell and push into the LFV. The flow rate of the pump is set to 600 mL/hour. (RFA, right femoral artery; RFV, right femoral vein; LFA, left femoral vein; LAD, left anterior descending coronary artery; GCV, great cardiac vein; and IV, intravenous.) The arrows indicate the direction of blood flow.

make a shunt from a femoral artery of the rat to an EPR cell positioned in an EPR cavity; the blood was returned to the rat by a continuation of the shunt from the EPR cell back to a femoral vein of the animal. In these experiments, the investigators demonstrated that introduction of iron, as ferric citrate, to the rat results in an increase in the circulating $[\text{Asc}^{\cdot -}]_{ss}$. This increase in $[\text{Asc}^{\cdot -}]$ correlates with other parameter of oxidative stress.

Canine in vivo/ex vivo *myocardial ischemia/reperfusion studies*
Free radical mediated oxidative stress is now thought to be a significant source of tissue damage during myocardial ischemia/reperfusion episodes. We have developed a means to monitor by EPR whole blood *ex vivo* from an open-chest canine method of myocardial ischemia/reperfusion [53]. Using this method we can monitor for the presence of Asc$^{\cdot -}$ in myocardial blood within ≈ 4–5 s from leaving the heart. By following the intensity of the Asc$^{\cdot -}$ signal versus time we can determine

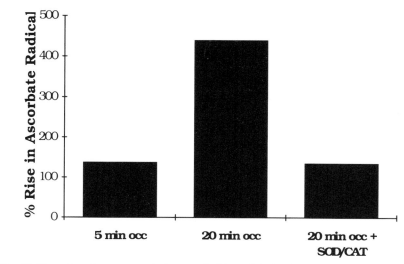

Fig. 10. These data demonstrate the increase in [Asc· ⁻] during ischemia/reperfusion episodes. The percent rise is the change in the area under the curves of plots of [Asc· ⁻] vs. time; the areas were determined using the Simpson integration method; the limits being from the beginning of reperfusion through the thirty minute time point [53]. Both the 5 min occlusion (+134 ± 73%) and the 20 min occlusion + SOD/CAT experiment (+136 ± 94%) were statistically different p < 0.005 than the 20 min occlusion (+440 ± 236%) experiments.

the changes in oxidative stress within the myocardium with various interventions during ischemia/reperfusion episodes [53].

These studies have used an open-chest canine model (≈20 kg) model of ischemia. General anesthesia is achieved with fentanyl droperidol. Briefly, a midsternal thoracotomy is performed and the heart exposed. A cannula is manipulated into the coronary sinus. Blood is withdrawn from the coronary sinus and passed through the EPR spectrometer, which is positioned next to the animal; we have refined this system so that the blood is scanned by EPR within ≈4–5 s of withdrawal from the coronary sinus, Figure 9. At the beginning of the experiment we administer 1 gram of vitamin C as an intravenous bolus, followed by an intravenous infusion (usually 3.8–15.2 mg/min) in order to attain a steady-state arterial concentration of ascorbate free radical. Arterial blood is also initially passed through the EPR spectrometer to demonstrate the steady-state arterial level, which is usually ≈14 nM. The venous level is usually ≈8 nM. The blood is periodically rescanned to further demonstrate that the arterial level has not changed (if it has changed, the IV infusion of ascorbate is adjusted as necessary). In spite of this, if the arterial level of AFR is shown to vary more than 15% during a study, that study is discarded. During the experiment the coronary venous blood is continuously scanned to determine ascorbate free radical signal intensity: the amplitude of the ascorbate radical signal is linearly proportional to the concentration of the radical, thus permitting real-time quantitative AFR determination, a demonstrated index of oxygen free radical generation.

In this method we have observed that 20 min of regional ischemia will increase the integrated $Asc^{\bullet-}$ signal intensity by over 400% upon reperfusion, Figure 10. A 5-min occlusion produced substantially less $Asc^{\bullet-}$; superoxide dismutase (SOD) and catalase (CAT) are able to blunt the 20-min reperfusion oxidative stress, bringing it to near the 5-min occlusion results.

The disadvantage of this method is that we lose information on the exact radicals being produced. However, the big advantage is that we are able to get a relative estimate on the total free oxidative flux, in real time. Because many types of radicals are produced during the oxidative cascade no one primary radical can be a reliable marker of the total radical flux. However, ascorbate, being at the bottom of the pecking order for oxidizing free radicals, can serve as a marker of the total free radical oxidative flux in a carefully controlled system.

Conclusion

Ascorbate is well known for its reducing properties. As such, it is an excellent antioxidant; it is thermodynamically at the bottom of the pecking order for oxidizing free radicals [3], thus we view it as the terminal small-molecule antioxidant [4]. Ascorbate protects cells from oxidative stress by scavenging free radicals and recycling other antioxidants, such as vitamin E. We have described here how using EPR, the ascorbate free radical can be used as a maker of oxidative stress.

Acknowledgments
This work was supported in part by The Center for Global and Regional Environmental Research at The University of Iowa as well as by NIH 1 PO1 HL 49264.

References

1. Lewin, S. (1976) *Vitamin C: Its Molecular Biology and Medical Potential*. Academic Press, New York.
2. Davies, M.B., Austin, J. and Partridge, D.A. (1991) *Vitamin C: Its Chemistry and Biochemistry*. Royal Society of Chemistry, Cambridge.
3. Buettner, G.R. (1993) The pecking order of free radicals and antioxidants: Lipid peroxidation, α-tocopherol, and ascorbate. *Arch. Biochem. Biophys.* 300: 535–543.
4. Buettner, G.R. and Jurkiewicz, B.A. (1993) Ascorbate free radical as a marker of oxidative stress: An EPR study. *Free Radic. Biol. Med.* 14: 49–55.
5. Creutz, C. (1981) The complexities of ascorbate as a reducing agent. *Inorg. Chem.* 20: 4449–4452.
6. Frei, B., England, L. and Ames, B.N. (1989) Ascorbate is an outstanding antioxidant in human blood plasma. *Proc. Natl. Acad. Sci. USA* 86: 6377–6381.

7. Halliwell, B. (1990) How to characterize a biological antioxidant. *Free Rad. Res. Comms.* 9: 1–32.
8. McCay, P.B. (1985) Vitamin E: Interactions with free radicals and ascorbate. *Ann. Rev. Nutr.* 5: 323–340.
9. Rees, S. and Slater, T.F. (1987) Ascorbic acid and lipid peroxidation: The cross-over effect. *Acta Biochim. Biophys. Hung.* 22: 241–249.
10. Niki, E. (1991) Vitamin C as an antioxidant. *World Rev. Nutr. Diet.* 64: 1–30.
11. Krinsky, N.I. (1992) Mechanism of action of biological antioxidants. *Proc. Soc. Exp. Biol. Med.* 200: 248–254.
12. Koppenol, W.H. and Butler, J. (1985) Energetics of interconversion reactions of oxy radicals. *Adv. Free Radical Biol.* 1: 91–131.
13. Kalyanaraman, B., Darley-Usmar, V.M., Wood, J., Joseph, J. and Parthasarathy, S. (1992) Synergistic interaction between the probucol phenoxyl radical and ascorbic acid in inhibiting the oxidation of low density lipoprotein. *J. Biol. Chem.* 267: 6789–6795.
14. Sharma, M.K. and Buettner, G.R. (1993) Interaction of vitamin C and vitamin E during free radical stress in plasma: An ESR study. *Free Radic. Biol. Med.* 14: 649–653.
15. Rose, R.C. and Bode, A.M. (1993) Biology of free radical scavengers: An evaluation of ascorbate. *FASEB J.* 7: 1135–1142.
16. Retsky, K.L., Freeman, M.W. and Frei, B. (1993) Ascorbic acid oxidation product(s) protect human low density lipoprotein against atherogenic modification. *J. Biol. Chem.* 268: 1304–1309.
17. Navas, P., Villalba, J.M. and Cordoba, F. (1994) Ascorbate function at the plasma membrane. *Biochim. Biophys. Acta* 1197: 1–13.
18. Davis, H.F., McManus, H.J. and Fessenden, R.W. (1986) An ESR study of free-radical protonation equilibria in strongly acid media. *J. Phys. Chem.* 90: 6400–6404.
19. Wardman, P. (1989) Reduction potentials of one-electron couples involving free radicals in aqueous solution. *J. Phys. Chem. Ref. Data.* 18: 1637–1755.
20. Buxton, G.V., Greenstock, C.L., Helman, W.P. and Ross, A.B. (1988) Critical review of rate constants for reactions of hydrated electrons, hydrogen atoms and hydroxyl radicals ($^{\cdot}OH/^{\cdot}O^-$) in aqueous solution. *J. Phys. Chem. Ref. Data* 17: 513–886.
21. Erben-Russ, M., Michel, C., Bors, W. and Saran, M. (1987) Absolute rate constants of alkoxyl radical reactions in aqueous solution. *J. Phys. Chem.* 91: 2362–2365.
22. Neta, P., Huie, R.E. and Ross, A.B. (1990) Rate constants for reactions of peroxyl radicals in fluid solutions. *J. Phys. Chem. Ref. Data* 19: 413–513.
23. Packer, J.E., Willson, R.L., Bahnemann, P. and Asmus, K.D. (1980) Electron transfer reactions of halogenated aliphatic peroxyl radicals: measurement of absolute rate constant by pulse radiolysis. *J. Chem. Soc. Perkin Trans. II.* 2: 296–299.
24. Tamba, M. and O'Neill, P. (1991) Redox reactions of thiol free radicals with the antioxidants ascorbate and chlorpromazine: Role in radioprotection. *J. Chem. Soc. Perkin Trans. II* 1681–1685.
25. Forni, L.G., Monig, J., Mora-Arellano, V.O. and Willson, R.L. (1983) Thiyl free radicals: Direct observations of electron transfer reactions with phenothiazines and ascorbate. *J. Chem. Soc. Perkin Trans. II* 961–965.
26. Simic, M.G. and Jovanovic, S.V. (1989) Antioxidation mechanisms of uric acid. *J. Am. Chem. Soc.* 111: 5778–5782.
27. Bielski, B.H.J. (1982) Chemistry of ascorbic acid radicals. *In*: Seib, P.A. and Tolbert, B.M. (eds): *Ascorbic Acid: Chemistry, Metabolism, and Uses.* Washington DC, American Chemical Society, pp 81–100.
28. Pelizzetti, E., Meisel, D., Mulac, W.A. and Neta, P. (1979) On the electron transfer from ascorbic acid to various phenothiazine radicals. *J. Am. Chem. Soc.* 101: 6954–6959.
29. Bielski, B.H.J., Cabelli, D.E. and Arudi, R.L. (1985) Reactivity of HO_2/O_2^- radicals in aqueous solution. *J. Phys. Chem. Ref. Data* 14: 1041–1100.
30. Cabelli, D.E. and Bielski, B.H.J. (1983) Kinetics and mechanism for the oxidation of ascorbic acid/ascorbate by HO_2/O_2^- radicals. A pulse radiolysis and stopped-flow photolysis study. *J. Phys. Chem.* 87: 1809–1812.
31. Nishikimi, M. (1975) Oxidation of ascorbic acid with superoxide anion generated by the xanthine-xanthine oxidase system. *Biochem. Biophys. Res. Comms.* 63: 463–468.
32. Buettner, G.R. (1986) Ascorbate autoxidation in the presence of iron and copper chelates. *Free Rad. Res. Comms.* 1: 349–353.

33. Buettner, G.R. (1990) Ascorbate oxidation: UV absorbance of ascorbate and ESR spectroscopy of the ascorbyl radical as assays for iron. *Free Rad. Res. Comms.* 10: 5–9.
34. Ross, A.B., Mallard, W.G., Hleman, W.P., Buxton, G.V., Huie, R.E. and Neta, P. (1994) NDRL-NIST Solution Kinetics Database: -Ver. 2.0. Gaithersburg; NIST.
35. Kahn, M.M.T. and Martell, A.E. (1967) Metal ion and metal chelate catalyzed oxidation of ascorbic acid by molecular oxygen. I. Cupric and ferric ion catalyzed oxidation. *J. Am. Chem. Soc.* 89: 4176–4185.
36. Khan, M.M.T. and Martell, A.E. (1967) Metal ion and metal chelate catalyzed oxidation of ascorbic acid by molecular oxygen. II. Cupric and ferric chelate catalyzed oxidation. *J. Am. Chem. Soc.* 89: 7104–7111.
37. Scarpa, M., Stevanto, R., Viglino, P. and Rigo, A. (1983) Superoxide ion as active intermediate in the autoxidation of ascorbate by molecular oxygen. *J. Biol. Chem.* 258: 6695–6697.
38. Williams, N.H. and Yandell, J.K. (1982) Outer-sphere electron-transfer reaction of ascorbate anions. *Aust. J. Chem.* 35: 1133–1144.
39. Winterbourn, C.C. (1993) Superoxide as an intracellular radical sink. *Free Radic. Biol. Med.* 14: 85–90.
40. Koppenol, W.H. (1993) A thermodynamic appraisal of the radical sink hypothesis. *Free Radic. Biol. Med.* 14: 91–94.
41. Foerster, G., Weis, W. and Staudinger, H. (1965) Messung der Elektronenspinresonanz an Semidehydroascorbinsäure. *Annalen der Chemie* 690: 166–169.
42. Stegmann, H.B., Schuler, P., Westphal, S. and Wagner, E. (1993) Oxidative stress of crops monitored by EPR. *Z. Naturforsch. C.* 48: 766–772.
43. Minetti, M., Forte, T., Soriani, M., Quaresima, V., Menditoo, A. and Ferrari, M. (1992) Iron-induced ascorbate oxidation in plasma as monitored by ascorbate free radial formation. *Biochem. J.* 282: 459–465.
44. Miller, D.M. and Aust, S.D. (1989) Studies of ascorbate-dependent, iron catalyzed lipid peroxidation. *Arch. Biochem. Biophys.* 271: 113–119.
45. Buettner, G.R. and Chamulitrat, W. (1990) The catalytic activity of iron in synovial fluid as monitored by the ascorbate free radical. *Free Radic. Biol. Med.* 8: 55–56.
46. Buettner, G.R., Motten, A.G., Hall, R.D. and Chignell, C.F. (1987) ESR detection of endogenous ascorbate free radical in mouse skin: Enhancement of radical production during UV irradiation following topical application of chlorpromazine. *Photochem. Photobiol.* 46: 161–164.
47. Jurkiewicz, B.A. and Buettner, G.R. (1994) Ultraviolet light-induced free radical formation in skin: An electron paramagnetic resonance study. *Photochem. Photobiol.* 59: 1–4.
48. Roginsky, V.A. and Stegmann, H.B. (1994) Ascorbyl radical as natural indicator of oxidative stress: Quantitative regularities. *Free Radic. Biol. Med.* 17: 93–103.
49. Timmins, G.S. and Davies, M.J. (1993) Free radical formation in murine skin treated with tumour promoting organic peroxides. *Carcinogensis* 14: 1499–1503.
50. Tomasi, A., Albano, E., Bini, A., Iannone, A.C. and Vannini, V. (1989) Ascorbyl radical is detected in rat isolated hepatocytes suspensions undergoing oxidative stress: and early index of oxidative damage in cells. *Adv. in the Biosciences* 76: 325–334.
51. Arroyo, C.M., Kramer, J.H., Dickens, B.F. and Weglicki, W.B. (1987) Identification of free radicals in myocardial ischemia/reperfusion by spin trapping with nitrone DMPO. *FEBS Lett.* 221: 101–104.
52. Nohl, H., Stolze, K., Napetschnig, S. and Ishikawa, T. (1991) Is oxidative stress primarily involved in reperfusion injury of the ischemic heart? *Free Radic. Biol. Med.* 11: 581–588.
53. Sharma, M.K., Buettner, G.R., Spencer, K.T. and Kerber, R.E. (1994) Ascorbyl free radical as a real-time marker of free radical generation in briefly ischemic and reperfused hearts. *Circulation Res.* 74: 650–658.
54. Pietri, S., Culcasi, M., Stella, L. and Cozzone, P.J. (1990) Ascorbyl free radical as a reliable indicator of free-radical-mediated myocardial ischemic and post-ischemic injury. *Eur. J. Biochem.* 193: 845–854.
55. Pietri, S., Seguin, J.R., D'Arbigny, P. and Culcasi, M. (1994) Ascorbyl free radical: A noninvasive marker of oxidative stress in human open-heart surgery. *Free Radic. Biol. Med.* 16: 523–528.
56. Minakata, K., Suzuki, O., Saito, S. and Harada, N. (1993) Ascorbate radical levels in human sera and rat plasma intoxicated with paraquat and diaquat. *Arch. Toxicol.* 67: 126–130.

57. Stark, J.M., Jackson, S.K., Rowlands, C.C. and Evans, J.C. (1988) Increases in ascorbate free radical concentration after endotoxin in mice. *In*: C. Rice-Evans and B. Halliwell (eds): *Free Radicals: Methodology and Concepts*. Richelieu, London, pp 201–209.

58. Sasaki, R., Kurokawa, T. and Tero-Kubota, S. (1982) Nature of serum ascorbate radical and its quantitative estimation. *Tohoku J. Exp. Med.* 136: 113–119.

59. Sasaki, R., Kurokawa, T. and Tero-Kubota, S. (1983) Ascorbate radical and ascorbic acid level in human serum and age. *J. Gerontology* 1: 26–30.

60. Sasaki, R., Kobayasi, T., Kurokawa, T., Shibuya, D. and Tero-Kubota, S. (1984) Significance of the equilibrium constant between serum ascorbate radical and ascorbic acids in man. *Tohoku J. Exp. Med.* 144: 203–210.

61. Sasaki, R., Kurokawa, T. and Shibuya, D. (1985) Factors influencing ascorbate free radical formation. *Biochem. Intern.* 10: 155–163.

62. Ohara, T., Sasaki, R., Shibuya, D., Asaki, S. and Toyota, T. (1992) Effect of omeprazole on ascorbate free radical formation. *Tohoku J. Exp. Med.* 167: 185–188.

63. Buettner, G.R. (1988) In the absence of catalytic metals ascorbate does not autoxidize at pH 7: ascorbate as a test for catalytic metals. *J. Biochem. Biophys. Meth.* 16: 27–40.

64. Laroff, G.P., Fessenden, R.W. and Schuler, R.H. (1972) The electron spin resonance spectra of radical intermediates in the oxidation of ascorbic acid and related substances. *J. Am. Chem. Soc.* 94: 9062–9073.

65. Buettner, G.R. and Kiminyo, K.P. (1992) Optimal EPR detection of weak nitroxide spin adduct and ascorbyl free radical signals. *J. Biochem. Biophys. Meth.* 24: 147–151.

66. Miller, D.M., Buettner, G.R. and Aust, S.D. (1990) Transition metals as catalysts of "autoxidation" reactions. *Free Radic. Biol. Med.* 8: 95–108.

67. Guzman Barron, E.S., DeMeio, R.H. and Klemperer, F. (1936) Studies of biological oxidations. Copper and hemochromogens as catalysts for the oxidation of ascorbic acid. The mechanism of the oxidation. *J. Biol. Chem.* 112: 625–640.

68. Borsook, H., Davenport, H.W., Jeffreys, C.E.P. and Warner, R.C. (1937) The oxidation of ascorbic acid and its reduction *in vitro* and *in vivo*. *J. Biol. Chem.* 117: 237–279.

69. Weissberger, A., LuValle, J.E. and Thomas, D.S. (1943) Oxidation processes. XVI. The autoxidation of ascorbic acid. *J. Am. Chem. Soc.* 65: 1934–1939.

70. Halliwell, B. and Foyer, C.H. (1976) Ascorbic acid, metal ions and the superoxide radical. *Biochem. J.* 155: 697–700.

71. Britigan, B.E., Pou, S., Rosen, G.M., Lilleg, D.M. and Buettner, G.R. (1990) Hydroxyl radical is not a product of the reaction of xanthine oxidase and xanthine. *J. Biol. Chem.* 265: 17533–17538.

72. Buettner, G.R. (1990) Use of ascorbate as test for catalytic metals in simple buffers. *Meth. Enzymol.* 186: 125–127.

73. Niki, E. (1990) Free radical initiators as source of water- or lipid-soluble peroxyl radicals. *Methods Enzymol.* 186: 100–108.

74. Willis, E.D. (1969) Lipid peroxide formation in microsomes, general considerations. *Biochem. J.* 113: 315–324.

75. Willis, E.D. (1969) Lipid peroxide formation in microsomes, the role of non-haem iron. *Biochem. J.* 113: 325–332.

76. Willis, E.D. (1966) Mechanisms of lipid peroxide formation in animal tissues. *Biochem. J.* 99: 667–675.

77. Baysal, E., Sullivan, S.G. and Stern, A. (1989) Prooxidant and antioxidant effects of ascorbate on tBuOOH-induced erythrocyte membrane damage. *Int. J. Biochem.* 21: 1109–1113.

78. Wagner, B.A., Buettner, G.R. and Burns, C.P. (1993) Increased generation of lipid-derived and ascorbate free radicals by L1210 cells exposed to the ether lipid edelfosine. *Cancer Res.* 53: 711–713.

79. Wagner, B.A., Buettner, G.R. and Burns, C.P. (1993) Free radical-mediated lipid peroxidation in cells: Oxidizability is a function of cell lipid bis-allylic hydrogen content. *J. Biol. Chem.* 33: 4449–4453.

80. Taylor, H.R., West, S.K., Rosenthal, F.S., Munoz, B., Newland, H., Abbey, H. and Emmett, E.A. (1988) Effect of ultraviolet radiation on cataract formation. *New Eng. J. Med.* 319: 1429–33.

81. Weiter, J.J. and Finch, E.D. (1975) Paramagnetic species in cataractous human lenses. *Nature* 254: 536–537.
82. Murakami, J., Okazaki, M. and Shiga, T. (1989) Near UV-induced free radicals in ocular lens, studies by ESR and spin trapping. *Photochem. Photobiol.* 49: 465–473.
83. Mori, A., Wang, X. and Liu, J. (1994) Electron spin resonance assay of ascorbate free radicals *in vivo*. *Methods Enzymology* 233: 149–154.
84. Wang, X., Liu, J., Yokoi, I., Kohno, M. and Mori, A. (1992) Direct detection of circulating free radicals in the rat using electron spin resonance spectrometry. *Free Radic. Biol. Med.* 12: 121–126.

Analysis of Free Radicals in Biological Systems
Favier et al. (eds)
© 1995 Birkhäuser Verlag Basel/Switzerland

Lipid hydroperoxide analysis by reverse-phase high-performance liquid chromatography with mercury cathode electrochemical detection

W. Korytowski[1,2] and A.W. Girotti[2]

[1]*Institute of Molecular Biology, Jagiellonian University, Krakow, Poland*
[2]*Department of Biochemistry, Medical College of Wisconsin, Milwaukee, WI 53226, USA*

Summary. Lipid hydroperoxides (LOOHs) are important intermediates generated during peroxidative degeneration of unsaturated lipids in cell membranes and other organized assemblies. Such degeneration typically occurs under conditions of oxidative stress. Since LOOHs are subject to rapid turnover in the presence of peroxidases, reductants, or metal ions, their levels may be very low and difficult to measure in complex systems such as peroxidizing cells or lipoproteins. This has prompted the development of new approaches for the high sensitivity/high specificity detection of biological LOOHs. One such approach is reverse-phase high-performance liquid chromatography with electrochemical detection using a renewable mercury drop cathode [HPLC-EC(Hg)]. Using HPLC-EC(Hg) under optimal mobile phase conditions, we have been able to achieve baseline separation of several cholesterol hydroperoxides (ChOOHs), not only from one another, but also from phospholipid hydroperoxide species (PLOOHs). ChOOHs and PLOOHs in model systems, e.g. photodynamically treated liposomes or erythrocyte membranes, could be readily resolved and quantified by this means, with detection limits of <0.5 pmol and <30 pmol, respectively. Moreover, with HPLC-EC(Hg) it was possible to determine the relative susceptibility of membrane ChOOHs and PLOOHs to selenoperoxidase-catalyzed reduction (detoxification). HPLC-EC(Hg) analysis of LOOHs has been extended to more complex systems, viz. (i) murine leukemia L1210 cells subjected to photooxidative stress; and (ii) oxidatively modified low density lipoprotein (LDL). Discrete peroxide families could be identified and quantified in each case. On the strength of these observations, and because of its low cost, ease of operation, and extraordinary accuracy/precision, HPLC-EC(Hg) is a highly recommended technique for ultrasensitive LOOH analysis.

Introduction

Since its introduction in the 1970s [1], electrochemical (amperometric) measurement has become increasingly more popular as a detection modality for liquid chromatography. This popularity is largely attributed to the versatility, high sensitivity, and relatively low cost of electrochemical (EC) detection. Of particular note is the versatility factor, which pertains to the broad spectrum of operating potentials that can be used with different EC electrodes [2]. This permits a wide range of redox-active compounds to be analyzed, and can afford varying degrees of selectivity for complex mixtures. A thin film of glassy carbon or gold-amalgamate is commonly used as an indicator electrode; this is typically coupled with a silver/silver chloride reference. The working range for these electrodes is nominally -0.5 V to $+1.0$ V *versus* Ag/AgCl. When operated at positive potentials (anodically), the electrodes can detect oxidizable analytes such as ascorbate and thiols; conversely, when operated at negative potentials (cathodically), they can detect reducible analytes such as quinones and peroxides. Used less frequently than anodic operation, cathodic operation is more problematic in terms of (i) O_2 interference, which intensifies as applied voltage becomes more negative; and (ii) loss of sensitivity with continued use, which is largely due to analyte deposits. The idea of using a renewable mercury drop electrode (polarographic approach) to circumvent these problems was introduced in 1971 [3]. Peroxide measurement was the proposed application at the time. Since then, there has been a surge of interest in developing new high performance techniques for analyzing biologically important organoperoxides such as lipid hydroperoxides (LOOHs). Although a few EC-based approaches have been described [4–8], most of these employ static thin layer electrodes, which, as already indicated, are subject to deterioration and loss of sensitivity when operated in the reductive mode. Cognizant of this, we have recently adapted mercury cathode EC detection to reverse-phase high-performance liquid chromatography, and have found it to be extraordinarily well suited for high sensitivity LOOH analysis [9]. This new approach is called HPLC-EC(Hg). In this report, we describe optimal HPLC-EC(Hg) running conditions and detection limits for cholesterol, cholesteryl linoleate, and phosphatidylcholine hydroperoxide standards, and illustrate how selected LOOHs in four different oxidized systems (liposomes, erythrocyte membranes, murine leukemia cells, and low density lipoprotein) can be identified and quantified by HPLC-EC(Hg).

Materials and methods

Materials

Cholesterol (Ch), cholesteryl linoleate (CL), ubiquinone-10 (CoQ_{10}), dicetylphosphate (DCP), reduced glutathione (GSH), glutathione reductase, Chelex-100, and various cell culture components (except serum) were from Sigma Chemical Co. (St. Louis, Missouri). Fetal calf serum was from Hyclone laboratories (Logan, Utah). Avanti Polar Lipids (Birmingham, Alabama) supplied the 1-palmitoyl-2-oleoyl-phosphatidylcholine (POPC) and 1-palmitoyl-2-linoleoyl-phosphatidylcholine (PLPC). HPLC-grade acetonitrile, methanol and isopropanol were obtained from Burdick and Jackson (Muskegon, Michigan). The photosensitizing dyes, chloroaluminum phthalocyanine tetrasulfonate ($AlPcS_4$) and merocyanine 540 (MC540), were from Porphyrin Products (Logan, Utah) and Eastman Kodak (Rochester, New York), respectively. Ciba-Geigy Corp. (Suffern, New York) provided the desfer-rioxamine (DFO). Phospholipid hydroperoxide glutathione peroxidase (PHGPX) was isolated from rat testes and purified as described [10]. Enzymatic activity was measured by coupled spectrophotometric assay, using NADPH, GSH, glutathione reductase, and photoperoxidized egg phosphatidylcholine [11]. Low density lipoprotein (LDL) was isolated from human plasma by flotation ultracentrifugation and stored at 4°C in phosphate-buffered saline, PBS (25 mM sodium phosphate/125 mM NaCl, pH 7.4) containing 1 mM EDTA. All aqueous solutions were prepared with glass-distilled, deionized water.

Hydroperoxide standards

The following cholesterol hydroperoxide (ChOOH) standards were prepared by $AlPcS_4$-sensitized photooxidation of cholesterol and isolated as described [8, 12]: 3β-hydroxy-5α-cholest-6-ene-5-hydroperoxide (5α-OOH); 3β-hydroxycholest-4-ene-6β-hydroperoxide (6β-OOH); and the

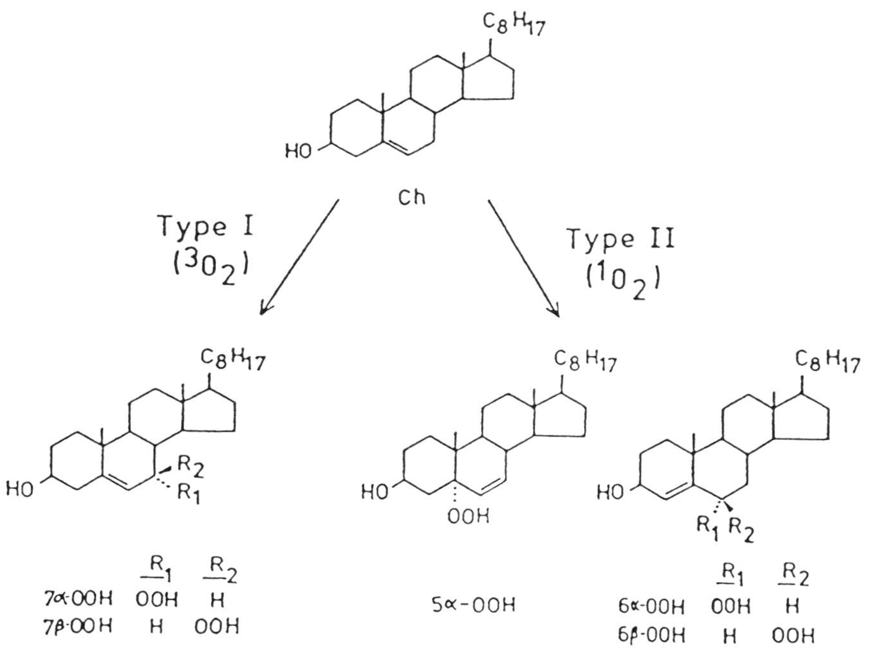

Fig. 1. Cholesterol hydroperoxides generated by Type I and Type II photochemistry. Peroxides are typically derived from ground state oxygen ($^3O_2$) in Type I (free radical) reactions, and from singlet oxygen ($^1O_2$) in Type II reactions. In the former case, only the major photoproducts are shown, i.e. 7α- and 7β-OOH. From Girotti [15].

epimeric pair, 3β-hydroxycholest-5-ene-7α-hydroperoxide and 3β-hydroxycholest-5-ene-7β-hydroperoxide (7α/7β-OOH). Photogeneration pathways and structures of these compounds are shown in Figure 1. 5α-OOH and 6β-OOH are singlet oxygen adducts [13–15]; 7α-OOH and 7β-OOH are typically produced by free radical reactions [16], but can also arise via 5α-OOH rearrangement to 7α-OOH, followed by epimerization [14, 17], as used here. Product assignments were confirmed by melting point determinations and by proton NMR [8]. Hydroperoxides of POPC (POPC-OOH) and PLPC (PLPC-OOH) were prepared by photooxidizing POPC/DCP (10:1, mol/mol) and PLPC/DCP (10:1, mol/mol) liposomes, respectively [9]. Cholesteryl linoleate hydroperoxides (CLOOH) were prepared by photooxidizing CL in chloroform/methanol (1:1, v/v). AlPcS$_4$ was used as the sensitizing dye in each case, and peroxides were isolated by means of semipreparative HPLC-EC(Hg) [9]. The major phosphatidylcholine hydroperoxide (PCOOH) family generated by mild photooxidation of L1210 leukemia cells was isolated as described [18].

Membrane preparation and photooxidation

Unilamellar liposomes ($\sim$100 nm average diameter) consisting of various combinations of Ch, POPC and DCP in Chelex-treated PBS were fabricated by an extrusion process [12, 19]. Unsealed ghost membranes were prepared by hypotonic lysis of freshly isolated human erythrocytes, followed by extensive washing with Chelex-treated PBS, and resuspension in this buffer. All membrane preparations were stored under argon at 4°C and used for experiments within a fortnight. Liposomes ($\sim$5 mM total lipid) or ghosts ($\sim$1 mg protein/ml) were photoperoxidized by irradiating in the presence of 5–10 μM AlPcS$_4$, a potent singlet oxygen generator [20]. Reactions were carried out in thermostatted beakers, using a 90-W quartz-halogen lamp (fluence rate $\sim$30 mW/cm^2). After predetermined light doses, samples were extracted with chloroform/methanol (2:1, v/v), and recovered lipid fractions were analyzed for peroxides by iodometric assay and/or by HPLC-EC(Hg).

Preparation and photooxidation of cultured cells

Murine leukemia L1210 cells from the American Type Culture Collection (Rockville, Maryland) were grown in suspension culture, using 1% serum/RPMI-1640 medium supplemented with selenium, growth factors, and antibiotics [21]. Viability was assessed by clonogenic assay [21]. Exponentially growing cells were used for all experiments. Immediately before irradiation, cells ($\sim$10^7/ml) were transferred to serum-free

RPMI medium, treated with sensitizing dye (typically MC540), and irradiated with cool-white light (fluence rate ~ 0.75 mW/cm^2). After a brief exposure (< 5 min), samples were extracted and lipid fractions analyzed by HPLC-EC(Hg); see [18] for additional details.

Iodometric assay

The LOOH content of standards and sufficiently peroxidized experimental samples was determined by iodometric assay [22]. Hydroperoxides were reduced anaerobically in the presence of excess iodide, with stoichiometric formation of triiodide. The latter was determined spectrophotometrically at 353 nm; quantitation was based on an extinction coefficient of 22.5 mM^{-1} cm^{-1} at this wavelength.

Chromatographic conditions

The HPLC-EC(Hg) instrumentation consisted of an Isco integrated HPLC system (Isco, Inc., Lincoln, Nebraska) interfaced with an EG&G-Princeton Model 420 electrochemical detector (Princeton, New Jersey). The detector was equipped with a renewable mercury drop electrode that was operated in the cathodic (reductive) mode. Analytical chromatography was carried out at 25–27°C, using a C18 Ultrasphere column (4.6×150 mm; 5 μm particles) from Beckman Instruments (San Ramon, California) and a C18 guard cartridge (4.6×15 mm) from Alltech Associates (Deerfield, Illinois). Three different premixed mobile phases were used, as indicated: (A) 83% methanol, 10% acetonitrile, and 7% aqueous solution containing 10 mM sodium acetate and 0.25 mM sodium perchlorate; (B) 79.5% methanol, 11.5% acetonitrile, and 9.0% aqueous solution containing 10 mM ammonium acetate and 0.25 mM sodium perchlorate; (C) 54% isopropanol, 23% acetonitrile, 14% methanol, and 9% aqueous solution containing 10 mM ammonium acetate and 0.25 mM sodium perchlorate. (Designated proportions are by volume.) Each solvent system was sparged continuously with high-purity argon ($> 99.98\%$; BOC Gases, Chicago, Illinois) that had been passed first through an OMI-1 oxygen scrubber (Supelco, Inc. Bellefonte, Pennsylvania) to reduce O$_2$ concentration to < 10 ppb, and then through a presaturating mobile phase scrubber. Typically, overnight purging, followed by column washing for 0.5 h was sufficient to achieve a background current in the range of 0.4–1.0 nA (depending on the applied potential), which was satisfactory for most analyses. The mobile phase was delivered isocratically at a flow rate of 1.0–2.0 ml/min against a back pressure of 1100–1500 psi, depending on the solvents used. Samples were dissolved in isopropanol and injected through a

10 μl loop. For high sensitivity settings (0.1–0.2 nA full scale), samples were chilled and sparged with high purity helium for 30 s prior to injection in order to minimize interference from O_2. A mercury drop was dispensed and equilibrated for 3–4 min before each sample injection, which typically reduced background noise to a satisfactory low level. Unless indicated otherwise, the operating potential of the mercury electrode was set at -300 mV *versus* a Ag/AgCl reference. Data collection, manipulation and storage was accomplished with an IBM 486 clone and Isco Chemresearch software.

Semipreparative HPLC-EC(Hg) of photooxidized cellular lipids was carried out on a Beckman Ultrasphere C18 column (10.0×250 mm; 5 μm particles), using Mobile Phase A. The flow rate was 4.0 ml/min.

Results

Chromatography of lipid hydroperoxide standards

We have devoted considerable effort to ChOOH analysis for the following reasons: (i) cholesterol is a prominent lipid in most eucaryotic cells, comprising 40–50 mol % of the plasma membrane; (ii) the peroxide population will be relatively small, since cholesterol is the only naturally occurring lipid that exists as a single molecular species; (iii) identification of different ChOOHs can provide valuable information about oxidative reaction mechanisms [23, 24]. By testing various solvent combinations, we found that Mobile Phase A or B (ternary mixtures) afforded excellent baseline separation of three isomeric ChOOH standards: 5α-OOH, 6β-OOH, and $7\alpha/7\beta$-OOH. (Phase A, the less polar of the two, resulted in sharper peaks and shorter retention times.) 6α-OOH could also be resolved under these conditions, but we have not yet isolated it in sufficient quantity to define its HPLC-EC(Hg) properties. Unlike 6α- and 6β-OOH, the rapidly eluting 7α- and 7β-OOH epimers did not resolve from one another. With Mobile Phase B, the capacity factor, k′ (defined as $(t - t_0)/t_0$, where t = analyte retention time, and t_0 = solvent breakthrough time) was found to be 7.4, 8.8, and 12.5 for $7\alpha/7\beta$-OOH, 5α-OOH, and 6β-OOH, respectively. Standard curves for the EC responses (peak areas) of the different ChOOHs were linear with amount of injected solute out to at least 1 nmol. Slopes of the response curves were as follows: 2.29 ± 0.01 ($7\alpha/7\beta$-OOH); 2.11 ± 0.01 (5α-OOH); 1.70 ± 0.05 (6β-OOH) (mean $\pm$ SE, n > 10). Thus, 5α-OOH and $7\alpha/7\beta$-OOH have nearly equal EC responses, which are $\sim 30\%$ greater than that of 6β-OOH. The detection limit (based on a signal-to-noise ratio of 3) was found to be 0.13, 0.20, and 0.31 pmol for $7\alpha/7\beta$-OOH, 5α-OOH, and 6β-OOH, respectively. It is important to note that these limits are far below the limit observed for UV (212 nm) absorbance by each of these compounds (~ 0.25 nmol).

POPC-OOH and PLPC-OOH families could also be resolved from one another, as well as from most of the ChOOH species, using Mobile Phase B. POPC-OOH could be partially separated into two components (presently undefined), while PLPC-OOH emerged as a single broad peak. The k' values were as follows: 14.2 and 15.3 for POPC-OOH peaks I and II; 10.0 for PLPC-OOH. We found that EC peaks were significantly sharper when ammonium acetate was included in the mobile phase. This was especially apparent for POPC-OOH; 0.8–1.0 mM ammonium acetate in the bulk phase was sufficient for maximal (albeit still partial) resolution of the two POPC-OOH peaks. Standard response curves for POPC-OOH and PLPC-OOH were linear with injected peroxide out to at least 1 nmol, with slopes of 1.10 ± 0.05 and 1.09 ± 0.04, respectively (mean $\pm$ SE, $n > 8$). Thus, both phospholipid hydroperoxides exhibit the same detection sensitivity, which is approximately one-half that of $7\alpha/7\beta$-OOH or 5α-OOH, and two-thirds that of 6β-OOH. The detection limit was estimated as 20 pmol for PLPC-OOH and 30 pmol for POPC-OOH. These values are ~ 100-fold greater than those for the ChOOHs, reflecting the much broader peaks and longer retention times of the phospholipid species. Despite these higher limits, EC detection of POPC-OOH and PLPC-OOH is still at least 10-times more sensitive than UV (212 nm) detection under the chromatographic conditions described.

The retention time of a CLOOH standard (eluting as a single peak) was found to be longer than 60 min with Mobile Phase B. Switching to a lower polarity system, Mobile Phase C, resulted in a considerably shorter retention time. The following HPLC parameters for CLOOH were established with this system: $k' = 4.08$; detection limit ~ 0.1 pmol.

The optimal working potential of the mercury cathode for any given hydroperoxide or hydroperoxide family was established by examining its hydrodynamic voltamogram under selected chromatographic conditions. For each of the LOOHs studied except CLOOH, EC peak height increased with negative applied potential until some limiting value was reached. This value was approximately -250 mV for 5α-OOH, 6β-OOH, and $7\alpha/7\beta$-OOH; and -300 mV for POPC-OOH and PLPC-OOH. The behavior of CLOOH was strikingly different in that maximal response was attained at the lowest applied potential (0 mV) and remained constant out to -300 mV. This translates into an analytical advantage, since working at relatively low negative potentials results in less interference from background noise and contaminating O_2.

Lipid hydroperoxides in photooxidized membranes

LOOHs generated by $AlPcS_4$-sensitized photooxidation of a model membrane, unilamellar Ch/POPC/DCP liposomes, were analyzed by

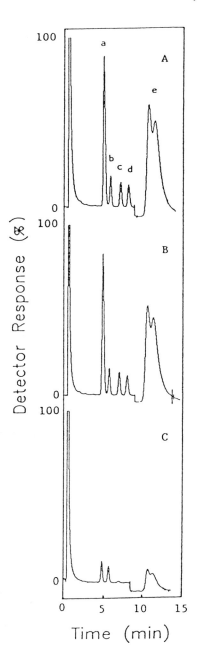

HPLC-EC(Hg). A lipid extract of the peroxidized membranes was chromatographed using Mobile Phase A. As shown in Figure 2A, the EC profile consisted of six major peaks; four of these (labeled a–d, 4–8 min range) were baseline-separated, and two (labeled e collectively; 10–13 min range) were only partially separated from one another. Using retention times of known species as references, we have assigned peaks, a, b, c, d, and e as $7\alpha/7\beta$-OOH, 5α-OOH, 6α-OOH, 6β-OOH, and POPC-OOH, respectively. Detection of 5α-OOH, 6α-OOH and 6β-OOH is unambiguous evidence for singlet oxygen involvement in the photoperoxidation reaction (Fig. 1; [24]). The prominent $7\alpha/7\beta$-OOH peak could signify free radical involvement as well [23, 24]; however, we ascribe its presence in this particular experiment to 5α-OOH rearrangement, which becomes more prominent as membranes become heavily photooxidized [12]. All of the peaks observed in Figure 2A disappeared after triphenylphosphine treatment, which is consistent with peroxide identity. In seeking additional confirmatory evidence, we treated photooxidized liposomes with GSH and two different selenoperoxidases, "classical" glutathione peroxidase (GPX) and phospholipid hydroperoxide glutathione peroxidase (PHGPX). Unlike GPX, which recognizes only relatively polar species such as H_2O_2 and fatty acid hydroperoxides, PHGPX reacts with a wide variety of peroxides, including lipid hydroperoxides in membrane environments [11, 25]. There was no significant change in the HPLC-EC(Hg) profile after incubating peroxidized membranes with GSH alone for 30 min (Fig. 2B). However, 30 min with GSH/PHGPX resulted in a large reduction of all peaks (Fig. 2C). By contrast, incubation with GSH and GPX (even at 10 times as many units as PHGPX) produced no change in the profile (not shown). Taken together, these results support our conclusions about peroxide identity in this system. In subsequent experiments on liposomes photooxidized as described in Figure 2, we compared the kinetics of PHGPX-catalyzed reduction for different peroxide species. A kinetic profile showed that all of the resident LOOHs decayed in apparent first-order fashion during GSH/PHGPX treatment, but some at significantly different rates than others. For example, 5α-OOH (the only tertiary peroxide) was reduced most slowly, and 6β-OOH most rapidly;

Fig. 2. HPLC-EC(Hg) of lipid hydroperoxides from photoperoxidized liposomes before and after GSH/PHGPX treatment. Unilamellar Ch/POPC/DCP liposomes (1:1:0.1 mol/mol) in Chelex-treated PBS (4.0 mM total lipid) were sensitized with 10 μM AlPcS$_4$ and irradiated with broad-band visible light for 3 h at 15°C, giving a total [LOOH] of ~2 mM. The liposomes were then diluted 10-fold in PBS containing 50 μM DFO. Lipids were extracted and analyzed by HPLC-EC(Hg) before incubation (A) or after a 30-min dark incubation at 37°C in the presence of 5 mM GSH alone (B) or 5 mM GSH plus 0.05 U/ml PHGPX (C). Mobile Phase A was used. Total LOOH injected: 3.7 nmol (A). Full scale detector sensitivity was 20 nA over the first 9 min and 5 nA thereafter. Hydroperoxides are denoted as follows: (a) $7\alpha/7\beta$-OOH; (b) 5α-OOH; (c) 6α-OOH; (d) 6β-OOH; (e) POPC-OOH.

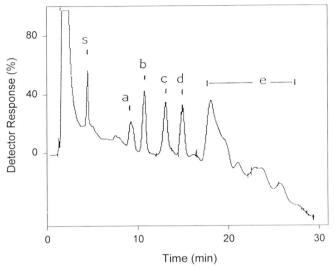

Fig. 3. HPLC-EC(Hg) of photoperoxidized erythrocyte ghosts. Ghost membranes (1.0 mg protein/ml in Chelex-treated PBS) were sensitized with 10 μM AlPcS$_4$ and irradiated for 75 min at 10°C. Extracted LOOHs were analyzed by HPLC-EC(Hg) using Mobile Phase B. Total peroxide injected: 6.09 nmol. Full scale detector sensitivity was 5 nA from 0–12 min, 2 nA from 12–17 min, and 1 nA thereafter. Peak identities are as follows: (s) photooxidized cis-11-eicosenoic acid methyl ester (internal standard); (a) 7α/7β-OOH; (b) 5α-OOH; (c) 6α-OOH; (d) 6β-OOH; (e) PLOOHs.

rate constants differed in the following order: 6β-OOH > POPC-OOH > 7α/7β-OOH ≫ 5α-OOH.

We have also examined the LOOH distribution in a photooxidized natural membrane, the human erythrocyte ghost. After exposure of ghost membranes to a single dose of white light in the presence of AlPcS$_4$, lipids were extracted and analyzed by HPLC-EC(Hg), using Mobile Phase B. A representative chromatogram for these membranes is shown in Figure 3, where one sees at least 10 EC peaks over a 30-min elution span. The earliest peak (∼4 min) represents a peroxidized eicosenoate ester, which was added as an internal standard to check stability of the electrode response. Referring to retention times of ChOOH standards with Mobile Phase B, we have assigned peaks a, b, c, and d as 7α/7β-OOH, 5α-OOH, 6α-OOH, and 6β-OOH, respectively. Mobile Phase B is more polar than A, which probably accounts for the longer retention of peaks a–d in Figure 3 than in Figure 2A. The group of peaks collectively denoted as fraction e (18–27 min) in Figure 3 is believed to represent phospholipid hydroperoxides on the following basis. When photooxidized ghosts were incubated with Ca^{2+} and phospholipase A$_2$, all the peaks in fraction e disappeared, whereas peaks a–d were unaffected (not shown). This is consistent with fraction e being phospholipid-derived, since phospholipase A_2 specifically catalyzes the hydrolysis of sn-2 linkages in phospholipids. When photooxidized ghosts were incubated with GSH/PHGPX under conditions similar to those described in Figure 2, there was a progressive loss of peaks a–e (not shown), confirming that all of the compounds detected were hydroperoxides. All peaks decayed in apparent first order fashion, the rate constants (h^{-1}) being as follows: PLOOH (7.9); 6β-OOH (3.0); 6α-OOH $= 7\alpha/7\beta$-OOH (1.8); 5α-OOH (0.9). Thus, in this system, as with liposomes, 5α-OOH stands out as the peroxide that is least reactive with GSH/PHGPX. Solubilization of the membranes with Triton X-100 caused all of the LOOHs to decay more rapidly during GSH/PHGPX treatment; however, their relative decay rates remained essentially the same, suggesting that low reactivity of 5α-OOH, for example, is not due to poor accessibility in the membrane environment, but rather low intrinsic reactivity. This might relate to the fact that 5α-OOH is a tertiary (inter-ring) peroxide, and as such might not be recognized by PHGPX as well as secondary peroxides such as 6β-OOH or $7\alpha/7\beta$-OOH. Photodynamically treated cell membranes can accumulate 5α-OOH at relatively high initial rates [12, 24]. Rapid formation on the one hand and slow enzymatic detoxification on the other could mean that 5α-OOH is potentially more cytotoxic than all of the other peroxides considered. Whether this proves to be the case remains to be seen.

Lipid hydroperoxides in photooxidized cells

The excellent results obtained with isolated membranes prompted us to extend our work to more complex systems. Murine leukemia L1210 cells subjected to a photooxidative insult were chosen as one example. Content of major lipids in these cells is as follows [18]: phospholipid (69%), triacylglycerol (27%), cholesterol (4%). MC540, an amphiphilic dye of therapeutic potential in connection with its antileukemic properties [26], was used as the photosensitizing agent in these experiments. Of special interest to us was whether LOOHs could be detected before any gross photokilling had occurred, i.e., in minimally damaged cells. This would require operation of the HPLC-EC(Hg) system at maximal or near-maximal sensitivity. Results of a typical experiment [9, 18] are shown in Figure 4. Using a detector sensitivity of 2.0 nA full-scale, we observed no EC signals for unirradiated cells (Fig. 4A) or cells that had been irradiated in the absence of MC540 (not shown). By contrast, a complex profile with at least 10 well-defined peaks was observed when an equal number of dye/light-treated cells was analyzed (Fig. 4B). All of these peaks disappeared when a photooxidized sample was treated with triphenylphosphine, consistent with peroxide identity. We have deter-

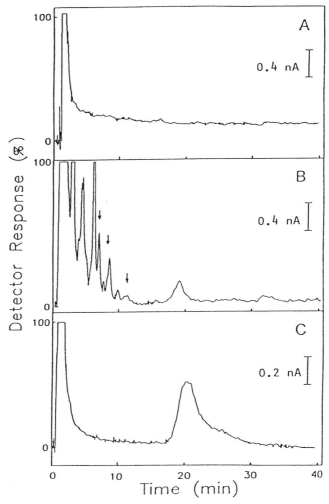

Fig. 4. Chromatograms of lipid hydroperoxides from photooxidized L1210 cells. Lipids were extracted and analyzed by HPLC-EC(Hg) with Mobile Phase B before (A) and after (B) exposing L1210 cells (10^7/ml) to ~ 0.11 J/cm^2 of cool-white fluorescent light in the presence of 25 μM MC540. Each injection corresponded to 2×10^7 cells ($\sim 280\,\mu$g total lipid). The sample in (B) contained 1.4 ± 0.2 nmol of total LOOH (mean $\pm$ SD, n = 3). Vertical arrows indicate retention times of cholesterol hydroperoxide standards run separately: $7\alpha/7\beta$-OOH (7.2 min); 5α-OOH (8.3 min); 6β-OOH (11.4 min). The 18–22 min fraction in (B) was isolated by semipreparative HPLC and rechromatographed (C): 575 pmol LOOH injected. From Korytowski et al. [9].

mined that the three peaks marked with arrows represent mostly 7-OOH, 5α-OOH, and 6β-OOH (in order of increasing retention time); this was based on sample spiking with authentic compounds and thin layer chromatography (TLC) of isolated peak fractions [18]. Material eluting as a broad peak at 18–22 min (Fig. 4B) was isolated in relatively large amount by semipreparative HPLC-EC(Hg). Re-examination by analytical HPLC-EC(Hg) revealed a single broad peak centered at ~20 min (Fig. 4C). We have determined that the 18–20 min fraction is hydroperoxide in nature by showing that it can be completely reduced by GSH/PHGPX, but not affected by GSH/GPX. That this fraction consists predominantly of phospholipid hydroperoxide (specifically PCOOH), was established on the following grounds: (i) It comigrates with authentic POPC-OOH or egg PCOOH on a TLC plate; (ii) Its EC response characteristics are identical to those of POPC-OOH or PLPC-OOH; (iii) It can be hydrolyzed by Ca^{2+}-dependent phospholipase A_2, giving stoichiometric amounts of fatty acid hydroperoxides, as determined by HPLC-EC(Hg). Since triacylglycerols account for a large proportion of cellular lipid (see above), some of the other EC peaks might represent triacylglycerol hydroperoxides, but we have no evidence for this yet. It is important to note that if these species are generated, their yields might be low, since triacylglycerols are located in the cytosol, whereas MC540 photoactivation occurs mainly in the plasma membrane [26]. The light fluence resulting in the peroxide profile shown in Figure 4B (~0.11 J/cm²) caused a relatively modest cell kill, ~30% loss of clonogenicity. Using HPLC-EC(Hg), we were able to detect LOOHs at a fluence of only 0.045 J/cmˢ, which killed only 10% of the cells. Thus, it appears that lipid peroxidation was an early event in dye-mediated phototoxicity, raising the possibility that it was linked to cell killing in some fashion, as recently proposed [21].

Lipid hydroperoxides in low density lipoprotein

We have also used HPLC-EC(Hg) to determine trace levels of LOOHs in human LDL [27]. Oxidative modification of the apoB-100 protein of LDL, a process mediated by lipid peroxidation, is believed to play a causal role in atherogenesis [28]. Since rampant free radical-mediated peroxidation of LDL might be triggered by preexisting LOOHs, it is important that these species be determined with high accuracy and sensitivity. We have empirically developed a quaternary solvent system (Mobile Phase C) which effects a good separation of all major LOOHs found in oxidized LDL, i.e. species derived from cholesteryl esters (the most abundant lipid family), phospholipids, cholesterol, and triacylglycerols. A great advantage of EC detection over, e.g. UV (234 nm) detection, was demonstrated for a cholesteryl ester standard, CLOOH. The single EC peak for CLOOH (retention time ~10 min) disappeared after triphenylphosphine treatment; however, with UV (234 nm), which detects conjugated dienes of fatty acyl peroxides as well as their similarly eluting hydroxy analogues, no change was observed. Therefore, as

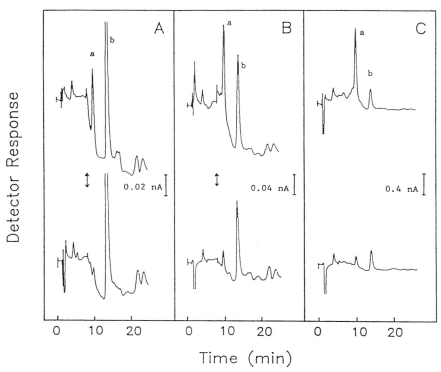

Fig. 5. LOOHs in freshly prepared and aged LDL: susceptibility to enzymatic reduction. A single preparation of LDL was analyzed by HPLC-EC(Hg) (Mobile Phase C) within 24 h after final preparative centrifugation (A), and after two weeks (B) or six weeks (C) storage at 4°C. LDL samples (1.0 mg protein/ml in Chelexed PBS/1 mM EDTA) were either analyzed directly (top traces) or after a 1-h incubation with GSH (5 mM)/PHGPX (10 mU/ml) at 37°C (bottom traces). Full scale sensitivity: (A) 0.5 nA over the first 7 min, then a switch to 0.1 nA (arrow); (B) 0.5 nA over the first 7 min, then a switch to 0.2 nA (arrow); (C) 2.0 nA throughout. (a) cholesteryl ester (9.9 min); (b) ubiquinone-10 (14.1 min). From Thomas et al. [27].

pointed out previously for other compounds [8, 9], EC is far superior to UV for specific peroxide detection (distinction of peroxides from hydroxides). With regard to sensitivity, the HPLC-EC(Hg) detection limit for CLOOH was found to be ~ 100 fmol (signal-to-noise ratio of 3). This is at least 200-times lower than the limit determined for UV (234 nm) detection. An EC response curve for CLOOH showed linearity out to at least 4.0 nmol. With the detector set at maximum sensitivity (0.1 nA full scale), we have been able to detect and quantify cholesteryl ester hydroperoxide (CEOOH) in fresh LDL prepared with the utmost care to avoid autoxidation. Pristine LDL (analyzed within 24 h after final centrifugation) exhibited a prominent CEOOH peak at ~ 10 min and a peak at ~ 14 min identified as ubiquinone-10 (CoQ_{10}) (Fig. 5A).

The CEOOH content of four different preparations of fresh LDL (representing different donors) was found to be 9.8 ± 2.2 pmol/mg protein (mean $\pm$ SD), which corresponds to ~ 1 CEOOH for every 200 LDL particles. (The antioxidant, e.g. α-tocopherol, status of these preparations was normal.) When pristine LDL was allowed to stand at $4°C$ in Chelexed PBS/1 mM EDTA, CEOOH content increased to ~ 25 pmol/mg protein after 2 weeks (Fig. 5B) and ~ 300 pmol/mg protein after six weeks, a 30-fold increase overall (Fig. 5C). Thus autoxidation took place, despite precautionary measures. It is important to note that the 6-week LOOH level was still far below the detection limit of a conventional iodometric or thiobarbituric acid assay.

We found that incubating freshly prepared LDL with GSH/PHGPX caused CEOOH to be completely reduced without affecting CoQ_{10} (Fig. 5A). Similar results were obtained with the aged preparations (Fig. 5B,C). Importantly, the GSH/PHGPX-treated LDL was found to be much more resistant to Cu^{2+}-induced peroxidation, as evidenced by a longer lag in the formation of thiobarbituric acid-reactive substances [27]. These findings – largely derived through the use of HPLC-EC(Hg) – have clearly established that preexisting LOOHs in LDL can act as powerful sensitizers of large-scale oxidative damage.

Discussion

Interest in biological lipid peroxidation has intensified in recent years. This is attributed to a growing awareness that lipid peroxidation may play an important role in a wide variety of disorders and pathological conditions, including autoimmune inflammatory states, ischemia-reperfusion injury, atherogenesis, and carcinogenesis [25]. Along with this interest has come the need for improving the methods used for analyzing peroxidation intermediates and products. Many investigators have focused on direct determination of fatty acid and lipid hydroperoxide intermediates [4–8]. When carried out carefully, direct measurement of hydroperoxides is subject to fewer artifacts and uncertainties than many other approaches used to assess lipid peroxidation, e.g., determination of conjugated dienes or aldehyde by-products such as malonaldehyde and 4-hydroxynonenal [24]. In biological systems, LOOHs generated as a result of some oxidative stress are present in complex mixtures typically representing many different lipid classes and different molecular species. In addition to the polydispersity, naturally occurring LOOHs may exist at very low steady-state levels due to metabolic detoxification or metal ion-mediated conversion to free radicals and other species [29]. These have been important considerations in the drive to develop high-resolution and high-sensitivity methods for LOOH analysis. Several different HPLC-based approaches have been described in recent years, along with a variety of detection modalities, including chemiluminescence [30, 31], fluorescence [32], visible absorbance [33], ultraviolet absorbance [34], and electrochemical [4–9] detection. All of these modalities except the latter two depend on some type of postcolumn reaction (LOOH-induced oxidation) in order to generate a signal. The non-requirement for postcolumn chemistry is a major advantage for EC detection, since potential problems associated with mobile phase incompatibility, mixing inaccuracy, reagent instability, or interfering contaminants are avoided. An additional advantage of EC detection that distinguishes it from ultraviolet or radiochemical detection is a high selectivity factor. The importance of selectivity can be illustrated for phospholipid hydroperoxides, which are often accompanied by hydroxy analogues (reduction products) in biological samples. Since phospholipid hydroperoxides and hydroxides are not resolved in any type of reverse-phase or normal-phase HPLC, differentiating them in systems relying on ultraviolet detection (e.g., 234 nm absorbance of conjugated dienes) or radiochemical detection (radioactivity of ^{14}C- or ^{3}H-labeled species) is virtually impossible. However, hydroxy derivatives of all LOOHs, including fatty acid, phospholipid, cholesterol, and cholesteryl ester hydroperoxides, are EC silent [7, 8], meaning that only the peroxides will be detected in complex mixtures containing both forms. Although similar specificity is possible with other detection methods [30–33], occasionally false or interfering signals arise, which lead to confusion or uncertainty in peroxide determinations. For example, in the well known chemiluminescence approach involving isoluminol as an indicator, peaks attributed to ubiquinols and troughs attributed to antioxidants such as ascorbate or α-tocopherol have been observed along with authentic peroxide peaks in chromatograms of plasma samples [31]. If coincident with or closely adjacent to peroxide signals, false signals such as these could obviously present problems. In our experience, none of the above mentioned compounds interferes with HPLC-EC(Hg). However, we have seen that the fully oxidized form of ubiquinol-10, i.e., ubiquinone-10 (CoQ$_{10}$), gives a strong EC signal at operating potentials used for HPLC-EC(Hg). Under the conditions described for LDL-LOOH analysis, CoQ$_{10}$ emerges well beyond all recognized peroxide species, including CEOOHs, and therefore does not pose a problem. We have not yet looked for CoQ$_{10}$ in cell samples; a system of much lower polarity than Mobile Phase B (cf. Fig. 4) would be necessary in order to elute CoQ$_{10}$ in a reasonable time (< 1 h). It is possible that other oxidized species that extract with LOOHs (e.g., other lipophilic quinones or peroxides of β-carotene and α-tocopherol) will also be detected by HPLC-EC(Hg).

In earlier work involving reductive mode HPLC-EC for LOOH analysis, we [8] and others [6, 7] used a thin-layer (glassy carbon) indicator electrode. Progressive loss of responsiveness due to electrolyte deposits poses a major problem with permanent electrodes of this type. This problem can become critical when large negative operating potentials are used. Quite often, samples need to be spiked with internal standards to correct for response losses. Electrodes can be rejuvenated by polishing, but this is tedious and the results are often disappointing. These difficulties were eliminated by switching to the renewable mercury drop system. The EG&G-Princeton accessory that we describe can be interfaced with any available EC controller unit. We have determined that under normal conditions, several samples can be injected consecutively on a single mercury drop (at least five for CLOOH) without significant loss of detector sensitivity ($<5\%$ from run-to-run). However, we typically dispense a new drop before each injection. Dissolved O_2 is the major source of interference in HPLC-EC(Hg). Although O_2 elutes far upstream and well ahead of most LOOHs, high sensitivity runs (<0.5 nA full scale) require that samples be carefully deoxygenated before injection. Scrupulous deoxygenation of the mobile phase is also necessary in order to minimize the equilibration time for each newly dispensed mercury drop. Under normal circumstances, background currents of <1 nA are attained within $2-3$ min during equilibration, which is far better than the best possible time with the glassy carbon electrode [8]. Unlike thin-film electrodes, the mercury cathode is easily renewed. This translates into a great advantage for chromatographic flexibility, since the mobile phase can be changed at will without concern about permanent loss of sensitivity. With a detection limit of <0.5 pmol for ChOOHs and ~0.1 pmol for CLOOH under specified conditions, mercury electrode EC is much more sensitive than glassy carbon EC [8], and probably more sensitive than all other HPLC-based detection modes.

Using peroxidized membranes, cells, and LDL as examples, we have demonstrated that HPLC-EC(Hg) is eminently well suited for analyzing LOOHs in complex biological systems. As it continues to be developed and refined, this technique should prove to be extremely valuable for assessing LOOH status in oxidatively stressed cells and tissues. There is a growing interest in other biological peroxides besides LOOHs, e.g. nucleic acid and protein peroxides [35, 36]. Studies involving these species should also benefit from the availability of HPLC-EC(Hg).

Acknowledgements
The expert technical assistance of Pete Geiger and helpful discussions with Jim Thomas are greatly appreciated. The following grant support is acknowledged: Grants CA49089 and HL47250 from the National Institutes of Health; Grant MCB-9106117 from the National Science Foundation; and KBN Grant 6P203-033-06.

References

1. Kissinger, P.T., Refshauge, C.J., Dreiling, R. and Adams, R.N. (1973) Electrochemical detector for liquid chromatography with picogram sensitivity. *Anal. Lett.* 6: 465–477.

2. Shoup, R.E. (1986) Liquid chromatography/electrochemistry. *In:* C. Horvath (ed.): *High-Performance Liquid Chromatography*, Vol. 4, Academic Press, New York, pp 91–194.
3. Mair, R.D. and Hall, R.T. (1971) Determination of organic peroxides by physical, chemical and colorometric methods. *In:* D. Swern (ed.): *Organic Peroxides*, Wiley Interscience, New York, p. 578.
4. Funk, M.O., Keller, M.B. and Lewison, B. (1980) Determination of peroxides by high-performance liquid chromatography with amperometric detection. *Anal. Chem.* 52: 773–774.
5. Funk, M.O., Walker, P. and Andre, J.C. (1987) An electroanalytical approach to the determination of lipid peroxides. *Bioelectrochem. Bioenerget.* 18: 127–135.
6. Yamada, K., Terao, J. and Matsushita, S. (1987) Electrochemical detection of phospholipid hydroperoxides in reverse-phase high-performance liquid chromatography. *Lipids* 22: 125–128.
7. Terao, J., Shibata, S.S. and Matsushita, S. (1988) Selective quantification of arachidonic acid hydroperoxides and their hydroxy derivatives in reverse-phase high-performance liquid chromatography. *Anal. Biochem.* 169: 415–423.
8. Korytowski, W., Bachowski, G.J. and Girotti, A.W. (1991) Chromatographic separation and electrochemical determination of cholesterol hydroperoxides generated by photodynamic action. *Anal. Biochem.* 197: 149–156.
9. Korytowski, W., Bachowski, G.J. and Girotti, A.W. (1993) Analysis of cholesterol and phospholipid hydroperoxides by high-performance liquid chromatography with mercury drop electrochemical detection. *Anal. Biochem.* 213: 111–119.
10. Maiorino, M., Gregolin, C. and Ursini, F. (1990) Phospholipid hydroperoxide glutathione peroxidase. *Methods Enzymol.* 186: 448–457.
11. Thomas, J.P., Maiorino, M., Ursini, F. and Girotti, A.W. (1990) Protective action of phospholipid hydroperoxide glutathione peroxidase against membrane-damaging lipid peroxidation. *J. Biol. Chem.* 265: 454–461.
12. Korytowski, W., Bachowski, G.J. and Girotti, A.W. (1992) Photoperoxidation of cholesterol in homogeneous solution, isolated membranes, and cells: comparison of the 5α- and 6β-hydroperoxides as indicators of singlet oxygen intermediacy. *Photochem. Photobiol.* 56: 1–8.
13. Schenck, G.O., Gollnick, K. and Neumuller, O.A. (1957) Zur photosensibilisieren Autoxydation der steroide. Darstellung von Steroid-hydroperoxyden mittels phototoxischer Photosensibilisatoren. *Ann. Chem.* 603: 46–59.
14. Kulig, M. and Smith, L.L. (1973) Sterol metabolism XXV. Cholesterol oxidation by singlet molecular oxygen. *J. Org. Chem.* 38: 3639–3642.
15. Girotti, A.W. (1992) Photosensitized oxidation of cholesterol in biological systems: reaction pathways, cytotoxic events, and defense mechanisms. *J. Photochem. Photobiol.* 13: 105–118.
16. Smith, L.L., Teng, J.I., Kulig, M.J. and Hill, F.H. (1973) Sterol metabolism XXIII cholesterol oxidation by radical-induced processes. *J. Org. Chem.* 38: 1763–1765.
17. Beckwith, A.L.J., Davies, A.G., Davison, I.G.E., Maccoll, A. and Mruzek, M.H. (1987) The mechanism of the rearrangements of allylic hydroperoxides: 5α-hydroperoxy-3β-hydroxycholest-6-ene and 7α-hydroperoxy-3β-hydroxycholest-5-ene. *J. Chem. Soc. Perkin Trans.* II: 815–824.
18. Bachowski, G.J., Korytowski, W. and Girotti, A.W. (1994) Characterization of lipid hydroperoxides generated by photodynamic treatment of leukemia cells. *Lipids* 29: 449–459.
19. Mayer, L.D., Hope, M.J. and Cullis, P.R. (1986) Vesicles of variable size produced by a rapid extrusion procedure. *Biochim. Biophys. Acta* 858: 161–168.
20. Bachowski, G.J., Ben-Hur, E. and Girotti, A.W. (1991) Phthalocyanine-sensitized lipid peroxidation in cell membranes: use of cholesterol and azide as probes of primary photochemistry. *J. Photochem. Photobiol.* 9: 307–321.
21. Lin, F., Thomas, J.P. and Girotti, A.W. (1992) Selenoperoxidase-mediated cytoprotection against merocyanine 540-sensitized photoperoxidation and photokilling of leukemia cells. *Cancer Res.* 52: 5282–5290.
22. Bachowski, G.J., Pintar, T.J. and Girotti, A.W. (1991) Photosensitized lipid peroxidation and enzyme inactivation by membrane-bound merocyanine 540: reaction mechanisms in the absence and presence of ascorbate. *Photochem. Photobiol.* 53: 481–491.

23. Smith, L.L. (1981) *Cholesterol Autoxidation*. Plenum Press, New York.
24. Girotti, A.W. (1990) Photosensitized lipid peroxidation in biological systems. *Photochem. Photobiol.* 51: 497–509.
25. Ursini, F., Maiorino, M. and Sevanian, A. (1991) Membrane hydroperoxides. *In:* H. Sies (ed.): *Oxidative Stress: Oxidants and Antioxidants.* Academic Press, New York, pp 319–336.
26. Sieber, F. (1987) Merocyanine 540. *Photochem. Photobiol.* 46: 1035–1042.
27. Thomas, J.P., Kalyanaraman, B. and Girotti, A.W. (1994) Involvement of preexisting lipid hydroperoxides in Cu^{2+}-stimulated oxidation of low density lipoprotein. *Arch. Biochem. Biophys.* 315: 244–254.
28. Steinberg, D., Parthasarathy, S., Carew, T.E., Khoo, J.C. and Witztum, J.L. (1989) Beyond cholesterol: modifications of low-density lipoprotein that increase its atherogenicity. *N. Engl. J. Med.* 320: 915–924.
29. Lin, F. and Girotti, A.W. (1993) Photodynamic action of merocyanine 540 on leukemia cells: iron-stimulated lipid peroxidation and cell killing. *Arch. Biochem. Biophys.* 300: 714–723.
30. Miyazawa, T. (1989) Detection of phospholipid hydroperoxides in human blood plasma by a chemiluminescence-HPLC assay. *Free Radical Biol. Med.* 7: 209–217.
31. Yamamoto, Y., Frei, B. and Ames, B.N. (1990) Assay of lipid hydroperoxides using high-performance liquid chromatography with isoluminol chemiluminescence detection. *Methods Enzymol.* 186: 371–380.
37. Akasaka, K., Ohrui, H. and Meguro, H. (1993) Normal-phase high-performance liquid chromatography with a fluorometric postcolumn detection system for lipid hydroperoxides. *J. Chromatogr.* 628: 31–35.
33. Mulhertz, A., Schmedes, A. and Holmer, G. (1990) Separation and detection of phospholipid hydroperoxides in the low nanomolar range by high-performance liquid chromatography/isothiocyanate assay. *Lipids* 25: 415–418.
34. Terao, J., Asano, I. and Matsushita, S. (1985) Preparation of hydroperoxy and hydroxy derivatives of rat liver phosphatidylcholine and phosphatidylethanolamine. *Lipids* 20: 312–317.
35. Cadet, J. and Berger, M. (1985) Radiation-induced decomposition of the purine bases within DNA and related model compounds. *Int. J. Radiat. Biol.* 47: 127–143.
36. Gebicki, S. and Gebicki, J. (1993) Formation of peroxides in amino acids and proteins exposed to oxygen free radicals. *Biochem. J.* 289: 743–749.

Analysis of Free Radicals in Biological Systems
Favier et al. (eds)
© 1995 Birkhäuser Verlag Basel/Switzerland

Determination of primary and secondary lipid peroxidation products: Plasma lipid hydroperoxides and thiobarbituric acid reactive substances

C. Coudray[1], M.J. Richard[1] and A.E. Favier[1,2]

Groupe de Recherche et d'Etude sur les Pathologies Oxydatives (GREPO), [1]Laboratoire de Biochimie C, Centre Hospitalier Régional de Grenoble, F-38043 Grenoble Cedex, France [2]Laboratoire de Biochimie Pharmaceutique, Faculté de Pharmacie, Université J. Fourier, F-38700 La Tronche, France

Summary. Lipid peroxidation is a complex process whereby unsaturated lipid undergoes reaction with molecular oxygen to yield lipid hydroperoxides. In most situations involving biological samples the lipid hydroperoxides are degraded to a variety of products including alkanals, alkenals, hydroxyalkenals, ketones, alkanes. Although attack by singlet oxygen on unsaturated lipid has been shown to give hydroperoxides by a nonradical process, the vast majority of situations involving lipid peroxidation proceeds through a free radical-mediated chain reaction initiated by the abstraction of a hydrogen atom from the unsaturated lipid by a reactive free radical, followed by a complex sequence of propagation reactions. The involvement of free oxygen radicals in the pathology of certain diseases explains the growing interest in the assay of lipid peroxides. Assay of polyunsaturated fatty acid degradation products is currently performed by measuring the so-called thiobarbituric acid-reactive substances, of which malondialdehyde is the best known. Because this assay is controversial, a second index of free radical attack would be useful to confirm the peroxidative process.

In this paper, we describe the determination of plasma lipid hydroperoxides and thiobarbituric acid reactive substances (TBARS). We proposed an improved enzymatic technique for assay of lipid hydroperoxides in biological fluids. The technique previously described by Heath and Tappel cannot be used in biological determinations. In fact, the presence of endogenous enzymes such as glutathione peroxidase and glutathione reductase in the sample interferes with the reaction and makes the results unreliable. Elimination of these endogenous enzymes by deproteinization before assaying for lipid hydroperoxides in the plasma gives simple, reliable, and reproducible measurements. The determination of TBARS is a widely used method for investigating overall lipid peroxidation. The TBARS assay is accomplished by mixing the sample with a TBA reagent in acid medium and placing in a boiling water bath. After extracting the TBARS by organic solvent, their optical density or fluorescence intensity is measured. TBARS assay detects both preexisting malondialdehyde (MDA) plus whatever substances give rise to MDA during the assay. Lipid hydroperoxides can decompose during heating in the presence of acid and metals and give rise to MDA and other aldehydes capable of interacting with TBA during the assay. Many researchers use this assay in their laboratories but the procedures used vary; the variability could arise from differences in sample volume, acid type, pH of medium, heating duration, blank undertaking and detection conditions. These variations render impossible the comparison of results between laboratories. The use of an assay kit for plasma TBARS assay would enable the method to be standardized. The results reported here indicate that the MDA-kit manufactured by SOBIODA (GRENOBLE, France) complies with criteria of good analytical practices. However, we concluded that no single method sufficiently meets analytical standards in all application to make it the choice (let alone universal) one. We thus emphasize the need to integrate different analytical approaches in the assessment of oxidant stress *in vivo*.

Introduction

Reactive oxygen species (ROS) are involved in many pathologies (Aids, cancer, ischaemia/reperfusion, chronic inflammation, kidney failure) and in ageing process [1–6]. Free radicals are also produced as a secondary event in human disease, in particular trauma and toxins [7]. Though it is extremely difficult to measure these transient, highly reactive species directly in biological fluids, determination of the degradation products of the peroxidized substances (lipid, protein, or DNA) is becoming the procedure of choice [8]. Attack of a reactive species such as ˙OH upon the side-chains of unsaturated fatty acids causes hydrogen atom abstraction. Hydrogen can be abstracted at different points in the side-chain, so that several isomeric lipid hydroperoxides can result. These hydroperoxides can then decomose into many low-weight-molecular compounds. The measurement of putative elevated end products of lipids peroxidation in human material is probably the evidence most frequently quoted in support of the involvement of ROS in tissue damage in human diseases [9–11]. Estimation of lipid peroxidation in blood samples has been based on the analysis of conjugated dienes absorption, lipid hydroperoxides (LHP) and malondialdehyde (MDA) content. Other aldehydes are also among the parameters commonly assayed [11]. Determination of each of these parameters has both advantages and limits. However, we think that at least two parameters have to be explored to confirm the presence of lipid peroxidation process.

The most direct approach for the assessment of lipid peroxidation is the quantification of the primary (hydroperoxide) products. These products are relatively stable and usually the result of both free radical attack and biological enzyme activity such as lipooxygenase. Their assay can give information on the degree of peroxidation. Methods of LHP assay are based on several principles. The reference method uses the oxidation of iodide by LHP [12]. This method is not very sensitive in the presence of atmospheric oxygen when applied to human plasma [13]. This technique is indeed the most commonly used one for LHP assay in pure lipid solutions. Several other techniques have also been proposed [14, 15], in particular the enzymatic method described by Heath and Tappel [16], and recently taken up again by Allen et al. [17]. Their technique involves a coupled glutathione peroxidase-glutathione reductase reaction. Currently, it is the technique most used in human biology. Though the initial method has satisfactory within-run and between-run precision, it lacks accuracy and specificity. The enzymes involved in the assay (glutathione peroxidase [GPx], glutathione reductase [GRx]) exist, indeed, in the endogenous state in the sample at concentrations that can distort the result. We thus define methods of treating the sample that would overcome the problem of endogenous enzymes and allow better standardization of the procedure in order to satisfy the analytical criteria required for the biological assay.

Malondialdehyde is one of the products of lipid peroxidation which appears to be produced in relatively constant proportion to lipid peroxidation, it is therefore a good indicator of the rate of lipid peroxidation *in vivo*. MDA can act as either a nucleophile or an electrophile and forms multimeric adducts [18]. Upon heating at low pH, MDA readily participates in addition reactions giving rise to a variety of condensation products. However, all of these reactions lack selectivity with the exception of low-molecular-weight aldehydes [19]. The reaction between MDA and TBA produces a red pigment with a high molar absorbtivity. The red MDA:TBA condensation product is a 1:2 adduct with a MDA moiety at the end of the molecule of TBA. This adduct is both pigmented and fluorescent, and the reaction usually requires low pH and elevated temperature. However, several variations of the reaction have been reported to facilitate MDA detection and analysis by the laboratories. Subsequently, the assay of products reacting with thiobarbituric acid has led recently to the development of various commerical kits, one in collaboration with our laboratory [9, 20]. These kits generally give a better standardization of the method and allow interlaboratory comparisons.

Materials and methods

Reagents

Tris (hydroxymethyl) aminomethane (Prolabo), ethylene diaminetetraacetic acid, metaphosphoric acid, glutathione peroxidase, glutathione reductase, NADPH2, reduced glutathione, tert-butyl hydroperoxide (+BHP), (Sigma), bovine albumin (Fluka). Solutions of glutathione, glutathione peroxidase, glutathione reductase, and NADPH2 were freshly prepared in Tris buffer (50 mM, pH 7.4).

Reagents kit

The reagents in the kit included solution 1 (thiobarbituric acid: TBA), solution 2 (Perchloric acid: HClO4) and a calibration solution consisting of 20 mM 1,1,3,3-tetraethoxypropane (TEP) in ethanol. The TBA-Acid working solution was prepared as follows: TBA/HClO4, 2/1, V/V. The stock standard solution was diluted with deionized water to obtain a concentration of 10 μmol/L. The working solution and standard solution were prepared fresh daily.

Additional chemicals

n-Butanol, fluorometric grade, (Merck, Darmstadt, Germany) was used for extraction. 2% w/v butylated hydroxytoluene (BHT) from Sigma (Sigma Chemical Co, via Coger, Paris, France) was prepared in absolute ethanol. The control serum used was lyophilized Probiocal AB 43 serum (BioMérieux, Lyon, France), reconstituted daily according to the manufacturer's instructions.

Apparatus

Spectrophotometry was performed using an Uvicon 860 (Kontron Instruments). Fluorescence was measured with a Perkin Elmer LS 50 Fluorometer (Perkin-Elmer Ltd., Bucks, UK.). The reaction was run in disposable 7 ml polypropylene screw cap tubes (Sobioda, Grenoble, France) which had previously been tested to verify that they released no transition metals or substances interfering with the fluorometry assay.

Blood collection

Blood was collected by venous puncture into 5 ml trace element-free heparinized vacum tubes. After centrifugation (1600 g) for 10 min, the supernatant plasma was removed carefully to avoid contamination with platelets or white blood cells and was stored as rapidly as possible after sampling at $-20°C$ until further analysis.

Principle of assays

Lipid hydroperoxides measurement
Glutathione peroxidase catalyses the reduction of LHPs (ROOH) in the presence of glutathione (GSH) to their corresponding alcohol:

$$2GSH + ROOH \xrightarrow{GPx} ROH + GssG + H2O$$

The reaction can be followed by regeneration of glutathione by coupled oxidation of NADPH2 in the presence of glutathione reductase:

$$GssG + NADPH2 \xrightarrow{GRx} 2GSH + NADP+$$

The decreased absorbance of NADPH2 at 340 nm is a direct function of the amount of LHP present in the reaction mixture. The LHP concentration is determined through a calibration curve of tert-butyl hydroperoxide as a standard.

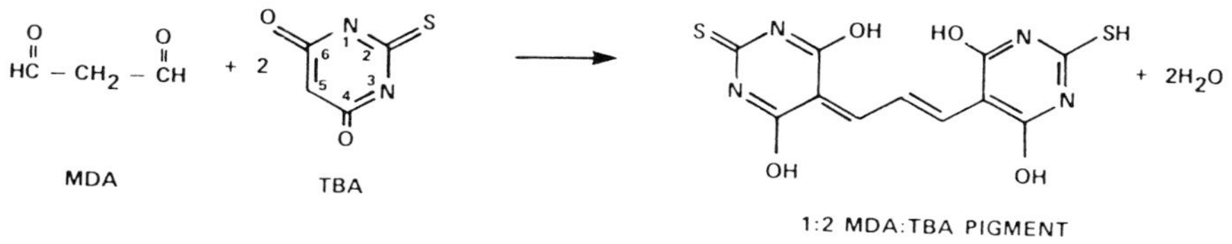

Fig. 1. Formation of the fluorescent red adduct between one MDA and two TBA via an acid-catalyzed nucleophile addition reaction (according to [18]).

TBA/MDA reaction

Malondialdehyde (MDA) is one of the several-low-molecular-weight end products formed predominantly via the decomposition of certain primary and secondary lipid peroxidation products. At low pH and elevated temperature, MDA readily participates in nucleophilic addition reaction with 2-thiobarbituric acid (TBA), generating a red, fluorescent 1:2 MDA:TBA adduct (Fig. 1), with high molar absorptivity at 532 nm or high fluorescent intensity at 552 nm when excited at 532 nm.

Results

Optimized procedure of LHP measurement

400 μl of plasma was treated with 400 μl of 6% metaphosphoric acid in plastic tubes. After shaking, and a contact time of 3 min, centrifugation was performed at 1000 g for 15 min. 500 μl of supernatant was neutralized with 100 μl of molar NaOH (in order to obtain a pH of 7). 700 μl of 150 mM Tris HCl buffer containing 2 mM EDTA, pH 7.5, was added to the preceding solution and the mixture incubated for 5 min at room temperature. The following solutions were then added: 100 μl of 2 mM NADPH2, 10 μl of 20 U/ml of glutathione peroxidase, and 100 μl of 5 mM reduced glutathione. Tubes were shaken and incubated for 10 min at room temperature. Absorbance of the solution (ABS1) was then read at 340 nm.

10 μl of 200 U/ml of glutathione reductase was added and the tubes were again shaken and incubated for 10 min at room temperature. Absorbance of the solution (ABS2) was again read at 340 nm. Calibration was performed with tBHP (0, 15, 30, 60, 120 μmoll/l) in a solution of 65 g/l of bovine albumin containing 0.9% NaCl and treated in a similar fashion.

The calibration curve was drawn as follows: the abscissa indicates the concentrations of tBHP as a function of the difference between the two absorbances (ABS1-ABS2), minus the difference from the absorbance of the blank. DABS = (ABS1-ABS2 of assay) (ABS1-ABS2 of blank). (For more details, see [21]).

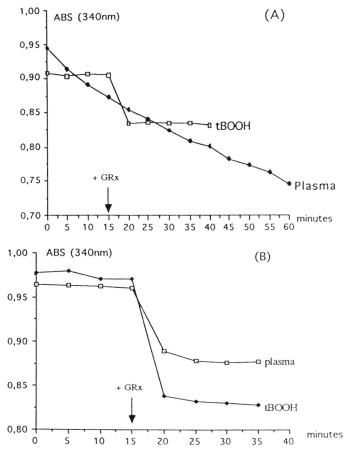

Fig. 2. Time-course of the absorbance of the reaction mixture without (A) and with (B) treatment of samples with metaphosphoric acid. (A) The assay was carried out using the Heath and Tappel method by monitoring the disappearance of NADPH2 during the second and third incubation steps on a plasma sample and on a solution of 120 μM tBHP. (B) As in (A) but after treatment of sample and standard solution with 6% metaphosphoric acid as described in the deproteinization section from [21].

It should be noted that inactivation of endogenous enzymes with 6% metaphorsphoric acid and neutralization of the supernatant with molar NaOH enabled us to achieve a stable absorbance in the first step of incubation as well as a clear termination of the reaction at the end of the 10 min incubation during the third stage of the reaction (Fig. 2).

Optimized procedure of TBARS assay
In a polypropylene test tube, 0.10 ml of assay specimen and 0.75 ml of kit working solution were mixed. In order to inhibit lipoperoxidation

when measuring native TBA reactants, $10\,\mu l$ of a 2% solution of butylated hydroxytoluene (BHT) was added. After vortexing tubes were tightly capped and placed in a controlled 95°C water bath for 60 min. They were then chilled in an ice bath. Reagent and assay blanks were left at room temperature. Two ml of butanol were added to each tube, the TBARS complex was extracted by shaking, and the phases were separated by centrifugation. Fluorometric determination of the TBA reactive substances (TBARS) complex in the n-butanol extract was done at excitation wavelength of 532 nm and at emission wavelength of 553 nm. The calibration was performed with TEP solutions $(0, 1, 2, 4, 6, 8\ \mu mol/1)$ as before.

Validation of the LHP assay

Detection limit
To evaluate the limit of detection, we used the protocol proposed by the French Clinical Biology Society (SFBC) "Validation of techniques" committee (ISB 1989, 15, Nr. 3). We measured the DABS of the blank (point 0 of the calibration range) 30 times in a single run. The calculation was made using the formula:

$$\text{limit of detection (LD)} = K \times (SDb/s),$$

where SDb is the standard deviation (0.0022) of the blank (n = 30), s is the initial slope of the curve (0.0008), and K is the square root of $1/nb + 1/ny \times 2.325$, where nb is the number of measurements on blanks (for routine use = 1), and ny is the number of measurements on samples (for routine use = 1), i.e. K = 3.3. Therefore LD = (0.0022/0.0008) × 3.3 = 9 μM.

Linearity
We carried out this test on increasing concentrations of tBHP (0 to 360 μmol/1). Figure 3 shows the mean of the deltaABS obtained for each point on the calibration curve, each measured three times. The curve is linear up to at elast 360 μmol/1 of hydroperoxide (r = 1). Plasma samples (n = 6) were also diluted by 50% and 25% in a solution of albumin (60 g/1), and LHP was assayed in both pure and diluted samples. Results were as follows: undiluted plasma, 66 ± 8.8 μM; 50% diluted plasma, 33.3 ± 4 μM; and 25% diluted plasma, 17 ± 2.8 μM.

Precision
Within-run precision was evaluated on known concentrations of tBHP and on plasma samples. We carried out 20 determinations on the same plasma sample and on solutions of 60 and 120 μmol/1 tBHP. The determinations were made on the same day and in the same run. Results

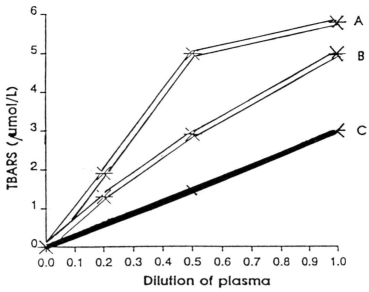

Fig. 3. Influence of assay specimen volume and their dilutions on linearity (Specimens were A = 0.5, b = 0.2; C = 0.1 ml) (From [9]).

expressed as deltaABS (difference between the first and second reading) are given in Table 1. Within-run precision was satisfactory, the CV being about 5%. Between-run precision: We determined the level of lipid hydroperoxides under the conditions already described day after day for 6 days on 10 different plasmas. The means of the six determinations of three of the plasmas studied are given in Table 1 in terms of deltaABS. Similarly, we carried out the determination at two points on the concentration range, 60 and 120 μmol/1 tBHP. Between-run precision

Table 1. Within-run (WR) and between-run (BR) precision of LHP and TBARS optimized methods

LHP (Delta $ABS_{340nm} \times 1000$)				TBARS (μmol/l)			
Medium	Precision	Mean $\pm$ SD	CV, %	Medium	Precision	Mean $\pm$ SD	CV, %
Plasma	WR	74 ± 4.7	6.3	Low Plasma	WR	2.12 ± 0.04	1.8
	BR	56 ± 3.4	6.1		BR	2.08 ± 0.07	3.3
Standard							
60 μmol/l	WR	64 ± 3.7	5.7	Normal plasma	WR	2.77 ± 0.06	2.1
	BR	67 ± 2.8	4.2		BR	2.76 ± 0.09	3.2
120 μmol/l	WR	101 ± 5.1	5.0	High plasma	WR	3.41 ± 0.11	3.2
	BR	102 ± 4.7	4.5		BR	3.38 ± 0.16	4.4

Table 2. Percent recovery of LHP and TBRAs obtained with the optimized methods

LHP (μmol/l)			TBARS (μmol/l)		
LHP added	LHP detected	Recovery, %	MDA added	MDA detected	Recovery, %
0	60	—	0	2.70	—
75	132	96	0.25 μmol/l	2.95	100
0	65	—	0.50	3.20	100
150	215	100	1.00	3.64	94
0	43	—	2.50	4.90	88
250	283	96	5.00	6.80	82

was good both for the standard solutions and for the plasma samples. The coefficients of variation of the plasma samples ranged from 4 to 7.5% (mean 6%).

Accuracy

We calculated the percentage of recovery of different concentrations of tBHP added to eight different plasma samples. The amount of hydroperoxides before and after addition of tBHP was measured for each sample. The percentage of tBHP recovered with this technique ranged from 84 to 108% (Tab. 2).

Stability of plasma LHP

We assayed three pools of plasma collected into lithium heparinate-containing tubes, fractionated, and stored at $+22$, $+4$, -20, and $-80°C$. Lipid hydroperoxides were assayed on D0, D1, D2, D4, D10, D15, and D30. The stability was satisfactory for several weeks at $-80°C$, for 10 days at $-20°C$, and for 2 days at $+4°C$. On the other hand, at room temperature stability was satisfactory only for a few hours. Thus, it is preferable to treat samples as rapidly as possible or store them at $-80°C$ for a maximum of 2 months.

Reference values

The reference range was determined on blood samples drawn from healthy volunteers into vacutainer tubes by the blood transfusion centre or from laboratory personnel. Seven ml glass tubes (Becton Dickinson),

Table 3. Reference values of plasma LHP and TBARS levels obtained using the optimized methods. Assays were performed on blood samples drawn into tubes containing lithium heparinate as an anticoagulant

LHP (μmol/l)				TBARS (μmol/l)			
n	Mean	SD	M $\pm$ 2SD	n	Mean	SD	M $\pm$ SD
30	54.0	9.01	36.0–72.0	32	2.51	0.24	2.03–2.99

either dry or containing lithium heparinate or EDTA K3 were used. After centrifugation, samples were analysed the day of collection under the conditions defined earlier. Results obtained with blood drawn into potassium EDTA were significantly lower than those obtained with serum or blood drawn into lithium heparinate (Tab. 3).

Analytical performance of TBARS assay

Detection limit
The detection limit (DL) was determined as described by Gatautis and Pearson [22]. A sample containing MDA at a concentration three to five times that of the reagent blank was measured 10 times and the detection limit was calculated with the formula:

$$DL = (2*SD*(c))/S,$$

where S is the mean of fluorescence measurements, SD is the corresponding standard deviation and c the concentration of the tested solution. The detection limit obtained with our kit was 0.11 μmol/1. This shows the excellent sensitivity of the proposed method. It is sufficient to enable the method to be applied, for example, with isolated human lipid fractions. Sensitivity decreased if the water used to prepare the standards and samples was not deionized.

The concentrations of the TBA-perchloric acid reagent played a fundamental role in determining the linearity of the method when used with plasma samples. Linearity was acceptable only for a sample/reagent ratio of 1/7.5 (V/V) (Fig. 3). This role was minimized when a standard solution was used. The correlation coefficient of the regression line (r = 0.9996; p = 0.0001) in the TEP range from 0 to 8 μmol/1 was excellent and thereby enabling TBARS to be assayed in most samples of human plasma. Linearity remained satisfying when TEP concentrations were 10 times higher (r = 0.982; p = 0.005).

Linearity
Linearity was established by using the correlation coefficient according to E.E.C. instructions (Additive to Directive 75/318/EEC, August 1989). The standard calibration solutions (1, 2, 3, 4, 5, 6 and 8 μmol/l) were determined in triplicate. Linear regressions and the correlation coefficient were then calculated. A similar study was carried out with a high concentration range (5, 20, 30 and 60 μmol/l).

Precision
In order to determine between-run and within-run precisions, aliquots of plasma from a control subject were frozen at $-20°C$ and were thawed only before analysis. Within-run precision was calculated from

15 assays done on the same day. Between-run precision was calculated from 15 assays done over a period of 30 days. Within-run precision and between-run precision over 30 days were acceptable (Tab. 1). During the assay period, plasma samples were stored at $-20°C$, with no freeze-thaw cycles, in trace element-free tubes. This guaranteed the stability of the plasma samples and precluded any *in vitro* lipoperoxidation.

Accuracy
Accuracy was measured by evaluating the recovery of standard additions. Known quantities of the 10 μmol/l standard solution were added to plasma from a healthy subject before adding the TBA-acid reagent. After homogenizing the sample, TBARS were measured as described above. The standard recovery was excellent when the final concentration of MDA was lower than 2.5 μmol/l (Tab. 2). There are however relatively few published data on this aspect. As seen above, the recovery of standard additions was satisfactory only when the plasma/MDA-acid ratio was 1/7.5 (V/V).

Physiological human plasma TBARS values
Normal values were established with 32 healthy subjects between 20 and 40 years of age (14 males, 18 females). The normal range (2.51 $\pm$ 0.25 μmol/l) determined in 32 normal subjects showed that there existed no sex-related difference for TBARS (Tab. 3).

Discussion

Lipoperoxidation is a biochemical process shared by a number of different phenomena, either physiologic (phagocytosis, mitochondrial respiration, platelet activation, etc.) or pathologic (atherosclerosis, ischemia, chronic infections, hemodialysis, etc.) [23]; and [1–6]. The detection of oxidation of polyunsaturated fatty acids *in vivo* and the assay of primary and secondary molecules released (whether or not responsible for tissue lesions) require the development of quantitative methods satisfying the fundamental analytical criteria of within- and between-run precision, sensitivity, specificity and accuracy. The thiobarbituric acid (TBA) test is the method most often used to quantify lipid peroxidation. Among the numerous fluorometric methodologies proposed, only a few studies have evaluated these analytical criteria.

In the absence of standard assay methods, laboratory investigation of free radicals is unsatisfactory at the present time. Indeed, many disorders are related either directly or indirectly to overproduction of free oxygen radicals and, consequently, accelerate the peroxidative process. These phenomena can be involved in the aetiology of certain disorders. The real impact of this overproduction is much debated and poorly clarified. Potential markers of the occurrence and extent of lipid peroxidation, though numerous, are limited by the practical requirements of a routine laboratory determination. However, assay of LHP and TBARS seems appropriate.

Various methods have been proposed for LHP assay using techniques that are either simple such as titration or very complex such as mass spectrometry. In this regard, the works of Khoda et al. [24] have shown that the coupled enzymatic technique using glutathione reductase as described by Heath and Tappel initially suffers from various interferences that make it unsuitable for assay of plasma LHP. In fact, this technique was originally developed on pure lipid products in simple media, and its application in complex biological media such as plasma can be compromised by endogenous enzymes that can interfere by participating in the reactions used in the assay. We therefore developed an improved protocol involving inactivation of the endogenous enzymes implicated in the continued decrease in absorbance during the second incubation period. In addition, the deproteinization could also eliminate other enzymes likely to use $NADPH2$ or $NADP+$ as a cofactor and thus interfere with the assay. Such enzymatic systems are known to be very numerous in biological media like plasma. The type of acid used for this inactivation is of prime importance. Inactivation of endogenous enzymes with 6% metaphosphoric acid and neutralization of the supernatant with molar NaOH enabled us to achieve a stable absorbance in the first step of incubation as well as a clear termination of the reaction at the end of the 10 min incubation during the third stage of the reaction.

Under the new experimental conditions, the different parameters of the assay were optimized. The limit of detection calculated according to the recommendations of the SFBC Validation Committee is 9 μmol/l. This relatively high detection limit can be lowered to 4 μmol/l by carrying out two readings instead of one for both samples and reagent blanks. The calibration curve is perfectly linear up to at least 360 μmol of LHP/l. The new protocol gives good within-run precision with a coefficient of variation of 6.3% as well as good between-run precision with a CV of 6%. In addition, the percentage of recovery of the different concentrations of tBHP added to the plasma is 97 $\pm$ 6% for concentrations up to 150 μmol/l and 87 $\pm$ 7% for concentrations of 250 μmol/l. This percentage of recovery is very satisfactory. The stability of plasma hydroperoxides is good at $-80°C$ up to 30 days, at $-20°C$ up to 10 days, while it is 2 days at $+4°C$. At room temperature, it is stabile only for a few hours. It is thus advisable to treat the plasma samples as rapidly as possible.

Lastly, the reference range of LHP was determined on blood samples collected from healthy volunteer donors. The mean obtained varied as a function of the anticoagulant used. EDTA tripotassium gave significantly lower values than those obtained with lithium heparinate or serum. EDTA is believed to be responsible for the degradation of a part of LHP when it is linked to a transition metal, in particular iron. It is thus recommended to assay LHP on blood samples drawn into lithium heparinate. This study shows clearly that elimination of endogenous enzymes from the reaction medium when assaying hydroperoxides in human plasma gives a simple, reliable, and easily reproducible procedure.

The measurement of TBRAs is still a more widely used test of lipid peroxidation. Its main advantage is its capacity to detect many kinds of peroxidation products and internadiates, but its specificity is rather low since various substances not related to the lipid peroxidation process could also react with the TBA during the test procedure [25]. Moreover, it is known that MDA molecules undergo self-condensation reactions, yielding polymers of variable molecular weight and polarity. They can be hydrolyzed in acid medium and are heat-labile. In order to insure satisfying linearity of the method, it is necessary to hydrolyze TEP directly in the presence of TBA so that the adduct can form as soon as the polymer is hydrolyzed [26].

Linearity measured in standard solutions was excellent in the assay range corresponding to plasma concentrations. At higher values, the regression coefficient was 0.982 up to 6 μmol/l. It was also found that the reaction was linear only in very defined analytical conditions [sample/reagent ratio 1/7.4 (V/V)], which can be explained by prior results. Thus, MDA can act as a nucleophilic or electrophilic compound. Its reactivity is such that it can non-specifically and covalently bind to various biological molecules in the samples (proteins, nucleic acids, aminophospholipids, etc.) [18, 27]. The response of the TBA test is thus both a function of the conditions of hydrolysis of these complexes (pH, type of acid, temperature) and the conditions of 1,2-MDA-TBA adduct formation [18, 28, 29]. In agreement with the results of Wong et al. [30], we have shown that there exists optimal conditions for adduct formation which require an optimum TBA concentration, as well as for the precipitation of plasma proteins or heating time. Also, in agreement with Wong et al. [30], we noted that it is more important to define these conditions for the samples than for the standards. When the method will be applied to tissue homogenates, it will be necessary to redefine optimal assay conditions. In addition, whenever the measured values are higher than 10 μmol/l (*in vitro* peroxidation studies, patients treated with bleomycin), it is preferable to dilute the sample. The pH of the acid mixture is 1, equivalent to optimal conditions of complex formation [31].

The detection limit of the technique is 0.11 μmol/l, which means that the method is applicable to TBARS assay in urine and lipoprotein fractions and cultured cells which contain small quantities of TBARS [9, 32].

The within- and between-run precisions of the method are satisfactory, with a coefficient of variation in the range of 1.8 to 4.4%. They are better than those reported by Yagi [33] varying from 6.5% in our experience to 12% for Conti et al. [34]. One of the advantages of using this kit for assaying biological samples when following patients is the excellent recovery. Some authors [29] have reported limitations of their fluorometric method, with recovery of additions in the range of 55 to 61%. Our study has shown that satisfactory recovery of standard additions requires very strict assay conditions, i.e., a sample/reagent ratio of 1/7.5 (V/V).

The specificity of the method was tested using the same reaction mixture with plasma from healthy subjects. The MDA-TBA complex was determined by high performance liquid chromatography (HPLC) with a method adapted from Knight [27]. The linear regression between the values measured with HPLC and those determined by fluorometry yielded a correlation coefficient of $r = 0.80$; $p = 0.0001$. These results show, thus, a good correlation between the two methods. The values generated by fluorometric analysis were significantly higher than those determined by HPLC. They confirm previous works showing that fluorometric methods give an overall view of peroxidation and that other aldehydes can also yield fluorescent complexes with TBA.

Finally, this simple, rapid, reproducible and sensitive assay is adapted to screening patients who may be subjected to oxidative stress. It can be adapted by laboratories for the routine monitoring of their patients. Fluorometry remains necessary to prevent possible interference by molecules such as glucose [31] or bilirubin [35]. As the lack of specificity of the TBA reaction could lead to misinterpretation of increased lipid peroxidation in studies of human disorders, the more-specific HPLC methods should be used to confirm *in vivo* lipid peroxidation. Aside from the need for special equipment and the limitation of detector sensitivity, considerable time is required. The ease of use of the MDA-kit, the ability to simultaneously process many derivatized samples, and the speed with which the derivative, once formed, can be quantified are practical reasons for supporting its use in screening and monitoring lipid peroxidation in human disorders.

In conclusion, if we are positive that oxidative stress occurs in most human diseases, our understanding of the role played by ROS in disease pathology is still weak, largely because of the lack of accurate methods applicable to human patients. Our present ignorance of the signifance of oxidant stress status in human health arises from the substantial gaps in our analytical capabilities and in our abilities to clearly interpret the data. For example, for following the general change in lipid peroxidation, it is appropriate to use a variety of methods for cross-checking purposes. This means that methods and hypotheses always need improvement. In this scenerio, the amelioration and the standardization of already existing methods should contribute to the advancement of research in this field.

Acknowledgements

This work was supported in part by Région Rhone-Alpes, Research Program G.B.M. 1991. The authors thank Catherine Mangournet, Colette Augert and Jacqueline Meo for technical assistance.

References

1. Halliwell, B. and Cross, C.E. (1991) Reactive oxygen species, antioxidants, and acquired immunodefeciency syndrome. *Arch Intern. Med.* 151: 29–31.
2. Ames, B.N. (1989) Endogenous oxidative DNA damage, aging and cancer. *Free Rad. Biol. Med.* 7: 121–128.
3. McCord, J.M. (1985) Oxygen derived free radicals in postischemic tissue injury. *N. Engl. J. Med.* 312: 159–163.
4. Flohé, L. (1988) Superoxide dismutase for therapeutic use: clinical experience, dead ends and hops. *Mol. Cell. Biochem.* 84: 123–131.
5. Richard, M.J., Arnaud, J., Jurkovitz, C., Hachache, T., Meftahi, H., Laporte, F., Foret, M., Favier, A. and Cordonnier, A. (1991) Trace elements and lipid peroxidation abnormalities in patients with chronic renal failure. *Nephron* 57: 10–15.
6. Harman, D. (1988) Free radicals in aging. *Mol. Cell. Biochem.* 84: 155–161.
7. Halliwell, B., Gutteridge, J.M.C. and Cross, C.E. (1992) Free radicals, antioxidants and human disease; where are we now? *J. Lab. Clin. Med.* 119: 598–620.
8. Pacifici, R.E. and Davies, J.A. (1991) Protein, lipid and DNA repair systems in oxidative stress: the free-radical theory of aging revisited. *Gerontology* 37: 166–180.
9. Richard, M.J., Portal, P., Meo, J., Coudray, C., Hadjian, A. and Favier, A. (1992a) Malondialdehyde kit evaluated for determining plasma and lipoprotein fractions that react with thiobarbituric acid. *Clin. Chem.* 38: 704–709.
10. Hunter, M. and Mohamed, J. (1987) Plasma antioxidants and lipid peroxidation products in Duchenne muscular dystrophy. *Clin. Chem. Acta.* 155: 123–132.
11. Gutteridge, J.M.C. and Halliwell, B. (1990) The measurement and mechanism of lipid peroxidation in biological systems. *TIBS* 15: 129–135.
12. Gebicki, J. and Guille, J. (1989) Spectrophotometric and high-performance chromatographic assays of hydroperoxides by iodometric technique. *Anal. Biochem.* 176: 360–364.
13. Cramer, G., Miller, J., Pendleton, R. and Lands, W. (1991) Iodometric measurement of lipid hydroperoxides in human plasma. *Anal. Biochem.* 193: 204–211.
14. Ohishi, N., Ohkawa, H., Miike, A., Tatano, T. and Yagi, K. (1985) A new assay method for lipid peroxides using a methylene blue derivative. *Biochem. Inter.* 10: 205–211.
15. Cathcart, R., Schwilers, E. and Ames, B. (1983) Detection of picomole levels of hydroperoxides using a fluorescent dichlorofluorescein assay. *Anal. Biochem.* 134: 111–116.
16. Heath, R. and Tappel, A.H. (1976) A new sensitive assay for the measurement of hydroperoxides. *Anal. Biochem.* 7: 184–191.
17. Allen, K., Hung, C. and Morin, C. (1990) Determination of picomole quatities of hydroperoxides by a coupled glutathione peroxidase and glutathione reductase and glutathione disulfide specific glutathione reductase assay. *Anal. Biochem.* 186: 108–111.
18. Janero, D. (1990) Malondialdehyde and thiobarbituric acid-reactivity as diagnostic indices of lipid peroxidation and peroxidative tissue injury. *Free Rad. Biol. Med.* 9: 515–540.
19. Nair, V. and Turner, G. (1984) The thiobarbituric acid test for lipid peroxidation: structure of the adduct with malondialdehyde. *Lipids* 19: 804–805.
20. Richard, M.J., Guiraud, P., Meo, J. and Favier, A. (1992b) High-performance liquid chromatographic separation of malondialdehyde-thiobarbituric acid adduct in biological materials (plasma and human cells) using a commercially available reagent. *J. Chromatogr.* 577: 9–18.
21. Hida, H., Coudray, C., Mangournet, C. and Favier, A. (1994) Improved enzymatic assay for plasma hydroperoxides: inactivation of interfering enzymes. *Ann. Biol. Clin.* 52 (9): 639–644.

22. Gatautis, V. and Pearson, K.H. (1987) Separation of plasma carotenoids and quantitation of beta carotene using HPLC. *Clin. Chem. Acta* 166: 195–206.
23. Halliwell, B. (1989) Free radicals, reactive oxygen species and human disease: a critical evaluation with special reference to atherosclerosis. *Br. J. Exp. Path.* 70: 737–757.
24. Kohda, K., Arisue, K. and Maki, A. (1982) The enzymatic determination of lipid hydroperoxides in serum. *Jpn. J. Clin. Chem.* 11: 306–313.
25. Halliwell, B. and Gutteridge, J.M.C. (1989) Lipid peroxidation: a radical chain reaction. *In*: *Free Radicals in Biology and Medicine*. Claredon Press, Oxford, pp 188–276.
26. Gutteridge, J.M.C. (1975) The use of standards for malondialdehyde. *Anal. Biochem.* 69: 518–526.
27. Knight, J., Pieper, R. and McClellan, L. (1988) Specificity of the thiobarbituric acid reaction: its use in studies of lipid peroxidation. *Clin. Chem.* 34: 2433–2438.
28. Nair, V., Cooper, C.S., Vietti, D.E. and Turner, G.A. (1986) The chemistry of lipid peroxidation metabolites: crosslinking reactions of malondialdehyde. *Lipids* 21: 6–10.
29. Hackett, C., Linley-Adams, M., Lloyd, B. and Walker, V. (1988) Plasma malondialdehyde: a poor measure of *in vivo* lipid peroxidation. *Clin. Chem.* 34: 208.
30. Wong, S., Knight, J., Hopfer, S., Wong, S.H.Y., Knight, J.A., Hopfer, S.M., Zahavia, O., Leach, C.N. and Sunderman, F.W. (1987) Lipoperoxides in plasma as measured by liquid-chromatographic separation of malondialdehyde-thiobarbituric acid adduct. *Clin. Chem.* 33: 214–220.
31. Bird, R.P. and Draper, H.H. (1984) Comparative studies on different methods of malondialdehyde determination. *Meth. Enzymol.* 105: 299–305.
32. Draper, H.H., Polennsek, L., Hadley, M. and McGirr, L.G. (1984) Urinary MDA as an indicator of lipid perxidation in the diet and tissue. *Lipids* 19: 836–843.
33. Yagi, K. (1976) A simple fluorometric assay for lipoperoxide in blood plasma. *Biochem. Res.* 15: 212–216.
34. Conti, M., Morand, P.C., Levillain, P. and Lemonnier, A. (1990) Methode simple et rapide de dosage du malondialdéhyde. *Acta. Pharm. Biol. Clin.* 5: 365–368.
35. Okawa, H., Ohishi, N. and Yagi, K. (1979) Assay for lipid peroxides in animal tissues by thiobarbituric acid reaction. *Anal. Biochem.* 95: 351–358.

Analysis of Free Radicals in Biological Systems
Favier et al. (eds)
© 1995 Birkhäuser Verlag Basel/Switzerland

Measurement of low-density lipoprotein oxidation

I. Hininger[1], A. David[2], F. Laporte[2], A.-M. Roussel[1], T. Foulon[2],
P. Groslambert[2] and A. Hadjian[3]

[1]*Laboratoire de Biochimie de la Faculté de Pharmacie, GREPO, Domaine de la Merci,
F-38706 La Tronche;*
[2]*Laboratoire de Biochimie A, and*
[3]*Laboratoire d'Enzymologie Centre Hospitalier Universitaire, BP 217,
F-38043 Grenoble Cédex 9, France*

Summary. Oxidative modification of low-density lipoprotein leads to enhanced uptake by macrophages and hence to the formation of atherosclerotic lesions. Low-density lipoprotein oxidizability can be determined *in vitro* by several methods. We describe here a rapid method for the isolation of low-density lipoprotein by density gradient ultracentrifugation. Cu^{2+}-catalyzed oxidation of low-density lipoprotein is then performed and the rate of conjugated diene formation is monitored continuously by the change in absorbance at 234 nm. This method provides useful information for the evaluation of individual susceptibility of low-density lipoprotein to oxidation and of the protection afforded by antioxidant molecules.

Introduction

Recent reports have clearly established the essential role of oxidatively-modified low-density lipoprotein (ox-LDL) in the etiology of atherosclerosis [for review see 1–3]. Ox-LDL is no longer recognized by the LDL receptor but binds with a high affinity to the macrophage scavenger receptor [4] and in some cases to the Fc receptor via immune complex formation with autoantibodies against ox-LDL [5]. The accumulation of cholesteryl esters into macrophages exposed to ox-LDL, leading to foam cell formation in the arterial intima, is thought to be one of the earliest events in atherogenesis. Many methods have been developed to monitor LDL oxidation *in vitro*: These were designed to elucidate the mechanism of the LDL modifications resulting in their recognition by the scavenger receptor, and also to study the metabolic pathways leading to protection against the oxidative process.

We will focus here on *in vitro* methods using isolated LDL. It should be kept in mind that LDL is also protected *in vivo* by water-soluble antioxidants and by other lipoprotein particles, especially high-density-lipoprotein [6]. Therefore, establishing a causal relationship between *in vitro* measurements of LDL oxidizability and predisposition to atherosclerotic disease is difficult.

LDL Structure and composition

LDL is derived from the conversion of VLDL through the lipoprotein lipase activity. This lipoprotein population is defined by a density within the range of 1.006 to 1.063 g/ml and can be further separated, either by density gradient ultracentrifugation [7] or by electrophoretic migration in a 2–20% acrylamide gel [8, 9] into three or more subclasses differing in size, density, antioxidant content and susceptibility to oxidative modification. In electron microscopy, LDL appears as a spherical particle with a mean diameter of 25 nm. Each particle is made up of a hydrophobic core containing mainly cholesteryl esters and triglycerides, surrounded by free cholesterol and a phospholipid monolayer. A molecule of apolipoprotein B, a 550 kDa glycoprotein, is linked by hydrophobic interaction to the polar head groups of the phospholipids. Moreover, LDL is loaded with lipid-soluble antioxidants, ubiquinol, carotenoids and mainly α-tocopherol, whose concentrations vary between individuals. A phospholipase A2 is also associated with the lipoprotein particle [10, 11].

Oxidative modifications of LDL

– Depletion of endogenous antioxidants is the first observable modification that takes place before lipid peroxidation. Depletion of α-tocopherol precedes the disappearance of carotenes [12].
– Lipid peroxidation takes place after all the antioxidants are consumed. This radical chain reaction is thought to be a common initiating step in the oxidative modification of LDL. Briefly, the initiation step consists of a hydrogen abstraction from a polyunsaturated fatty acid, mainly linoleic acid and arachidonic acid. The resulting carbon-centered radical is stabilized by a rearrangement of the double bond to form a conjugated diene. Under aerobic conditions, this radical combines with an oxygen molecule, to form a lipid peroxyl radical, $ROO^{\bullet}$. Abstraction of a hydrogen atom from another polyunsaturated fatty acid by the peroxyl radical leads to the propagation stage. In the decomposition step, the lipid peroxide generates a variety of breakdown products including aldehydes.
– Phospholipids hydrolysis: a large amount (up to 50%) of the phosphatidylcholine is hydrolyzed into lysophosphatidylcholine through the action of the PAF acetyl hydrolase associated with LDL.
– Protein damage: apolipoprotein B undergoes a series of modifications: on the one hand, oxygen-derived radicals induce nonenzymatic peptide bond disruption [13], and the resulting fragmentation of the protein alters the secondary and tertiary structure of the protein. This can be measured by circular dichroism spectropolarimetry as a reduc-

tion of helicity of the apoprotein [14]. On the other hand, covalent binding of aldehyde products from fatty acids to ε-amino group of the lysine residues precludes the recognition of the lipoprotein by the native LDL receptor. The resulting neutralization of positive charge, alters the electrophoretic mobility of the protein.

Effects of naturally occurring antioxidants

α-Tocopherol is the most abundant antioxidant in LDL. Each particle contains from 3 to 12 moles of α-tocopherol. It has been clearly established that vitamin E supplementation *in vivo* correlates with a greater resistance of LDL to *in vitro* oxidation [15], whereas no correlation was found between LDL oxidizability and the tocopherol content of the lipoprotein in nonsupplemented subjects [16, 17]. The protective effect against lipid peroxidation depends upon the nature of the radical-generating system used [18]. Furthermore, the ability of α-tocopherol to prevent LDL oxidation is dependent on the polyunsaturated fatty acid content of the lipoprotein [19]: these authors have shown that the amount of tocopherol carried by LDL is related to the amount of PUFA per particle. Other antioxidants are much less abundant (0.3 mole of β-carotene and 0.1 mole of ubiquinol-10/mole LDL), so that the molar ratio of the antioxidants to the PUFA is very low, about 1/100, indicating that LDL should be protected in the blood by water-soluble antioxidants, namely ascorbate which is able to recycle vitamin E by reduction of the tocopheroxyl radical [20, 21], and probably by HDL when the LDL particle is transferred into the intracellular space of the vascular wall [6].

Isolation of LDL

Isolation of LDL for oxidation experiments should be performed with appropriate precautions in order to minimize lipid peroxidation. Blood samples and subsequent operations must be carried at 4°C and quickly processed in the presence of a metal chelator and/or in a nitrogen atmosphere. Kleinveld et al. [17] reported that saccharose (final concentration 0.6%) prevented LDL aggregation and supplemented plasma can be stored at −80°C for at least 1 month without any change in the oxidizability indices determined by the conjugated diene method. Ethylenediamine tetraacetic acid (EDTA) is commonly used because it acts as an anticoagulant and as an antioxidant. Butylated hydroxytoluene (BHT) was shown to noticeably affect the lag time in oxidation experiment when added to the plasma samples [17]. Some researchers used heparin in place of EDTA as anticoagulant to avoid the time-consuming dialysis [18, 22, 23].

Attempts to reduce the duration of the centrifugation step has initiated much research. A wide variety of single-run or two-step sequential ultracentrifugation protocols has been proposed, but a decisive improvement was provided by the emergence of tabletop ultracentrifuges capable of reaching 400 000 g within a few minutes. Nevertheless, two pitfalls have to be avoided:

(1) The protocol used should preserve the endogenous antioxidants [17];
(2) Contamination of the lipoprotein with plasma antioxidants and especially plasma protein should be minimized. Whatever the method chosen, it is important to ascertain the absence of albumin in the LDL preparation.

Generally, KBr is used to bring the solutions to the desired density. Alternatively, a method using deuterium oxide in place of KBr was recently proposed [24], thus avoiding LDL alterations induced by high ionic strength.

Quantitation of LDL

Due to the variability of LDL composition and size, the best measure seems to be the determination of apolipoprotein B, since one mole of protein is associated with each LDL particle [16]. Nevertheless, for practical or cost arguments, total protein or cholesterol contents are frequently determined.

Measurement of the resistance of LDL to oxidation

LDL oxidation is a complex process and therefore can be measured by a variety of techniques. Each method required two elements: the first one producing the oxidant radical and the second one measuring the oxidation products.

(1) Oxidant generation

Cellular systems. Endothelial cells (aortic or umbilical), smooth muscle cells, monocytes or neutrophils in the presence of catalytic amounts of metal ions or appropriate stimuli have all been shown to be capable of oxidizing LDL *in vitro* [25–27]. Although oxidants produced in these conditions are thought to be more relevant to the physiological process, the amount of oxidant produced per time unit differs greatly owing to the viability of the cells and the variability of the cell response, therefore affecting the reproducibility of the results.

Non-cellular systems. Metals: it is generally recognized that LDL oxidation by ferrous ion is very slow as compared to the efficiency of Cu^{2+}. It has been reported by Esterbauer et al. [28] that Cu^{2+} strongly binds to LDL to at least two distinct sites. It is suggested that the binding of Cu^{2+} to the LDL is essential for the initiation of lipid peroxidation and this has been proven by the inhibitory effect of EDTA, which if present in sufficiently high concentration, prevents binding of Cu^{2+} to LDL [29]. It is therefore important to carefully remove EDTA from LDL fraction.

This can be done either by extensive dialysis or by gel filtration.

The molar ratio of Cu^{2+} to LDL rather than the Cu^{2+} concentration exerts an influence on the lag phase duration and on the rate of diene formation. In our experience, 3 to 5 nanomole of Cu^{2+} for 0.05 to 0.15 g apo-B seems to be optimal.

Cu^{2+}-mediated oxidation requires the presence of trace amounts of preformed hydroperoxides or the reduction of the metal ion [18].

Azo-initiators: another way of generating oxidizing radicals is the thermal decomposition of azo reagents [30]. These compounds decomposed at a temperature-controlled rate to produce radicals which react with an oxygen molecule to yield peroxyl radicals. These peroxyl radicals are capable of abstracting a hydrogen atom from unsaturated fatty acids. In a recent publication, Frei and Gaziano [18] have shown that the kinetic of diene formation depends on the system used to produce radicals, indicating that LDL could be oxidized by different mechanisms.

The chemical process involved in LDL oxidation is not clearly understood. Lynch and Frei [22] have shown that exposure to O_2^- or H_2O_2 in the absence of Cu^{2+} does not result in oxidative modifications of LDL whereas superoxide dismutase partially inhibited Cu^{2+}-dependent LDL oxidation. Moreover, hydroperoxides were undetectable in their native LDL preparation and it has been reported that hydroxyl scavengers does not inhibit cell-mediated LDL oxidation [31]. It could be hypothesized that the extremely reactive hydroxyl radical is formed in the immediate vicinity or in the LDL particle and is therefore unavailable for the water-soluble scavenger.

(2) Detection and quantitation of the lipid peroxidation products

Measurements of lipid hydroperoxides, thiobarbituric acid reactive substances or other aldehydes, fluorescent products of lipid peroxidation and conjugated dienes are frequently used. The latter is easy to perform, is carried out without previous lipid extraction and can be quantified in absolute values. The increase of the 234 nm diene absorbtion reflects the formation of the conjugated double bonds. Kinetic data obtained by

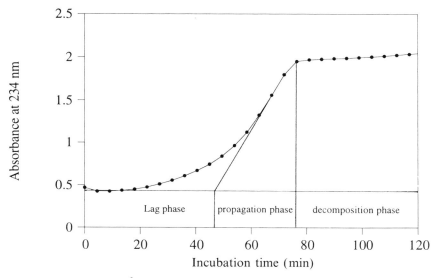

Fig. 1. Kinetic of the Cu^{2+}-induced formation of conjugated dienes. The LDL fraction (0.15 mg apo B in 1 ml PBS) was incubated in the presence of 3 μmole/l CuCl$_2$ at 37°C, and the absorbance at 234 nm was measured every 4.5 min.

monitoring the change in the absorption at 234 nm shows three distinct phases (Fig. 1):

– a lag time defined as the interval between the intercept of the tangent to the curve with the initial absorbance axis. During the lag phase the endogenous LDL antioxidants disappear and only minimal lipid peroxidation occurs [16].

The duration of the lag phase is drastically affected by a variation of the temperature and it is therefore essential that measurements are made at constant temperatures. As reported by Puhl et al. [32], an increase in the temperature from 25° to 30°C, for example, reduces the lag phase by about 30%, probably by affecting the rate of hydroperoxyl radical production.

– A propagation period defined as a rapid increase in diene formation. The maximal rate of diene production can be deduced from the maximal slope of the curve using a $\varepsilon_{234} = 29\,500$ M$^{-1} \cdot$ cm^{-1} [12].
– A decomposition phase beginning as the diene formation reaches a plateau. The maximal amount of diene formed is calculated from the difference between the initial and final absorbances. After this time a first decrease is observed by some authors (probably due to the decomposition of conjugated dienes), followed by a slow increase again. This second increase results from the UV absorption of some decomposition products of the dienes in the 234 nm range. [16].

Materials and methods

Principle
LDL isolated by a two-step, short-run ultracentrifugation is oxidized by the Cu^{2+}-mediated reaction and diene formation is monitored by the increase of the absorbance at 234 nm.

Isolation of low-density lipoprotein

Collection and treatment of the samples
After overnight fasting of test subjects, blood samples are withdrawn by venipuncture and collected into evacuated tubes containing EDTA (0.15 mg/ml, final concentration). Tubes are centrifuged at 1400 g for 15 min at 4°C. Generally, plasma is used the same day for low-density lipoprotein (LDL) isolation.

LDL isolation
The separation of plasma lipoproteins requires two sequential short-run ultracentrifugations in order to minimize contamination by adherence of albumin and other plasma proteins. All steps in the LDL isolation procedure are carried out at 4°C. 1.3 ml of plasma is adjusted to the

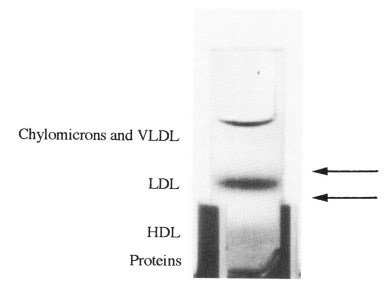

Fig. 2. Separation of plasma lipoproteins by density-gradient ultracentrifugation. Lipoproteins were prestained by addition of 15 µl of Sudan Black (5 g/l in ethyleneglycol) and centrifugation was carried out as described in the text. The arrows designate the slicing positions.

density of 1.220 g/ml by adding 0.426 g of solid KBr. 0.6 ml of the density-adjusted plasma is then layered under 1.4 ml of the isotonic saline solution (d = 1.006 g/ml) in each of two 2 ml polyallomer tubes (Beckman). Underlayering of the plasma sample is made by use of a syringe with a long needle. The tubes are then sealed by heating the plastic with the Beckman device and centrifuged at 430 000 g for 45 min at 4°C. At the end of the run the tubes are carefully removed from the rotor. VLDL are at the top, LDL at the center and HDL and proteins at the bottom of the tube (Fig. 2). The LDL band is readily visible, due to the presence of endogenous carotenoids. To recover the LDL fraction, the tubes are sliced as shown in Figure 2. To minimize the contamination of the LDL fraction with albumin, we proceed to a second run which results in the "flotation" of the LDL. In practice, the LDL preparation collected after the first run is adjusted to a density of 1.063 g/ml by adding solid KBr.

The KBr mass (m) to be added is calculated as follows: $m(g) = 0.042 \times V$, where V is the volume of LDL (in ml) collected after the first ultra-centrifugation. The preparation is then divided into two fractions of equal volume and each fraction is adjusted to 1 ml with the KBr solution (d = 1.063 g/ml) in a thick-walled polycarbonate centrifuge tube (Beckman Instruments). The tubes are then centrifuged for 1 h at 430 000 g at 4°C. Under these conditions albumin is at the bottom of the tube and the LDL fraction which floats on the top is easily recovered by pipetting. Albumin contamination measured on LDL fraction was under the detection limit of the radial immunodiffusion assay (25 mg/l).

Removal of EDTA

PBS containing chloramphenicol (0.1 g/l) as a bacteriocide is purged in a stream of nitrogen for 15 min. To remove EDTA and KBr each collected fraction is diluted with 0.5 volume of PBS and then dialysed against 1 l of oxygen-free PBS and kept at 4°C in darkness under stirring for 2 h. The buffer is changed twice (the one solution for 2 hours and the other one for overnight) to remove EDTA. The EDTA content in the bath and in the LDL fraction was undetectable using Erichrom Black $T + Mg^{2+}$ at the end of the dialysis.

Alternatively, EDTA and salt can be removed by gel filtration chromatography [32].

This EDTA-free LDL stock solution is used for subsequent oxidation studies. The LDL can be stored at 4°C in darkness, a sterile evacuated glass tube for no longer than 5 h. The apo B content is determined by an immunoprecipitation method. When 1.2 ml of a normolipidemic plsama is centrifuged, the final LDL contains about 0.5 mg apo B.

The cholesterol content of the LDL sample can be measured instead of apo B and correlated with the apo B measurement [32].

Oxidation of LDL

Standard procedure

Our procedure is adapted from the method described by Esterbauer et al. [12]. The EDTA-free LDL is diluted in oxygen-saturated PBS buffer pH 7.4 (without chloramphenicol) to give 1 ml containing 0.15 mg of apo B. Oxidation is initiated by the addition of 10 μl of 0.3 mM CuCl$_2$. This solution is gently mixed and then transferred into a 1 ml quartz cuvette. The reference cuve contains 2.5 mM CuCl$_2$ to compensate for the high initial absorbance. The absorbance at 234 nm is recorded every 4.5 min for 180 min at 37°C.

The spectrophotometer is connected to a computer for data acquisition.

Application of the method in biochemistry and clinical chemistry

Because oxidative modification of LDL is likely to be an important step in the initiation and progression of atherosclerosis, this method was used to study the role of several substances in preventing or in modifying LDL:

Estrogens, independently of alteration of the lipid levels, exert an antioxidant effect as demonstrated in postmenopausal women, thus preventing cardiovascular diseases [33]. Previous *in vitro* studies have shown that estrogens inhibited the subsequent modification of LDL by copper ions, monocytes or endothelial cells [34, 35].

It was also reported that gas phase oxidants of cigarette smoke, which represents a risk factor for coronary artery disease, lead to an increased susceptibility of LDL to oxidative modification *in vitro* [36].

Furthermore, the use of this method allows us to study the role of antioxidants such as α-tocopherol or carotenoids present in LDL in preventing LDL oxidation [28].

It would be of interest to investigate other parameters such as the antioxidant defence systems of blood plasma [37, 38], and to evaluate the *in vivo* preformed lipid hydroperoxide in LDL. In this respect, a vast array of methods have been developed and also very different results have been obtained, depending on the sensitivity or on the susceptibility of the techniques to various interferences. A protocol has recently been described [23], which allows the detection of nanomolar amounts of lipid peroxides by chemiluminescence after separation by HPLC. Depending on the overall goal of the study, it is prudent to use more than one method to measure the susceptibility of LDL to oxidation.

References

1. Steinberg, D., Parthasarathy, S., Carew, T.E., Khoo, J.C. and Witztum, J.L. (1989) Beyond cholesterol – Modifications of low-density lipoprotein that increase its atherogenicity. *New Engl. J. Med.* 320: 915–924.

2. Steinbrecher, U.P., Zhang, H. and Lougheed, M. (1990) Role of oxidatively modified LDL in atherosclerosis. *Free Rad. Biol. Med.* 9: 155–168.
3. Witztum, J.L. and Steinberg, D. (1991) Role of oxidized low density lipoprotein in atherogenesis. *J. Clin. Invest.* 88: 1785–1792.
4. Krieger, M. (1992) Molecular flypaper and atherosclerosis: Structure of the macrophage scavenger receptor. *Trends Biochem. Sci.* 17: 141–146.
5. Salonen, J.T., Ylä-Herttuala, S., Yamamoto, R., Butler, S., Korpela, H., Salonen, R., Nyyssönen, K., Palinski, W. and Witztum, J.L. (1992) Autoantibody against oxidised LDL and progression of carotid atherosclerosis. *Lancet* 339: 883–887.
6. Mackness, M.I., Abbott, C. and Durrington, P.N. (1993) The role of high-density lipoprotein and lipid-soluble antioxidant vitamins in inhibiting low-density lipoprotein oxidation. *Biochem. J.* 294: 829–834.
7. De Graaf, J., Hak-Lemmers, H.L.M., Hectors, M.P.C., Demacker, P.N.M., Hendriks, J.C.M. and Stalenhoef, A.F.H. (1991) Enhanced susceptibility to *in vitro* oxidation of the dense low density lipoprotein subfraction in healthy subjects. *Arterioscler. Thromb.* 11: 298–306.
8. Krauss, R.M. and Burke, D.J. (1982) Identification of multiple subclasses of plasma low density lipoproteins in normal humans. *J. Lipid Res.* 23: 97–104.
9. Williams, P.T., Vranizan, K.M. and Krauss, R.M. (1992) Correlations of plasma lipoproteins with LDL subfractions by particle size in men and women. *J. Lipid Res.* 33: 765–774.
10. Stafforini, D.M., Prescott, S.M. and McIntyre, T.M. (1987) Human plasma platelet-activating factor acetylhydrolase. Purification and properties. *J. Biol. Chem.* 262: 4223–4230.
11. Karabina, S.-A.P., Liapikos, T.A., Grekas, G., Goudevenos, J. and Tselepis, A.D. (1994) Distribution of PAF-acetylhydrolase activity in human plasma low-density lipoprotein subfractions. *Biochim. Biophys. Acta* 1213: 34–38.
12. Esterbauer, H., Striegl, G., Puhl, H. and Rotheneder, M. (1989) Continuous monitoring of *in vitro* oxidation of human low density lipoprotein. *Free Rad. Res. Comms.* 6: 67–75.
13. Fong, L.G., Parthasarathy, S., Witztum, J.L. and Steinberg, D. (1987) Non enzymatic oxidative cleavage of peptide bonds in apoprotein B 100. *J. Lipid Res.* 28: 1466–1477.
14. Herak, J.N. (1993) Physical changes of low-density lipoprotein on oxidation. *Chem. Phys. Lipids* 66: 231–234.
15. Abbey, M., Nestel, P.J. and Baghurst, P.A. (1993) Antioxidant vitamins and low-density lipoprotein oxidation. *Am. J. Clin. Nutr.* 58: 525–532.
16. Esterbauer, H., Gebicki, J., Puhl, H. and Jürgens, G. (1992) The role of lipid peroxidation and antioxidants in oxidative modification of LDL. *Free Rad. Biol. Med.* 13: 341–390.
17. Kleinveld, H.A., Hak-Lemmers, H.L.M., Stalenhoef, A.F.H. and Demacker, P.N.M. (1992) Improved measurement of low-density-lipoprotein susceptibility to copper-induced oxidation: application of a short procedure for isolating low-density lipoprotein. *Clin. Chem.* 10: 2066–2072.
18. Frei, B. and Gaziano, J.M. (1993) Content of antioxidants, preformed lipid hydroperoxides, and cholesterol as predictors of the susceptibility of human LDL to metal ion-dependent and independent oxidation. *J. Lipid Res.* 34: 2135–2145.
19. Thomas, M.J., Thornburg, T., Manning, J., Hooper, K. and Rudel, L.L. (1994) Fatty acid composition of low-density lipoprotein influences its susceptibility to autooxidation. *Biochemistry* 33: 1828–1834.
20. Jialal, I. and Grundy, S.M. (1991) Preservation of the endogenous antioxidants in low density lipoprotein by ascorbate but not probucol during oxidative modification. *J. Clin. Invest.* 87: 597–601.
21. Kagan, V.E., Serbinova, E.A., Forte, T., Scita, G. and Packer, L. (1992) Recycling of vitamin E in human low density lipoproteins. *J. Lipid Res.* 33: 385–397.
22. Lynch, S.M. and Frei, B. (1993) Mechanisms of copper- and iron-dependent oxidative modification of human low density lipoprotein. *J. Lipid. Res.* 34: 1745–1753.
23. Sattler, W., Mohr, D. and Stocker, R. (1994) Rapid isolation of lipoproteins and assessment of their peroxidation by high-performance liquid chromatography postcolumn chemiluminescence. *Method. Enzymol.* 233: 469–489.
24. Hallberg, C., Hadén, M., Bergström, M., Hanson, G., Pettersson, K., Westerlund, C., Bondjers, G., Östlund-Lindqvist, A.-M. and Camejo, G. (1994) Lipoprotein fractionation in deuterium oxide gradients: a procedure for evaluation of antioxidant binding and susceptibility to oxidation. *J. Lipid Res.* 35: 1–9.

25. Henriksen, T., Mahoney, E.M. and Steinberg, D. (1981) Enhanced macrophage degradation of low density lipoprotein previously incubated with cultured endothelial cells: Recognition by receptors for acetylated low density lipoproteins. *Proc. Natl. Acad. Sci. USA* 78: 6499–6503.
26. Parthasarathy, S., Steinbrecher, U.P., Barnett, J., Witztum, J.L. and Steinberg, D. (1985) Essential role of phospholipase A2 activity in endothelial cell-induced modification of low density lipoprotein. *Proc. Natl. Acad. Sci. USA* 82: 3000–3004.
27. Scaccini, C. and Jialal, I. (1994) LDL modification by activated polymorphonuclear leukocytes: a cellular model of mild oxidative stress. *Free Rad. Biol. Med.* 16: 49–55.
28. Esterbauer, H., Dieber-Rotheneder, M., Waeg, G., Striegl, G. and Jürgens, G. (1990) Biochemical, structural, and functional properties of oxidized low-density lipoprotein. *Chem. Res. Toxicol.* 3: 77–92.
29. Steinbrecher, P., Parthasarathy, S., Leake, D., Witztum, J. and Steinberg, D. (1984) Modification of low density lipoprotein phsopholipids. *Proc. Natl. Acad. Sci. USA* 81: 3883–3887.
30. Wainer, D.D.M., Burton, G.W., Ingold, K.U. and Locke, S. (1985) Quantitative measurement of the total, peroxyl radical-trapping antioxidant capability of human blood plasma controlled peroxidation. The important contribution made by plasma proteins. *FEBS Lett.* 187: 33–37.
31. Wilkins, G.M. and Leake, D.S. (1990) Free radicals and low-density lipoprotein oxidation by macrophages. *Biochem. Soc. Trans.* 18: 1170–1171.
32. Puhl, H., Waeg, G. and Esterbauer, H. (1994) Methods to determine oxidation of low-density lipoproteins. *Method. Enzymol.* 233: 425–441.
33. Sack, M.N., Rader, D.J. and O'Cannon, R. (1994) Oestrogen and inhibition of oxidation of low-density lipoproteins in postmenopausal women. *Lancet* 343: 269–270.
34. Mazière, C., Auclair, M., Ronveaux, M.-F., Salmon, S., Santus, R. and Mazière, J.-C. (1991) Estrogens inhibit copper and cell-mediated modification of low density lipoprotein. *Atherosclerosis* 89: 175–182.
35. Rifici, V.A. and Khachadurian, A.K. (1992) The inhibition of low-density lipoprotein oxidation by 17-β estradiol. *Metabolism* 41: 1110–1114.
36. Frei, B., Forte, T.M., Ames, B.N. and Cross, C.E. (1991) Gas phase oxidants of cigarette smoke induce lipid peroxidation and changes in lipoprotein properties in human blood plasma. *Biochem. J.* 277: 133–138.
37. Frei, B., Stocker, R. and Ames, B.N. (1988) Antioxidant defenses and lipid peroxidation in human blood plasma. *Proc. Natl. Acad. Sci. USA* 85: 9748–9752.
38. Miller, N.J., Rice-Evans, C., Gopinathan, V., Davies, M.J. and Milner, A. (1993) A new automated method for estimating plasma antioxidant activity and its application to the investigation of antioxidant status in premature neonates. *In*: F. Corongiu, S. Banni, M.A. Dessi and C. Rice-Evans (eds): *Free Radicals and Antioxidants in Nutrition*, Richelieu Press, London, pp 153–168.

Analysis of Free Radicals in Biological Systems
Favier et al. (eds)

Determination of 8-oxo-purines in DNA by HPLC using amperometric detection

T. Douki, M. Berger, S. Raoul, J.-L. Ravanat and J. Cadet

CEA; Département de Recherche Fondamentale sur la Matière Condensée, SESAM/LAN, F-38054 Grenoble Cédex 9, France

Summary. High-performance liquid chromatography coupled with electrochemical detection is a sensitive assay for 8-oxo-purines derivatives. This technique was applied to the measurement of the rate of formation of 8-oxo-dGuo in photooxidized isolated DNA following either enzymatic digestion or acidic hydrolysis. The assay was also used to assess the excision of 8-oxo-Gua from DNA upon incubation with the Fapy DNA glycosylase protein. It also allowed the determination of the respective yield of 8-oxo-dGuo and 8-oxo-dAdo upon γ-radiolysis of DNA in aqueous aerated solution.

Introduction

Oxidation reactions occurring within cells as the result of endogenous and/or exogenous processes may lead to several classes of DNA damage including base lesions, oligonucleotide strand breaks, abasic sites and DNA-protein crosslinks [1]. The detection of oxidized nucleobases within DNA has received increasing attention during the last decade. The precise quantitation of base lesions is required for the assessment of their biological role, including mutagenesis, lethality and repair. The measurement of a specific damage within DNA or in biological fluids can also be used as an indicator of "oxidative stress". Two main approaches can be considered for the measurement of modified nucleobases within DNA [2]. The first one involves the whole DNA. It is mainly based on the use of immunological assays and indirect methods, including sedimentation and gel sequencing techniques, which require the conversion of the modified bases into abasic sites. The other approach requires the release of the modified bases, nucleosides or nucleotides from DNA, followed by their chromatographic separation coupled with a sensitive and specific detection. Major results have been obtained by using HPLC-radioactive [32P] post labelling assay [3, 4], gas chromatography separation coupled with mass spectrometry detection [5] and HPLC analysis associated with various detection techniques [2].

HPLC coupled to electrochemical detection has been widely used for the measurement of 7,8-dihydro-8-oxo-2'-deoxyguanosine (8-oxo-dGuo) [6]. In the present work, two possible assays aiming at detecting 8-oxo-dGuo within DNA as a base or a nucleoside following either acidic hydrolysis or enzymatic digestion, respectively, are compared. The HPLC-EC assay was also applied to the determination of the action of the formamidopyrimidine DNA glycosylase (Fpg), a DNA repair enzyme of several guanine oxidative lesions. The extension of HPLC-EC assay to 7,8-dihydro-8-oxo-2'-deoxyadenosine (8-oxo-dAdo) is illustrated by the measurement of its formation within DNA exposed to gamma rays in aerated aqueous solution.

Materials and methods

Chemicals

Guanine, calf thymus DNA and 70% w/w solution of hydrogen fluoride in pyridine (HF/Pyr) were obtained from Sigma (St. Louis, Missouri). 7,8-Dihydro-8-oxoguanine (8-oxo-Gua) was purchased from Chemical Dynamics (South Plainfield, New Jersey) whereas 2'-deoxyadenosine (dAdo) and 2'-deoxyguanosine (dGuo) were obtained from Pharma-Waldorf (Geneva, Switzerland). Methylene blue was purchased from Aldrich (Milwaukee, Wisconsin). 7,8-Dihydro-8-oxo-2'-deoxyguanosine and 7,8-dihydro-8-oxo-2'-deoxyadenosine were prepared by catalytic hydrogenolysis of the corresponding 8-benzyloxy purine nucleoside derivatives [7].

HPLC-EC system

The high performance liquid chromatography system (HPLC) consisted of a model 2150 LKB pump (Pharmacia LKB Biotechnology, Uppsala, Sweden) equipped with a Rheodyne model 7125 loop injector (Berkeley, California) and an Interchrom HC18-25F octadecylsilyl silica gel column (250 × 4.6 mm I.D.) (Interchim, Montluçon, France). The eluent was a mixture of 50 mM sodium citrate (pH 5) and methanol, in a [87:13] v/v and [83:17] v/v ratio for the detection of 8-oxo-dGuo and 8-oxo-dAdo, respectively. The electrochemical detection was performed by amperometry using a model LC-4B/LC-17A(T) apparatus (Bioanalytical Systems, West Lafayette, Indiana) using two glassy-carbon electrodes in parallel. The working electrode was set at a potential of +650 mV (with respect to an $Ag^0/AgCl$ reference electrode) for the detection of 8-oxo-Gua and 8-oxo-dGuo. The potential was raised to +850 mV for the 8-oxo-dAdo assay. Unmodified nucleosides were simultaneously monitored by a Gilson Model 111b UV detector (Gilson, Middleton, Wisconsin) set at 280 nm.

Enzymatic hydrolysis of DNA

10 μL of P1 10X buffer (300 mM pH 5.3 sodium acetate, 1 mM ZnSO$_4$) and 10 μL (10 U) of a nuclease P1 solution (Boehringer, Mannheim, Germany) were added to 100 μL of the DNA solution. The sample was held at 37°C for 2 h. Then, the dephosphorylation of the resulting nucleotides was achieved by addition of 12 μL of 10X alkaline phosphatase buffer (500 mM pH 8 Tris-HCl, 1 mM EDTA) and 1 μL (1 U) of alkaline phosphatase solution (Boehringer, Mannheim, Germany). After incubation for 1 h at 37°C, the resulting solutions were stored at −20°C prior to HPLC analysis.

Acidic hydrolysis of DNA

The DNA sample was freeze-dried in a 1.5 mL polypropylene Eppendorf vial (Hamburg, Germany). HF/Pyr (50 μL) was added and the solution was homogenized by vortexing for 10 s prior to keeping at 37°C in a water bath. After 45 min, the solution was poured under stirring into a suspension of 150 mg of calcium carbonate in 2 mL of water. After neutralization, the resulting calcium fluoride and the excess of calcium carbonate were spun down by centrifugation. The supernatant was collected and the solid residue rinsed with 200 μL of water. The two aqueous solutions were mixed and freeze-dried. The resulting residue was dissolved in 500 μL of water prior to HPLC analysis.

Methylene blue mediated photosensitization of DNA

Calf thymus DNA (50 μg) was dissolved in 1 mL of water containing 10 μg of methylene blue. The solution was homogenized by stirring in the dark and subsequently photolyzed for increasing periods of time with a 100 W tungsten lamp fitted with a heat filter (10 mm of circulating water) and a 590 nm cut-off filter (Kodak no. 23A). After irradiation, the volume of the solution was reduced to 200 μL *in vacuo*. To this, a 3 M solution of sodium acetate (20 μL) and 500 μL of cold ethanol were added. After cooling overnight at −20°C, the resulting solution was centrifuged for 8 min at *12000xg*. The resulting DNA pellet was dissolved in 100 μL of water and each sample was split into two aliquot fractions. The first one was used for enzymatic digestion and the second aliquot was freeze-dried prior to HF/pyridine hydrolysis.

Fpg protein mediated DNA repair

A DNA solution (1 mL, 100 μg) was exposed to visible light for 30 min in the presence of 20 μg of methylene blue. The DNA was recovered by cold ethanol precipitation and resuspended in 150 μL of water. An aliquot fraction (75 μL, 50 μg) was digested (*vide supra*) by incubation with nuclease P1 and alkaline phosphatase and analyzed for 8-oxo-dGuo by HPLC-EC. The remaining DNA solution (75 μL, 50 μg) was mixed with 25 μL of Fpg buffer $4 \times$ (200 M potassium phosphate, 400 mM KCl, pH 7.5) and further incubated for 30 min at 37°C after addition of 2 μL (0.5 μg) of the Fpg solution. The reaction was stopped by ethanol precipitation of both the enzyme and DNA. The supernatant was collected, evaporated to dryness and the resulting residue was analyzed by HPLC-EC. The DNA pellet was hydrolyzed by incubation with nuclease P1 and alkaline phosphatase prior to HPLC-EC analysis. A control experiment was carried out with non oxidized DNA.

γ-Radiolysis of aerated solution of DNA

A 200 μg · mL^{-1} solution (5 mL) of calf thymus DNA was placed under constant air bubbling in a pool where a ^{60}Co source provided 50 Gy · min^{-1}. After increasing periods of exposure, 400 μL of the solution was collected, and 40 μL of 3 M sodium acetate pH 4.5 aqueous solution together with 1 mL of cold ethanol were added. After homogenization, the samples were kept overnight at -30°C. The DNA was recovered by centrifugation for 30 min in a cold centrifuge (5°C). The samples were resuspended in 100 μL of water and digested into nucleosides (*vide supra*).

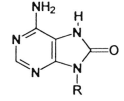

R=H: 8-oxo-Gua
R=2-deoxyribose: 8-oxo-dGuo

R=H: 8-oxo-Ade
R=2-deoxyribose: 8-oxo-dAdo

Fig. 1. Structure of 8-oxo-derivatives of guanine and adenine.

Results and discussion

It is now well established that 7,8-dihydro-8-oxo-2'-deoxyguanosine (8-oxo-dGuo) (Fig. 1) can be electrochemically detected in the oxidation mode [6, 8]. The availability of this sensitive assay has allowed the use of 8-oxo-dGuo as a biomarker of the effects of oxidative stress on DNA in both cultured cells and tissues [4, 8–10]. The amount of 8-oxo-dGuo present in biological fluids, like plasma and urine, has also been proposed to be an index of oxidative stress [10, 11]. The latter assay has recently been improved by immunoaffinity chromatography based on the use of specific antibodies raised against 8-oxo-Gua and its nucleoside derivatives, allowing a very efficient extraction of the lesions from the sample prior to its quantitation [12].

Amperometric detection of 8-oxo purine nucleobases and nucleosides

The hydrodynamic voltammogram (HPLC-EC response for decreasing potentials with constant amount of product) of 8-oxo-dGuo exhibited a half-oxidation wave potential around 600 mV (with respect to an $Ag^0/AgCl$ reference electrode). The corresponding value for the base derivative 8-oxo-Gua was found to be 550 mV. These relatively low values make the assay for 8-oxo-dGuo and 8-oxo-Gua quite specific. In particular, the unmodified bases and nucleosides are not detected at these potentials. The calibration curves, determined with an oxidation potential of 650 mV, were found to be linear over a range of 0.6 to 50 pmole for both the modified nucleobase and the 2'-deoxynucleoside. The detection limits were 0.02 and 0.03 pmole, for 8-oxo-Gua and 8-oxo-dGuo, respectively. The 8-hydroxylated analog of 2'-deoxyadenosine, 7,8-dihydro-8-oxo-2'-deoxyadenosine (8-oxo-dAdo), is also electroactive, but a higher potential (+850 mV) is required for its detection [13]. The formation of 8-oxo-purines within DNA can thus be efficiently monitored by measuring the related nucleosides subsequent to enzymatic digestion. It should be mentioned that several other DNA lesions can also be detected by HPLC-EC, including 5-hydroxy-2'-deoxycytidine [14], and cyclic adducts resulting from the reaction of guanine with malondialdehyde [15] and several unsaturated aldehydes [16].

Measurements in DNA

Measurement of oxidized purine bases in DNA may be carried out following either acidic hydrolysis or enzymatic digestion. 8-Oxo-purines are monitored by electrochemical detection whereas normal bases and nucleosides are detected by UV spectrometry (Fig. 2). The latter analy-

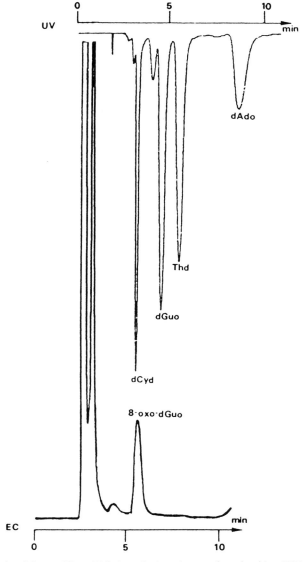

Fig. 2. Analysis of 8-oxo-dGuo (EC detection) and normal nucleosides (UV detection) in
DNA exposed to γ radiations in aerated aqueous solution.

sis is used for the precise quantitation of the amount of DNA injected,
which is inferred from the peaks corresponding either to thymidine or
2′-Deoxyguanosine. 2′-Deoxycytidine cannot be used because it is eluted
very rapidly, together with other fast eluting compounds. On the other
hand, 2′-deoxyadenosine must be avoided since it is eluted as a broad

peak. Thymidine was used for calibration in methylene blue photosensitization experiments since guanine is the main target of the latter process within DNA.

Formation of 8-oxo-dGuo in DNA upon methylene blue photosensitization

DNA samples were photooxidized by methylene blue in the presence of visible light [17]. A first aliquot fraction was digested into nucleosides, using a classical method involving two steps. First, DNA was digested into nucleoside 5'-phosphates by nuclease P1. The nucleosides were subsequently obtained by alkaline phosphatase treatment. The resulting solutions were analyzed for 8-oxo-dGuo. The second aliquot fraction of photooxidized DNA was hydrolyzed by hydrogen fluoride stabilized in pyridine (HF/Pyr), to release the DNA bases and 8-oxo-Gua, which was further measured by HPLC-EC analysis [18]. HF/Pyr has been successfully used to hydrolyze the N-glycosidic bond without modifying the base moiety of several oxidized nucleosides, including disastereoisomers of 5,6-dihydro-5,6-dihydroxy-thymidine [19] and 7,8-dihydro-4-hydroxy-8-oxo-2'-deoxyguanosine [20]. This mild acidic hydrolysis assay has also been applied to the measurement of several nucleobase lesions within DNA, including ultraviolet-induced (6-4) photoproducts [21], and the adducts of unsaturated aldehydes to 2'-deoxyguanosine [16].

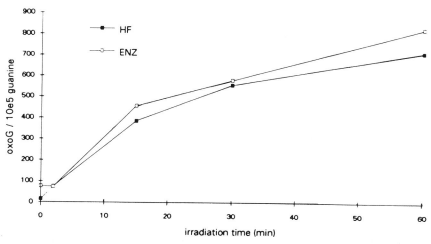

Fig. 3. Dose-response curves of formation of 8-oxo-guanine in DNA upon methylene blue photosensitization. Results obtained by HPLC-EC assay following: (ENZ) enzymatic digestion (release of 8-oxo-dGuo); (HF): acidic hydrolysis using HF/pyridine (release of 8-oxo-Gua).

The extent of formation of 8-oxo-dGuo residues in photooxidized DNA was determined following enzymatic digestion. This was compared with the yield of 8-oxo-Gua obtained by HF/Pyr hydrolysis followed by EC detection. In both cases, the level of modification was precisely determined by monitoring both the amount of lesions and normal bases in the HPLC profile. The dose-response curves (Fig. 3) were found to be very similar, confirming that 8-oxo-Gua is quantitatively released from DNA by acidic hydrolysis, as already observed at the nucleoside level [18]. However, the modification rates determined by HPLC-EC analysis following HF/Pyr hydrolysis are slightly lower than for the corresponding enzymatically digested DNA samples. An explanation for this result was provided by complementary experiments showing that small amounts of isolated 8-oxo-Gua are trapped during the hydrolysis process. This is likely to be due to the high amount of solid salts produced during the neutralization step. Autoxidation of 2'-deoxyguanosine during the 3 h incubation at 37°C can also partly account for the higher value of 8-oxo-dGuo rate found in enzymatically hydrolyzed samples of DNA.

Excision of 7,8-dihydro-8-oxoguanine residues from DNA by the Fpg protein
The *Escherichia coli* formamidopyrimidine DNA glycosylase (Fpg) has been isolated and initially characterized as an enzyme able to excise purine ring opening products [22–24]. 8-Oxo-Gua has been recently

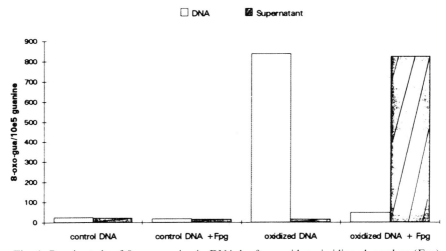

Fig. 4. Repair study of 8-oxo-guanine in DNA by formamidopyrimidine glycosylase (Fpg). Amount of 8-oxo-Gua detected in the supernatant fraction and of 8-oxo-dGuo from enzymatically digested DNA.

shown to be an excellent substrate for the glycosylase activity of Fpg [18, 25, 26]. As shown in Figure 4, only a small part of the original amount of 8-oxo-dGuo was left in DNA after incubation with Fpg. The results of the analysis of the supernatant obtained after incubation with the Fpg protein are consistent with the almost quantitative release of 8-oxo-Gua from DNA. This result confirms that the Fpg protein behaves like a glycosylase, by cleaving the N-glycosydic bond of the 7,8-dihydro-8-oxoguanine residues in DNA. It should be added that no 7,8-dihydro-8-oxoadenine was detected in the supernatant fraction upon treatment by the Fpg protein of DNA containing 8-oxo-dAdo (data not shown). This is in agreement with recent observations showing the lack of any detectable Fpg mediated release of 8-oxo-Ade from a synthetic oligonucleotide carrying a 8-oxo-dAdo residue at a specific site [27]. It should be noted that the use of Fpg may be an interesting alternative for the specific release of 8-oxo-Gua from DNA prior to its measurement by HPLC-EC. This could avoid autoxidation since 8-oxo-Gua is almost quantitatively released from double stranded DNA in 15 min under optimal conditions.

Formation of 8-oxo-dGuo and 8-Oxo-dAdo upon γ-radiolysis of aerated aqueous solution of DNA
The HPLC-EC assay was used to monitor the formation of 8-oxo-dAdo and 8-oxo-dGuo within DNA exposed to γ-radiation in aerated aqueous solution (Fig. 5). 8-oxo-dAdo was found to be generated in an

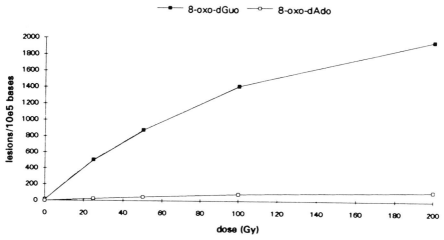

Fig. 5. Dose-dependent formation of 8-oxo-dGuo and 8-oxo-dAdo within DNA upon γ-radiolysis in aerated aqueous solution. The measurement was carried out by HPLC-EC following enzymatic digestion.

approximately 10-fold lower yield than 8-oxo-dGuo. However, 8-oxo-dAdo could also be a good marker of oxidative stress to DNA since its basal level is under the detection limit of the HPLC-EC assay (<one 8-oxo-dAdo for 10^5 dAdo) in calf thymus DNA. In contrast, the value for 8-oxo-dGuo was found to be 14 lesions for 10^5 guanine residues. Consequently, even a minor increase in the amount of 8-oxo-dAdo under conditions generating oxidizing species (mainly °OH radicals) could be detected. This may be partly explained by a lower sensitivity of adenine and 2'-deoxyadenosine to autoxidation. This was shown by the observation of an eight fold increase in the 8-oxo-dGuo level when DNA was solubilized in 10 mL of water and precipitated by addition of cold ethanol. Under identical conditions, the amount of 8-oxo-dAdo remained under the detection limit of the HPLC-EC assay.

Artefactual oxidation of DNA samples

As already mentioned, the formation of artefactual oxidation products is a key problem in the 8-oxo-dGuo assay, especially when the method is applied to cellular DNA [28]. Such autoxidation reactions can occur at several steps of the analysis. First, 8-oxo-dGuo can be produced from guanine residues when either DNA or digested nucleosides are left in aqueous solution. This artefactual production, often due to the presence of traces of transition metal generating °OH radicals or related reactive species through Fenton-like reactions [29], can be decreased by addition of metal ion chelators, like EDTA [30, 31]. Antioxidants, including glutathione [29], 8-hydroxyquinone [31], histidine [32] and butylated hydroxytoluene [33] can also be added to lower the artefactual formation of 8-oxo-dGuo. However, some antioxidants such as 5-aminosalicylic acid and ascorbate [29] can behave as prooxidants by enhancing Fenton-like reactions in reducing oxidized transition metal ions. Artefactual oxidation may also occur because of the chemicals used in the extraction of DNA from tissue or cell culture. For instance, the use of phenol, shown to be a prooxidant [30, 32], has been questioned. However, similar amount of 8-oxo-dGuo was measured in the same cellular DNA upon extraction involving either the use of phenol in association with antioxidants and EDTA or two other methods excluding the latter solvent [30]. Consequently, it seems that phenol extraction can be used prior to 8-oxo-dGuo analysis, if caution is taken [34, 35].

References

1. Cadet, J. (1994) DNA damage caused by oxidation, deamination, ultraviolet radiation and photoadded psoralens. *In:* K. Hemminki et al. (eds): *DNA Adducts: Identification and*

Biological Significance, Lyon International Agency for Research on Cancer; 1994. IARC Scientific Publications, No. 125, 245–276.

2. Cadet, J. and Weinfeld, M. (1993) Detecting DNA damage. *Anal. Chem.* 65: 675a–682a.
3. Randerath, K., Reddy, M.V. and Gupta, R.C. (1981) Post-labelling test for DNA damages. *Proc. Natl. Acad. Sci. USA* 78: 6126–6129.
4. Cadet, J., Odin, F., Mouret, J., Polverelli, M., Audic, A., Giacomoni, P., Favier, A. and Richard, M.-J. (1992) Chemical and biochemical postlabeling methods for singling out specific oxidative DNA lesions. *Mutat. Res.* 275: 343–354.
5. Fuciarelli, A.F., Wegher, B.J., Gajewski, E., Dizdaroglu, M. and Blakely, W.F. (1989) Quantitative measurement of radiation-induced base products in DNA by gas chromatography-mass spectrometry. *Radiat. Res.* 119: 219–231.
6. Floyd, R.A., Watson, J.J., Wong, P.K., Altmiller, D. H. and Rickard, R.C. (1986) Hydroxyl free radical adduct of deoxyguanosine: sensitive detection and mechanism of formation. *Free Rad. Res. Commun.* 1: 163–172.
7. Lin, T.-S., Cheng, J.-C., Ishiguro, K. and Sartorelli, A.C. (1985) 8-substituted guanosine and 2'-deoxyguanosine derivatives as potential inducers of differentiation of Friend erythroleukemia cells. *J. Med. Chem.* 28: 1194–1197.
8. Kasai, H., Crain, P.F., Kuchino, Y., Nishimura, S., Ootsuyama, A. and Tanooka, H. (1986) Formation of 8-hydroxyguanine moiety in cellular DNA by agents producing oxygen radicals and evidence for its repair. *Carcinogenesis* 7: 1849–1851.
9. Schraufsätter, I., Hyslop, P.A., Jackson, J.H. and Cochrane, C.G. (1988) Oxidant-induced DNA damage of target cells. *J. Clin. Invest.* 82: 10401–10405.
10. Fraga, C.G., Shigenaga, M.K., Park, J.-W., Degan, P. and Ames, B.N. (1990) Oxidative damage to DNA during ageing: 8-hydroxy-2'-deoxyguanosine in rat organ DNA and urine. *Proc. Natl. Acad. Sci. USA* 87: 4533–4537.
11. Shigenaga, M.K., Gimeno, C.J. and Ames, B.N. (1989) Urinary 8-hydroxy-2'-deoxyguanosine as a biological marker of *in vivo* oxidative DNA damage. *Proc. Natl. Acad. Sci. USA* 86: 9697–9701.
12. Park, E.-M., Shigenaga, M.K., Degan, P., Korn, T.S., Kitzler, J.W., Wehr, C.M., Kolachana, P. and Ames, B.N. (1992) Assay of excized oxidative DNA lesions: isolation of 8-oxo-guanine and its nucleoside derivatives from biological fluids with a monoclonal antibody column. *Proc. Natl. Acad. Sci. USA* 89: 3375–3379.
13. Berger, M., Anselmino, C., Mouret, J.-F. and Cadet, J. (1990) High performance liquid chromatography-electrochemical assay for monitoring the formation of 7,8-dihydro-8-oxoadenine and its related 2'-deoxynucleoside. *J. Liquid Chromatog.* 13: 929–932.
14. Wagner, J.R., Hu, C.-C. and Ames, B.N. (1992) Endogenous oxidative damage of deoxycytidine in DNA. *Proc. Natl. Acad. Sci. USA* 89: 3380–3384.
15. Goda, Y. and Marnett, L.J. (1991) High-performance liquid chromatography with electrochemical detection for determination of the major malondialdehyde-guanine adduct. *Chem. Res. Toxicol.* 4: 520–524.
16. Douki, T. and Ames, B.N. (1994) An HPLC-EC assay for $1,N^2$-propano adducts of 2'-deoxyguanosine with 4-hydroxynonenal and other α,β-unsaturated aldehydes. *Chem. Res. Toxicol.* 7: 511–518.
17. Floyd, R.A., West, M.S., Eneff, K.L. and Schneider, J.E. (1989) Methylene blue plus light mediates 8-hydroxyguanine formation in DNA. *Arch. Biochem. Biophys.* 273: 106–111.
18. Ravanat, J.-L., Berger, M., Boiteux, S., Laval, J. and Cadet, J. (1993) Excision of 7,8-dihydro-8-oxoguanine from DNA by the Fpg protein. *J. Chim. Phys.* 90: 871–879.
19. Polverelli, M., Berger, M., Mouret, J.-F., Odin, F. and Cadet, J. (1990) Acidic hydrolysis of the N-glycosidic bond of deoxyribonucleic acid by hydrogen fluoride stabilized in pyridine. *Nucleosides Nucleotides* 9: 451–452.
20. Cadet, J., Ravanat, J.-L., Buchko, G.W., Yeo, H.C. and Ames, B.N. (1994) Singlet oxygen DNA damage: chromatographic and mass spectrometric analysis of damage products. *Methods Enzymol.* 234: 79–88.
21. Douki, T., Voituriez, L. and Cadet, J. (1994) Measurements of the pyrimidine (6-4) photoproducts by a mild acidic hydrolysis – HPLC fluorescence assay. *Chem. Res. Toxicol.* 8: 244–253.
22. Boiteux, S., O'Connor, T.R. and Laval, J. (1987) Formamidopyrimidine-DNA glycosylase of *Escherichia coli*: cloning and sequencing of the *fpg* structural gene and overproduction of the protein. *EMBO J.* 6: 3177–3183.

23. Boiteux, S., O'Connor, T.R., Lederer, R., Gouyette, A. and Laval, J. (1990) FPG protein, a DNA glycosylase which excises imidazole ring-opened purines and nicks DNA at apurinic/apyrimidinic sites. *J. Mol. Biol.* 265: 3916–3922.

24. Boiteux, S., Gajewski, E., Laval, J. and Dizdaroglu, M. (1991) Substrate specificity of the *Escherichia coli* Fpg protein (Formamidopyrimidine-DNA glycosylase): excision of purine lesions in DNA produced by ionizing radiation or photosensitisation. *Biochemistry* 31: 106–110.

25. Tchou, J., Kasai, J., Shibutani, S., Ching, M.H., Laval, J., Grollman, A.P. and Nishimira, S. (1991) 8-Oxoguanine (8-hydroxyguanine) DNA glycosylase and its substrate specificity. *Proc. Natl. Acad. Sci. USA* 88: 4690–4694.

26. Boiteux, S. (1993) Properties and biological functions of the NTH and FPG proteins of *Escherichia coli*: two DNA glycosylates that repair oxidative damage in DNA. *J. Photochem. Photobiol. B: Biol.* 19: 87–96.

27. Tchou, J., Bodepudi, V., Shibutani, S., Antoshechkin, I., Miller, J., Grollman, A.P. and Johnson, F. (1994) Substrate specificity of Fpg protein. *J. Biol. Chem.* 269: 15318–15324.

28. Floyd, R.A., West, M.S., Eneff, K.L., Schneider, E.L., Wong, P.K., Tingey, D.T. and Hogsett, W.E. (1990) Conditions influencing yield and analysis of 8-hydroxy-2′-deoxyguanosine in oxidatively damaged DNA. *Anal. Biochem.* 188: 155–158.

29. Fischer-Nielsen, A., Poulsen, H.E. and Loft, S. (1992) 8-Hydroxyguanosine *in vitro*: effects of glutathione, ascorbate, and 5-aminosalicylic acid. *Free Rad. Biol. Medec.* 13: 121–126.

30. Claycamp, H.G. (1992) Phenol sensitization of DNA subsequent oxidative damage in 8-hydroxyguanine assays. *Carcinogenesis* 13: 1289–1292.

31. Harris, G., Bashir, S. and Winyard, P.G. (1994) 7,8-Dihydro-8-oxo-2′-deoxyguanosine present in DNA is not simply an artefact of isolation. *Carcinogenesis* 15: 411–413.

32. Takeuri, T., Nakajima, M., Ohta, Y., Mure, K., Takeshita, T. and Morimoto, K. (1992) Evaluation of 8-hydroxydeoxyguanosine, a typical oxidative DNA damage, in human leukocytes. *Carcinogenesis* 15: 1519–1523.

33. Adachi, S., Zeisig, M. and Möller, L. (1995) Improvements in the analytical method for 8-hydroxydeoxyguanosine in nuclear DNA. *Carcinogenesis* 16: 253–258.

34. Claycamp, H.G. and Ho, K.-K. (1993) Background and radiation-induced 8-hydroxy-2′-deoxyguanosine in γ-irradiated *Escherichia coli*. *Int. J. Radiat. Biol.* 63: 597–607.

35. Shigenaga, M.K., Aboujaoude, E.N., Chen, Q. and Ames, B.N. (1994) Assay of oxidative DNA damage biomarkers: 8-oxo-2′-deoxyguanosine and 8-oxo-guanine in nuclear DNA and biological fluids by high-performance liquid chromatography with electrochemical detection. *Methods Enzymol.* 234: 16–32.

Analysis of Free Radicals in Biological Systems
Favier et al. (eds)
© 1995 Birkhäuser Verlag Basel/Switzerland

Measurement of one of the features of apoptosis: DNA fragmentation

J. Mathieu, S. Ferlat, D. Ferrand, S. Platel, B. Ballester, J.C. Mestries, Y. Chancerelle and J.F. Kergonou

Unité de Radiobiochimie, Centre de Recherches du Service de Santé des Armées, 24, Avenue du Maquis du Grésivaudan, BP 87, F-38702 La Tronche Cédex, France

Summary. Unlike necrosis, apoptosis was observed to proceed in an orderly and controlled manner that appeared to follow a defined program. DNA fragmentation is one of the features of apoptosis in lymphocytes. Most chemical and physical treatments capable of inducing apoptosis are also known to evoke oxidative stress. Dexamethasone can induce apoptosis and trigger a quantitative increase in lipid peroxidation. This work shows two means to measure with flow cytometry DNA fragmentation in mouse thymocytes after dexamethasone treatment. Alternatively, electrophoretic analysis of thymocyte lysate DNA is performed.

Introduction

Apoptosis (derived from ancient Greek for "the falling leaves from trees or petals from flowers") is the name that has been given by Kerr et al. [1] to the process of physiological cell death. Unlike necrosis, physiological cell death was observed to proceed in an orderly and controlled manner that appeared to follow a defined program. The apoptosis process has been shown to proceed via a number of discrete steps, such as membrane blebbing, chromatin condensation and DNA fragmentation.

Apoptosis-inducing stimuli are diverse, some of which are important in pathology and include radiation, hyperthermia, calcium influx, glucocorticoids and cytotoxic agents [2]. Oxygen free radicals appear to be involved in cellular injury caused by hyperoxygenation, ischemia/reperfusion, inflammation, exposure to ionizing radiation and aging [3] and can induce apoptosis [4]. Most chemical and physical treatments capable of inducing apoptosis are also known to evoke oxidative stress. Both ionizing and ultraviolet radiation are capable of inducing apoptosis, and generate reactive oxygen intermediate (ROI) such as H_2O_2 and $HO^\cdot$ [5]. As reported by Forrest et al. [6] and Lennon et al. [7], H_2O_2 can induce apoptotic death in mouse thymocytes and in other cell types. Severe oxidative injury produces necrosis in target cells, consistent with the suggestion by Duvall and Wyllie [8] that the severity of insult determines the form of cell death. In a recent study, Langley et al. [9] examined the possibility that apoptosis can be triggered by agents that mimic different types of molecular damage that could be produced by ionizing radiation [10, 11]. These authors showed that tert-butyl hydroperoxide (t-BOOH) induced DNA fragmentation to the same extent as ionizing radiation. t-BOOH was used as a model for organic peroxides that are produced during irradiation [12]. As reported by Hockenbery et al. [13], other agents, such as dexamethasone can induce apoptosis and trigger a quantitative increase in lipid peroxidation within a 4-h treatment, which was much earlier than the effect upon cell viability. Reactive oxygen intermediates seem to be involved in dexamethasone-induced apoptotic cell death. In fact, antioxidants such as N-acetylcysteine can reduce this apoptotic cell death.

On the one hand, this work shows two means to measure with flow cytometry one of the features of apoptosis, i.e., DNA fragmentation subsequent to endonuclease activation, in mouse thymocytes after treatment with a glucocorticoid, dexamethasone [14]. Endonucleolysis, which accompanies apoptosis induced by dexamethasone, is assessed by two different flow cytometric methods. The first method is based on the DNA content measurement of ethanol-fixed and permeabilized cells, following extraction from these cells of the degraded, low MW DNA. Another method uses exogenous DNA polymerase to label *in situ* the DNA strand breaks with biotin-16,2′-deoxy-uridine-5′-triphosphate (biotin-dUTP).

Alternatively, electrophoretic analysis of thymocytes lysate DNA is performed. Most techniques used to assess apoptosis occurring in cells involve the quantitative or qualitative analysis of endonuclease-induced internucleosomal DNA fragmentation.

DNA fragmentation is an early irreversible step in apoptotic process arising from the activation of an endonuclease that cleaves the DNA into nucleosome-sized fragments of approximately 200 base pairs. The gel electrophoresis of DNA solution displays a characteristic ladder pattern of discontinuous DNA fragments [15], often considered as the biochemical hallmark of apoptosis. Electrophoretic analysis of DNA size is a widely used method to probe apoptosis.

Materials and methods

Thymocyte isolation

BalbC female mice (4–8 weeks old) are sacrificed with ether and their thymuses removed and placed in ice-cold PBS. Single-cell suspensions are prepared by pressing the organs followed by passage through a Falcon filter and a 25-gauge needle. The cells are washed once and resuspended in medium. Viable cells are determined by their ability to exclude trypan blue.

Dexamethasone treatment

Mouse thymocytes (2×10^6/ml) are exposed to 1 μM of dexamethasone in culture medium (RPMI 1640 medium supplemented with 2 mM L-glutamine, 50 μM 2-mercaptoethanol, 100 U/ml penicillin, 100 μg/ml streptomycin and 10% heat-inactivated foetal calf serum: Gibco/BRL). The cells are then incubated at 37°C under an atmosphere of 5% CO_2 in air.

Flow cytometry

DNA content analysis by using Propidium iodide (PI)

In order to record DNA histograms, about 10^6 cells are pelleted, ressuspended in PBS and fixed rapidly in 2 ml ice-cold ethanol for at least 1 h. Cells are centrifuged from the fixative, resuspended in 1 ml PBS, 50 μg/ml propidium iodide (PI), 100 μg/ml RNAase and incubated at 37°C for 15 min. The excitation wavelength in the cytometer was 488 nm and red (>620 nm) fluorescence was recorded [16, 17].

The flow cytometer is an Epics Profile 2 (Coulter) equipped with an argon-ion laser tuned to 488 nm. Histograms and cytograms are transferred to an IBM compatible PC on which data were analysed and figures prepared using software from Coulter.

In situ labeling of DNA strand breaks or in situ nick translation assay (ISNT)

The following method is a modification of those described by Gold et al. [18] and Gorczca et al. [19].

Reagents. 1% PFA solution of 1% formaldehyde: stock solution of PFA is prepared at 8% and stored at -20°C. (8 g PFA/100 ml distilled water, heating at 60°C, drop NaOH (10 N) in this solution until the liquid becomes clear. After filtration, store at -20°C. 1% PFA is freshly prepared in PBS, check pH (7.4 $-$ 7.6). Stock solution of formalin (Sigma) contains 4% of formaldehyde and is freshly diluted to 1% solution.

Nick-translation buffer consisting of: 50 mM Tris (pH 7.8), 10 μg/ml BSA, 10 mM β-mercapto-ethanol and 2.5 mM $MgCl_2$.

dATP, dGTP, dCTP (initial concentration = 100 mmoles/l) (Boehringer Mannheim, Germany). Three nucleotide concentrations are tested: 0.05, 0.2 nmol and 0.5 nmol/μl in nick translation buffer. Biotin-16-dUTP (b-dUTP: initial concentrations = 1 mmol/l or 1 nmol/μl), (Boehringer Mannheim, Germany). Three b-UTP concentrations are tested: 0.05, 0.2 nmol and 0.5 nmol in nick translation buffer. Prepare a solution consisting of the four nucleotides used at the same concentration: for each sample, one μl of this solution is used.

ISNT protocol. After dexamethasone treatment, cells were gently homogenized, harvested and centrifuged (400 g, 10 min, 4°C). After, the following centrifugations are done in the same conditions.

– Fixation: After lavage in 3 ml PBS, the pellet is resuspended and cells are fixed for 15 min on ice in 1% freshly prepared PFA or formaldehyde solution diluted in PBS.
  Store the cells at 4°C at least 30 min.
– Permeabilisation: After lavage with 3 ml PBS, the supernatant is discarded and the pellet is resuspended with addition of 1 ml of PBS. Drop ice-cold 70% ethanol into the cell suspension with gentle mixing. It is possible to store the cells at −20°C for several days.

After two lavages with 3 ml of PBS, the tube is carefully dripped. Cells· (5×10^5 in 20 μl) were incubated in small reaction tubes (microtube Eppendorf 0.5 ml) containing 25 μl of the reaction mixture in nick-translation buffer. This consisted of: 20 μl cell suspension, 1 μl of nucleotide solution, 0.2 μl of DNA polymerase I (one unit), 3.8 μl nick translation buffer in a final volume of 25 μl. A time-incubation study is performed to determine the maximum of DNA polymerase activity at 37°C from 45 min to 6 h in a shaking-water bath.

After two lavages in PBS, indirect labeling is carried out using streptavin-phycoerythrine (10 μl/10^6 cells; Dako) or streptavidin-FITC (2 μl/10^6 cells; Dako) in the ISNT assay. To stop the reaction, 4 μl of 0.5 M EDTA (Sigma) are added and the final volume of the reaction is 100 μl. After 30 min at +4°C, and two lavages, the pellet is resuspended in PBS (250 μl). The excitation wavelength in the cytometer is 488 nm and orange (575 nm) or green (525 nm) fluorescence is recorded.

DNA electrophoretic analysis

Reagents.

– Lysis buffer consisting of: 10 mM EDTA (pH 8), 50 mM tris-HCl (pH 8) containing 0.5% N-lauroylsarcosine.
– TE Buffer (10 mM Tris-HCl, pH 8 and 0.1 mM EDTA).
– TBE Buffer (Stock solution 10X) consisting of: 2 mM EDTA (pH 8), 89 mM Tris, 89 mM boric acid.

Cell preparation and DNA extraction. Mouse thymocytes (5×10^6/ml) are induced to undergo apoptosis, as described above, by exposure to dexamethasone for 16 h. Thymocytes incubated in complete medium are used as negative control.

Following treatment-induced apoptosis, thymus cells are washed with PBS buffer and pelleted by centrifugation at 200 × g for 10 min at room temperature. Cell pellets are gently resuspended in 500 ml of sterile lysis buffer and incubated for 2 h with proteinase K (0.1 mg/ml) in a 60°C water-bath.

In order to eliminate overlapping bands of RNA comigrating with the 180–200 base pairs of DNA fragments, RNAase (0.1 mg/ml; previously heat-treated to remove contaminating deoxyribonuclease activity) is added and samples are incubated for an additional hour at 60°C.

DNA for gel electrophoresis is generally extracted with the classical phenol/chloroform method. In our study, crude DNA preparations are extracted using an original centrifugal device, i.e., Ultrafree-Probind Filters (Millipore), providing the removal of protein from nucleic acid solution. Each sample is collected into ultrafree-probind inset and centrifuged at 12 000 g for approximately 1 min. Fragments of double stranded DNA lower than 10 Kb can only be recovered in the filtrate. DNA recovered is precipitated with 1:10 volume 3 M Na acetate (pH 4.8) and 2 volumes of ice cold 100% ethanol and kept for 2 h at −70°C.

Nucleic acid pellets are obtained by centrifugation (13 000 g; 30 min; +4°C), air dried, resuspended in TE buffer and stored at +4°C. DNA concentration is determined by measurement of OD at 260 nm, the ratio 260/280 allows to evaluate protein contamination.

Electrophoresis. Before electrophoresis, loading buffer is added in a 1:5 ratio, each sample is heated at 65°C for 10 min, in a dry bath. Electrophoresis is performed on a 2% agarose gel containing 1 mg/ml ethidium bromide. Approximately 10 mg of DNA is loaded into each well and electrophoresis is carried out at 80 V in TBE buffer for 2–3 h. DNA in the gel is visualized under UV light.

Results

As shown by Figure 1, glucocorticoid induced-apoptosis is observed by the measurement of DNA fragmentation by flow cytometry after labelling of DNA by propidium iodide: the fluorescence of apoptotic cells appeared lower than that of diploid cells, apoptotic cells are also called hypodiploid cells (sub-Go population). In these experiments, cells are treated by dexamethasone for 16 h. Cytograms A (control cells) and B (dexamethasone-treated cells) (Fig. 1) show two parameters: forward scatter (FS: it is an evaluation of cell size) and DNA content measured by the fluorescence of propidium iodide (FL3). Diploid cells and apoptotic cells appear in box 3 and in box 2, respectively. We can observe in cytogram B a decrease of both DNA content (FL3) and in cell size (FS). Histogram C (control cells) and histogram D (dexamethasone-treated cells) represent only one parameter: DNA content (FL3). A similar decrease of DNA content can be observed. As shown by Figure 2, the time-kinetic study of dexamethasone treatment allows to detect

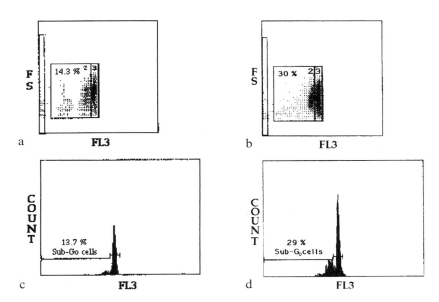

Fig. 1. DNA content analysis: control cells (A and C) and dexamethasone-treated cells (B and D). Thymocytes are treated during 4 h in culture medium.

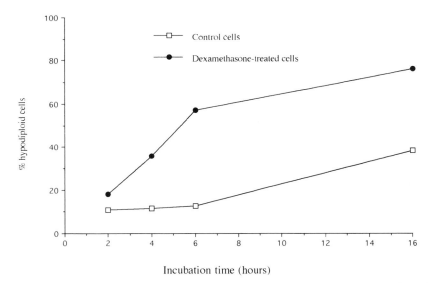

Incubation time (hours)

Fig. 2. This diagram shows a representative experiment of a time-kinetic study of DNA content analysis in control cells and in 1 μM dexamethasone-treated cells. At 2, 4, 6 and 16 h after addition of glucocorticoid, DNA fragmentation was determined as the percentage of hypodiploid cells.

apoptotic cells 2 h after addition of glucocorticoid, and a significant increase is observed at 4 h of incubation with dexamethasone: only a fraction of cell population is hypodiploid. Then, there is an increase of cell population in apoptosis after dexamethasone-treatment of cells and DNA damage also appear in control cells. In cells freshly harvested from normal mouse, no DNA fragmentation can be observed (data not shown).

To assess apoptosis, DNA strand-break measurements are performed using ISNT assays. Incubation of fixed and permeabilized thymocytes with *E. coli* DNA polymerase and four triphosphodeoxynucleotides (dNTPs) resulted in incorporation of the labelled nucleotide (dUTP). To determine the maximum of dNTP incorporation with one unit of DNA polymerase for 3 h, three experiments with different dNTP concentrations (0.05–0.5 nM) are performed with a population of apoptotic cells from thymocytes treated with 1 μM dexamethasone for 16 h or with control cells incubated without dexamethasone for 16 h. Figure 3a shows a representative experiment allowing to observe that the best dNTP incorporation is obtained with 0.2 nM dNTPs. A time-kinetic study of dUTP incorporation (Fig. 3b) in the two groups of cells

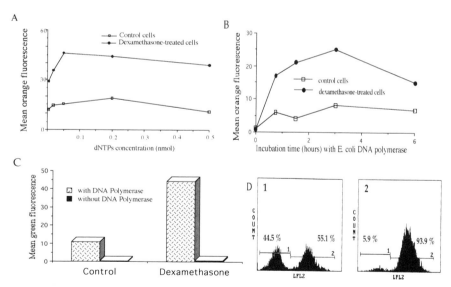

Fig. 3. ISNT assays: (A) effect of dNTP concentration on ISNT assay: mean orange fluorescence represents the mean of fluorescence intensity of the cells labelled with b-dUTP associated with phycoerythrin-streptavidin. The cells are incubated in culture medium for 16 h with or without dexamethasone. (B) Kinetic study with *E. coli* DNA polymerase (one unit). (C) ISNT assay with or without *E. coli* DNA polymerase, ISNT is performed on cells incubated in culture medium for 16 h. (D) A representative result of ISNT histograms (cell count against fluorescence of b-dUTP associated with phycoerythrin-streptavidin), 1: control cells and 2: dexamethasone-treated cells after an incubation of 16 h.

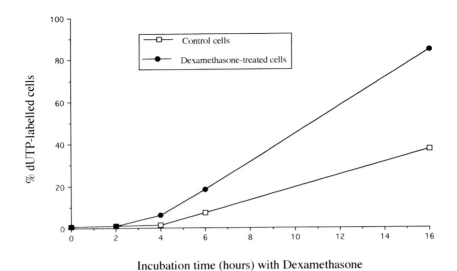

Incubation time (hours) with Dexamethasone

Fig. 4. DNA strand-breaks measured with ISNT assay: This diagram shows a representative experiment of a time-kinetic study of ISNT analysis in control cells and in 1 μM
dexamethasone-treated cells. At 2, 4, 6 and 16 h after addition of glucocorticoid, DNA
fragmentation was determined as the percentage of d-UTP-labelled cells.

described above, is done to find the best time of incubation at 37°C with
0.2 nM dNTPs. As shown by Figure 3b, in these experiments, the
maximum of dNTP incorporation is obtained at 3 h of incubation. For
the further experiments, therefore, we use to perform ISNT assays,
0.2 nM dNTPs and an incubation of 3 h with *E. coli* DNA polymerase.

Figure 3c shows that DNA strand-break detection is dependent of *E.
coli* DNA polymerase added in the incubation medium: in fact, without
this exogenous enzyme, no DNA strand-breaks can be observed. As
shown by Figure 4, in this experiment, thymocytes are incubated during
16 h: histograms A and B represent control cells and dexamethasone-
treated cells, respectively. We can observe a strong increase of DNA
strand-breaks in histogram B. Apoptotic cells already appear in control
cells after 16 h incubation; in fact, apoptosis is a physiological process
of death in thymocytes. After 4 h incubation with dexamethasone, few
DNA strand-breaks can be detected, albeit DNA damage begins to
appear as shown above. Without exogenous DNA polymerase in the
medium, no strand-breaks are observed.

To confirm DNA fragmentation, DNA electrophoresis is performed.
Figure 5 shows electrophoresis of thymocyte DNA extracted with
ultrafree probind filters. In dexamethasone-treated cells, the characteristic ladder pattern of discontinuous DNA fragments is exhibited, while
no DNA fragments are detected in control cells. This confirms the

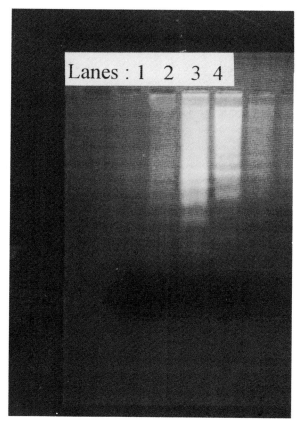

Fig. 5. Electrophoresis of thymocyte DNA extracted with ultrafree probind filters. Lanes 1 and 2, control cells. Lanes 3 and 4, thymocytes with 1 μM dexamethasone for 16 h. Lanes 3 and 4 exhibit the characteristic ladder pattern of discontinuous DNA fragments.

previous results obtained with flow cytometry in dexamethasone-treated cells but electrophoresis method cannot detect DNA fragmentation in control cells.

Discussion

The first method is based on measurement of cellular DNA content. In this approach, the partially degraded DNA, a result of activation of an endonuclease in apoptotic cells, is extracted from ethanol-fixed cells prior to their measurement and apoptotic cells are identified on the DNA frequency histograms or bivariate FS/DNA cytograms as the cells with a fractional DNA content. The disadvantage of this method is that when DNA degradation, in a given cell, is not much advanced, most DNA in that cell is still of high MW and, thus, is not extracted prior to measurement. This causes an overlap between the live and apoptotic cell populations which leads to underestimation of the latter. The correlation of DNA fragmentation with loss of apparent DNA fluorescence has also been demonstrated by Afanasyev et al. [20] by inducting internucleosomal DNA fragmentation in normal ethanol-fixed thymocytes with micrococcal nuclease. This flow cytometric method is easy, rapid, and the fixed samples can be prepared and analyzed at a later time.

In the second assay, formaldehyde-fixed and permeabilized cells are incubated in the presence of exogenous DNA polymerase and biotin-dUTP. DNA strand breaks are labelled with biotin-dUTP under these conditions, and apoptotic cells, which contain a large number of such lesions, can be discriminated based on intense labelling. Since, unlike the first method, the cross-linking agent formaldehyde is used as a fixative. Here, the extraction of DNA from apoptotic cells during the procedure is minor. As reported by Gorczyca et al. [19], the methodology of ISNT presents the advantage to detect DNA fragmentation on a single-cell level. In contrast, measurement of DNA content alone may not always be an adequate criterion of apoptosis because of possible changes in conformation of nuclear chromatin which may affect DNA accessibility to the dyes and the stoichiometry of its staining [21]. There are certain advantages of ISNT method over other assays of apoptosis [22]. This method directly detects DNA breaks (free 3′ hydroxyl ends of the DNA strand, to which dNTPs are being enzymatically added), rather than relying on more indirect evidence of DNA damage. Thus, activation of endonuclease, which is a very specific and well-characterized step of apoptosis, is actually being measured.

For example, the incorporation of the precursor under conditions of *in situ* nick translation is evidence of extensive DNA breaks, but cannot, *per se*, be proof of apoptosis. Other probes, therefore, should be used in parallel, such as analysis of cell morphology, plasma membrane assays or agarose gel electrophoresis, to identify the mode of cell death.

In this study, to assess DNA fragmentation, electrophoresis is performed. DNA purification is improved with ultrafree-probind filters. This device provides high recoveries of purified nucleic acid, using a special synthetic membrane which binds protein preferentially over double-stranded DNA. It is an alternative technique to phenol/chloroform DNA extraction, thus avoiding the use of toxic solvents. It is also a rapid method to recover in one centrifugation step a purified DNA solution containing only low weight (< 10 Kb) DNA.

Nevertheless, ethidium bromide staining might not be sensitive enough to determine low amounts of DNA fragments. This implies a sufficient percentage of cells undergoing apoptosis, to visualize the

DNA ladder. To overcome this lack of sensitivity, other procedures of DNA staining have been described [23].

As flow cytometry does not require destruction of the cells, it allows a more extensive study of the different steps leading to apoptosis, whereas cell lysate DNA electrophoresis only reveals one of the latest nuclear events, thereby this technique is currently used to confirm apoptosis detected by other methods [16, 17, 20].

References

1. Kerr, J.F.R., Wyllie, A.H. and Currie, A.R. (1972) Apoptosis: a basic biological phenomenon with wide ranging implications in tissue kinetics. *Br. J. Cancer* 26: 239–257.
2. Barr, P.J. and Tomei, L.D. (1994) Apoptosis and its role in human disease. *Bio/Technology* 12: 487–493.
3. Bulkley, G.B. (1983) The role of oxygen free radicals in human disease processes. *Surgery* 94: 407–411.
4. Buttke, T.M. and Sandstrom, P.A. (1994) Oxidative stress as a mediator of apoptosis. *Immunol. Today* 15: 7–10.
5. Halliwell, B. and Gutteridge, J.M.C. (1990) Ionizing and ultraviolet radiation generate ROI such as H_2O_2 and OH·. *Methods Enzymol.* 186: 1–85.
6. Forrest, V.J., Kang, Y.H., McClain, D.E., Robinson, D.H. and Ramakrishnan, N. (1994) Oxidative stress-induced apoptosis prevented by trolox. *Free Radiation Biol. Med.* 16: 675–684.
7. Lennon, S.V., Martin, S.J. and Cotter, T.G. (1991) Dose-dependent induction of apoptosis in human tumour cell lines by widely diverging stimuli. *Cell Prolif.* 24: 203–214.
8. Duval, E. and Wyllie, A.H. (1986) Death and the cell. *Immunol. Today* 7: 115–119.
9. Langley, R.E., Palayoor, S.T., Coleman, C.N. and Bump, E.A. (1993) Modifiers of radiation-induced apoptosis. *Radiat. Res.* 136: 320–326.
10. Bump, E.A. and Brown, J.M. (1990) Role of glutathione in the radiation response of mammalian cells *in vitro* and *in vivo*. *Pharmacol. Ther.* 47: 117–136.
11. Von Sonntag, C. (1987) *In: The Chemical Basis of Radiation Biology*. Taylor & Francis, London.
12. Raleigh, J.A. (1987) Radiation peroxidation in model membranes. *In:* T.L. Walden, Jr. and H.N. Hughes (eds): *Prostaglandin and Lipid Metabolism in Radiation Injury*. Plenum Press, New-York, pp 3–28.
13. Hockenberry, D.M., Oltval, Z.N., Yin, X.M., Millman, C.L. and Korsmeyer, S.J. (1993) Bcl-2 functions in an antioxidant pathway to prevent apoptosis. *Cell* 75: 241–251.
14. Cohen, J.J. and Duke, R.C. (1984) Glucocorticoid activation of a calcium-dependent endonuclease in thymocyte nuclei leads to cell death. *J. Immunol.* 132: 38–42.
15. Wyllie, A.H. (1980) Glucocorticoid-induced thymocyte apoptosis is associated with endogenous endonuclease activation. *Nature* 284: 555–556.
16. Ormerod, M.G., Collins, M.K.I., Rodriguez-Tarduchy, G. and Robertson, D. (1992) Apoptosis in interleukin-3-dependent haemopoietic cells. *J. Immunol. Methods* 153: 57–65.
17. Nicoletti, I., Migliorati, G., Pagliacci, M.C., Grignani, F. and Riccardi, C.A. (1992) A rapid and simple method for measuring thymocyte apoptosis by propidium iodide staining and flow cytometry. *J. Immunol. Methods* 139: 271–279.
18. Gold, R., Schmied, M., Rothe, G., Zischler, H., Breitschopf, H., Wekerle, H. and Lassmann, H. (1993) Detection of DNA fragmentation in apoptosis: application of *in situ* nick translation to cell culture systems and tissue sections. *J. Histochem. Cytochem.* 41: 1023–1030.
19. Gorczyca, W., Melamed, M.R. and Darzynkiewicz, Z. (1993) Apoptosis of S-phase HL-60 cells induced by DNA topoisomerase inhibitors: detection of DNA strand breaks by flow cytometry using the *in situ* nick translation assay. *Toxicol. Lett.* 67: 249–258.

20. Afanasyev, N.V., Korol, B.A., Matylevich, N.P., Pechatnikov, V.A. and Umansky, S.R. (1993) The use of flow cytometry for the investigation of cell death. *Cytometry* 14: 603–609.
21. Darzynkiewicz, Z., Traganos, F., Kapuscinski, J., Staiano-Coico, L. and Melamed, M.R. (1984) Accessibility of DNA *in situ* to various fluorochromes: relationship to chromatin changes during erythroic differentiation of Friend leukemia cells. *Cytometry* 5: 355–363.
22. Darzynkiewicz, Z., Bruno, S., Del Bino, G., Gorzyca, W., Hotz, M.A., Lassota, P. and Traganos, F. (1992) Features of apoptoic cells measured by flow cytometry. *Cytometry* 13: 795–808.
23. Facchinetti, A., Tessarollo, I., Mazzocchi, M., Kingston, R., Collavo, D. and Biasi, G. (1991) An improved method for the detection of DNA fragmentation. *J. Immunol. Methods* 136: 125–131.

Analysis of Free Radicals in Biological Systems
Favier et al. (eds)

Measurement of plasma sulfhydryl and carbonyl groups as a possible indicator of protein oxidation

P. Faure and J.-L. Lafond

Laboratoire de Biochimie C, CHU Albert Michallon, BP217, F-38043 Grenoble Cedex 9, France

Summary. Free radicals species can modify proteins as well as lipids. *In vitro* study of protein oxidation involves the evaluation of amino acid degradation, bityrosine production as well as structural modifications of proteins. The protein electrophoresis is relevant to determine protein aggregation and/or protein fragmentation. The change of protein solubility is linked to the modifications of their secondary and tertiary structure. Direct evidence of protein conformational changes secondary to free radical attack can be obtained with X-ray diffraction studies.

However, few methods are available in clinical chemistry to evaluate *in vivo* protein free radical attack. The measurement of sulfhydryl and carbonyl are an interesting way to evaluate the free radical attack of plasma protein. These measurements involve a spectrophotometric method using respectively for thiol and carbonyl groups 5,5'-dithio-bis(2-nitrobenzoic acid) and 2,4-dinitrophenylhydrazine.

Introduction

The implications of free radicals in protein and lipid damage in biological systems is becoming important. These modifications are suggested to occur in the course of many degenerative diseases such as diabetes mellitus, cancer or chronic renal failure. For instance exposure of lipids to free radicals and the resulting lipid peroxidation products, leads to vascular lesions or accelerated aging and inflammation. Moreover, the exposure of proteins to oxygen free radicals (OFR) also causes gross changes leading to the structural modification of their primary, secondary or tertiary assembly. Consequently, their enzymatic or hormonal activities can be modified, their susceptibility to spontaneous or induced fragmentation can be increased, and their immunological reactivity can be changed. These modifications are also involved in pathologic processes such as aging, inflammation or vascular lesions [1, 2, 3]. Despite considerable interest in this area, neither the mechanism(s) by which proteins are damaged nor their role in numerous diseases are well established and major questions still remain regarding protein oxidation at different steps of the evolution of chronic diseases. This may be due to difficulties in monitoring *in vivo* protein modifications by OFR. Therefore, we describe two methods that are relevant to study plasma protein modifications: the determination of thiol (SH) and carbonyl groups (C=O). But first, to better understand the mechanisms of protein damage secondary to free radical attack, it is important to catalogue the major modifications following the exposure of proteins to OFR. Generally, these modifications are obtained by *in vitro* systems on isolated proteins but one can speculate that similar modifications also occur during *in vivo* production of free radicals.

General modifications following protein exposure to free radicals

Loss of amino acids

Although all amino acids (AA) are susceptible to modifications secondary to free radical attack, some of them are more sensitive. The primary AA to undergo free radical modification are tryptophan, tyrosine, cysteine [4] and histidine [5]. The different species of free radical have different effects on AA. For instance, if a protein is exposed to $OH + O_2^{\cdot-}$ the loss of amino acids is not increased as compared to $\cdot OH$ alone. Moreover, the use of different chemical scavengers of $\cdot OH$ in system producing different free radical species, reveals that $\cdot OH$ is the primary radical responsive for amino acid modifications. The exception is cysteine residue which is both damaged by OH and $O_2^{\cdot-}$ resulting in a reduction of sulfhydryl groups.

Reaction between tryptophan and $\cdot OH$ results in the destruction of its aromatic groups, measured by a decrease in spontaneous fluorescence (280 nm exc., 340–350 nm em.). This observation can also be obtained with phenylalanine and tyrosine which have aromatic groups. Hence the interpretation of a protein auto fluorescence decrease must take into account the accessibility of AA to photon excitation [6] (i.e., depending on the secondary and tertiary structures of the protein). After free radical attack, tyrosine forms bityrosine. In this process, tyrosyl radicals react with other tyrosyl radicals or with tyrosine residues to form several stable biphenolic compounds. The 2, 2, biphenol bityrosine appears to be the major product of this reaction. Bityrosine production can be assessed by fluorescence at 325 nm excitation and 410–420 nm emission in comparison with authentic bityrosine or bityramine [4]. Generally, amino acid modifications caused by $\cdot OH$ alone or by $\cdot OH + O_2^{\cdot-}$ progressively affect the overall electrical charge of the protein leading to new isoelectric focusing bands. The alterations of AA provide a key to an understanding of the modification of the primary and secondary structure of proteins and their increased susceptibility to fragmentation following an exposure to OFR.

Protein aggregation and fragmentation

Exposure of protein to high ˙OH flux induces a generalized aggregation as shown by SDS-PAGE or acid precipitation studies. In contrast, exposure to both ˙OH and $O_2^{˙-}$ generally produces a disperse pattern of lower molecular weight fragmented proteins. To reveal the nature of the molecular bonds responsible for protein aggregation, comparison of non denaturing PAGE and SDS + PAGE + dithiothreitol can be performed. Such experiments show that most of the protein aggregates induced by ˙OH can be attributed to new intermolecular S–S bonds. Moreover, few aggregation products can be assigned to non covalent interactions. A denaturing electrophoresis (SDS PAGE + urea) performed on a protein exposed to both ˙OH + $O_2^{˙-}$ compared to a non denaturing electrophoresis, reveals that the aggregation products are conglomerates of protein fragmented together by non covalent attractions such as hydrophobic or ionic bonds. When proteins are exposed to hydroxyl radical and then incubated with proteolytic systems they are degraded almost 50 times faster than untreated proteins [7]. In contrast, superoxide ($O_2^{˙-}$) does not cause fragmentation of the protein in presence of cell proteolytic systems. Degradation in erythrocyte extracts seems to be accomplished via a ATP and Ca^{2+} independent pathway and, inhibitor profiles indicate that it could involve a metalloprotease and serine protease [8]. Then oxidative denaturation of proteins may increase the effective mechanisms for intracellular proteolysis and have important implications for the regulation of protein turnover [9, 10].

Modification of secondary and tertiary structure

The alteration of the primary structure of proteins results in gross modifications of secondary and tertiary structure [11]. These changes can be evaluated by the solubility of the proteins [12]. In high salt concentration the repulsion of hydrophobic groups by the solvent cage is maximized. In the native state the hydrophobic residues are shielded from the aqueous environment, contrary to the denatured state where hydrophobic residues are not protected. In these conditions, a denatured unfolded protein exhibits decreased solubility at its isoelectric point in high salt buffer because its exposed hydrophobic groups cluster together and cause precipitation. The study of protein solubility is relevant in estimating its secondary and tertiary structure modifications as a dose-dependent relation can be observed between the free radical production and the loss of protein solubility. This method only gives an indirect evidence of the conformational changes of a protein contrary to the X-ray diffraction study which is a direct approach to examine protein conformation. However, the X-ray study requires a large amount of pure material which is often difficult to obtain from biological samples.

Sulfhydryl groups in proteins

Sulfhydryl reactivity

Sulfhydryl groups can be schematically divided into two groups:

– non-protein thiols which are represented by very low molecular weight compounds as aminoacids (cysteine) or glutathione.
– high molecular weight thiols are exclusively represented by protein thiols. In this chapter we will only focus on some aspects of the latter. Readers interested in chemical aspects of thiols can refer to the literature [13, 14].

Particular properties of sulfhydryl in proteins

In proteins sulfhydryl groups play several important roles, some of which have been known for a long time:

– conformational structure;
– catalyst in the reactive site of numerous enzymes: For example, thiol is the active group of thiol protease. Standardized reagents are used for measuring creatine kinase activity in biological fluid containing SH donor as N acetyl cysteine;
– binding of the substrate for some enzymes;
– binding of some subunits;
– conformational change linked with allosteric processes;
– protective role against free radical effects.

Reactivity of protein thiol with reagents is linked to their accessibility: some react very quickly, some slowly, and others react only with denatured proteins. In enzymes, substrates can inhibit the reactivity of some thiol groups to a certain extent.

Sulfhydryl groups and metals

Metal ions seem to exert various and paradoxical effects on protein thiols. Stabilization of sulfhydryl groups can be enhanced by complexation with metal ions. The consequent ring formation gives an extra stability to the S-metal compounds. Zinc is often implicated in this chelate formation. For example, it seems that spermatozoid chromatin is protected, against early decondensation process, by the high content of zinc in the seminal fluid. This method probably protects thiol groups in histones by complexation with sulfhydryl groups [15].

Sulfhydryl groups are relatively unstable, unlike S–S which are more stable, in any case thiol groups are not oxidized spontaneously in the absence of metal ions. Moreover, other factors influencing oxidation, such as pH, temperature, oxygen, and metal ions are potent catalysts for this degradation.

Protein thiol and free radicals

The radioprotective action of low molecular weight thiols is well known. Glutathione, the major non-protein sulfhydryl compound, plays an important role in the protection of cells toward the ionizing radiation. Protein thiol seems to be involved in this protection, and probably takes part in the repair process as well [16].

The cell toxicity of some drugs such as Adriamicin and Menadione (Vitamine K derivatives) is probably due to oxygen radical generation, during metabolism. Alterations of protein thiol in cell or organelles membranes can alter Ca^{2+} efflux with an increase in cytosolic free Ca^{2+} level. This efflux precedes cell death [17].

The susceptibility of a thiol group to oxidation depends on the type of radical generated and how they are generated. Hydroxyl radical ($^{\cdot}OH$) seems to be able to react with protein SH groups, particularly when they are generated by radiolysis, near these SH sites. But OH is unable to reach protein SH when generated in the medium, by metal/H_2O_2 system. More diffusible species such as the hydroperoxyl and superoxide radicals, can be rapidly scavenged by protein thiol. Numerous authors have speculated that protein thiol could be an immediate reservoir which can protect against free radical attack [18]. Glutathione seems to be the only thiol that is able to prevent peroxidation.

Relation with nitric oxide ($^{\cdot}NO$)

Thiol seems to enhance $^{\cdot}NO$ production or $^{\cdot}NO$ protection: Properties of $^{\cdot}NO$ donors such as trinitrine are known and used in clinical medicine, even though the therapeutic action is not fully understood. The clinicians have observed that the drug lowers its efficiency when it has been used for long-time therapy. It was also noted that a low molecular weight molecule such as N acetyl cysteine (NAC) can counter this side-effect. Knowing that the action of trinitrine is linked to its capacity to liberate ˙NO, one can now speculate about the effect of thiol molecule. Nitroprussiade which liberates ˙NO is not affected by a decreased reactivity [19]. But other authors pointed out that NAC has hypotensive action without any relation to ˙NO, and that it does not scavenge $O_2^{˙-}$, as SOD [20].

Thiols seem to react with ˙NO metabolites: Peroxynitrite, a very potent oxidant, is probably produced during ˙NO metabolism, and can react with protein thiol at a rate 2600 times faster than H_2O_2. Using electrophoresis, some authors have shown that this action does not generate any intermolecular disulfide bounds. Therefore, one can speculate that SH is rather oxidized to sulfinic $(R–SO_2^-)$ and sulfonic $(R–SO_3^-)$ acids [21].

Thiol can also react with ˙NO derivative oxidants to form nitrosothiol derivatives which are more stable than ˙NO itself. It is postulated that these compounds, especially in plasma proteins, could be considered as stable ˙NO reservoir [22].

Analytical procedures

Numerous methods have been developed (amperometric, gas chromatographic, etc.) in order to determine the thiol group [13, 14]. Here, one of the colorimetric methods will be described.

For instance, aromatic disulfurs oxidize the thiols [23]. In this group, a water-soluble compound is used: 5,5'-dithiobis(2-nitrobenzoic acid) (DTNB) (Fig. 1) [13]. This reagent is reduced by thiol and gives a yellow derivative which absorbs around 412 nm (Fig. 2). This reagent is the most frequently used for colorimetric assay of thiol groups. There are some limitations in using this compound, however.

It is unsuitable for hemo protein measurements, because these proteins absorb in the same wavelength.

The chromophores liberated by the reaction are intensely colored and tend to oxidize.

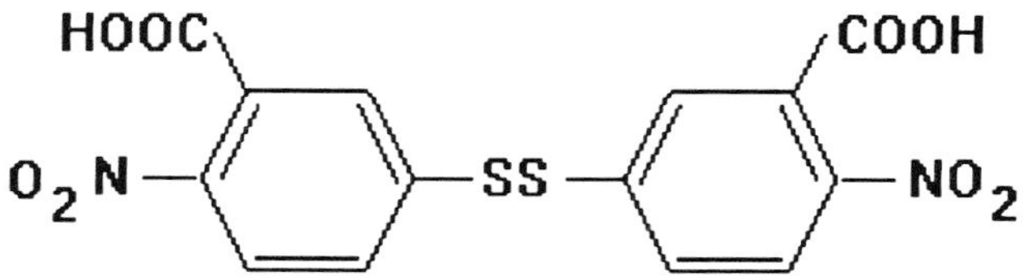

Fig. 1. 5,5'-dithiobis(2-nitrobenzoic acid).

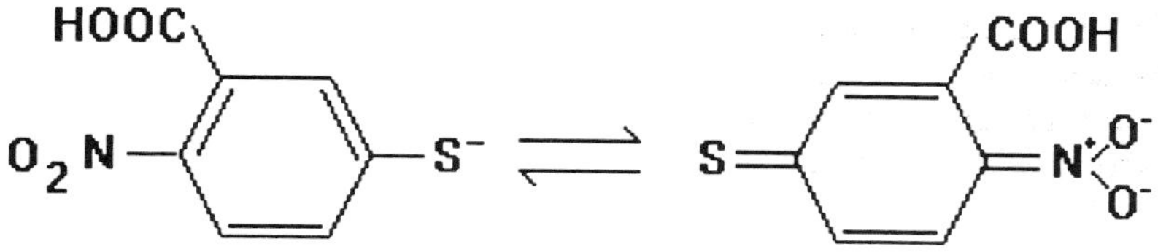

Fig. 2. *p*-nitrothiophenol anion.

The stability of Ellman's reagent is not very good, especially when it is exposed to natural light. DTNB is very sensitive to the action of light near 325 nm and is reduced to TNB. Ascorbate is unable to protect DTNB. To lower any damage to the reagent, it is necessary to operate in a dark room and to keep the reagents and the assay tubes in the dark [24].

An added problem, which is not specific to DTNB, is the lack of standardization.

Protein thiol can be evaluated using a kinetic reaction, allowing the reaction to continue for 1 h, and using a large amount of DTNB. Under these conditions, it is possible to observe a polyphasic absorbance corresponding to fast- and slow-reacting thiol groups [23].

The methods to distinguish protein thiol from no protein thiol consist of:

– protein precipitation with trichloracetic or perchloric acids, but the best one seems to be salicylsulphonic or metaphosphoric acids, since they present antioxidant properties [13];
– dialysis or gel chromatographies;
– using different pH and concentration of DTNB, fast (non protein thiol) and slow reactive (protein thiol) can be distinguished [25].

Other authors have slightly modified the methods in order to resolve specific problems. For example, the presence of H_2O_2 in the sample assay affects the reaction so catalase must be added before thiol analysis with Ellman reagent can be conducted [26].

The thiol groups buried in the protein can be determined after protein denaturation. This can be achieved with a high concentration of urea, or with sodium dodecyl sulfate. This method can also be applied to measure protein thiol in cells or in tissues [27].

Carbonyl groups in proteins

Introduction of carbonyl groups in amino acids residues is a hallmark for oxidative modification and can lead to structural modifications of proteins and increased susceptibility to proteolytic degradation. The mechanism of the introduction of carbonyl groups into proteins involves a metal ion catalyzed oxidation as described previously [28]. Consequently, assay for the carbonyl content of proteins is a suitable indicator for the evaluation of metal catalysed oxidation. We present here a colorimetric reaction using 2,4-dinitrophenylhydrazine. Other methods using radio-labeled reagents are available, in particular the reaction with tritiated borohydride. This is the most sensitive method, but it involves the utilization of a laboratory equipped to handle radio-labeled material. Thus, the results of the two methods are equal, and when one has sufficient sample, the 2,4-dinitrophenylhydrazine method provides a nonradiochemical assay for carbonyl groups in protein. The reaction involves the formation of Shiff bases as follows:

$$\text{Protein–C=O} \longrightarrow \text{Protein=N–NH–2,4-DNP} + H_2O$$

$$\text{Protein–C=N–Protein} \longrightarrow \text{Protein=N–NH–2,4-DNP} + H_2O$$

The first assays were performed without the extraction of the excess of 2,4-dinitrophenylhydrazine because of the possible loss of protein. This rendered the method rather insensitive and subsequent studies have shown that the reagent could be extracted without a significant loss of proteins [29].

Technical aspects

Carbonyl groups measurement

Reagents
– HCl, 2 M;
– 2,4-dinitrophenylhydrazine, 10 mM, in HCl 2 M;
– Trichloroacetic acid, 20% (w/v);
– Guanidine 6 M, with potassium phosphate 20 mM adjusted to pH 2.3 with trifluoroacetic acid.

Procedure
The blood sampling must be performed with heparinized tubes and a minimum of 500 μl of plasma is required (200 μl for the blank, 200 μl for the reaction, and 100 μl to determine the protein concentration). Remove the plasma from blood sample by centrifugation (3000 g, 15 min) and pipet into 1.5 ml centrifuge tubes. Aliquot the plasma sample for the blank and for the reaction and dilute each aliquot with 800 μl deionized water. Precipitate each protein solution with trichloroacetic acic (2 vol for 1 vol of plasma), centrifugate 10 min (3000 g) and discard the supernatant. Add 1 ml of 10 mM 2,4-dinitro-

phenylhydrazine in 2 M HCl and allow to stand at 37°C for 50 min, with vortexing every 15 min. Parallelly use 1 ml of 2M HCl as blank solution and allow it to stand at the same temperature. Add 1 ml of 20% trichloroacetic acid, centrifuge the tubes (3000 g) for 15 min, and discard the supernatant. Wash the pellet in 1 ml ethanol/ethyl acetate solution (1:1) to remove free reagent: allow the sample to stand 10 min before centrifugation and discard the supernatant each time. This procedure must be repeated until a clear supernatant is obtained. To be able to read the colorimetric reaction, you have to dissolve the precipitated protein in 1 ml guanidine solution. Plasma is generally dissolved in 30 min at 37°C. The insoluble material must be removed by centrifugation. Obtain the O.D. of the reaction and of the blank at 380 nm and calculate the carbonyl content (after subtracting with the blank value) using the molar absorption coefficient of 22 000/M/cm. Use the 100 μl plasma to determine the protein concentration of the plasma sample. Protein recovery is generally excellent but it may be checked by its measure at 280 nm and calculated using a standard curve of bovine albumin also treated with guanidine.

Results
The precision of the method shows that the CV of the within run (n = 30) is 4.5% and the CV of the between run is 7%. The reference value of the laboratory, established in 30 apparently healthy adults (age, 25 to 48 years) is 0.35 ± 0.08 μmol/g of proteins.

Total sulfhydryl groups measurement using Ellman's reagent

Reagents
– 0.05 M and 0.2 M pH 8 phosphate buffer;
– Bovine serum albumin (BSA) 10 g/l stock solution;
– N acetyl cysteine (NAC) 1 mM stock solution;
– Ellman's reagent: DTNB 10 mM in 0.2 M phosphate buffer.

Procedure
Blood collection and preparation are shown for carbonyl assay. Plasma is diluted 10 times before analysis. Standard curves are made from BSA or NAC stock solutions. In a plastic cuvette add:

– 500 μl of water for blank, dilutes solution of BSA or NAC stock or, dilution of plasma.
– 750 μl of 0.05 M phosphate buffer.

Gently mix and measure the first O.D. at 412 nm. Then add, in all the cuvettes, 250 μl of freshly prepared Ellman's reagent and after mixing

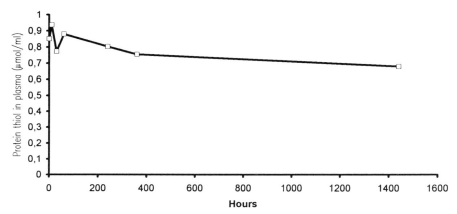

Fig. 3. Evolution of protein thiol in plasma at ambient temperature (0 to 4 h) and at −20°C.

observe a 15 mn delay with assay tubes in the dark. Then read the second O.D. at 412 nm.

Results

Results are based on the following data and expressed in μmol/ml of thiol:

– BSA molecular weight: 69 000
– Sulfhydryl content of BSA: 0.66 per mole [13].

CV of the within run for this method is 6.01% (n = 30) and our reference value (established with 30 blood donors) is $0.78 \pm 0.12\ \mu$mol/ml (values are around 0.6 μmol/ml [13]).

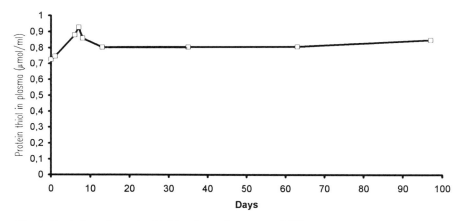

Fig. 4. Evolution of protein thiol in plasma frozen at −80°C.

Note: In order to minimize variation in protein content of plasma it seems better to express the results as nmol/g protein. With these units our reference value is 12.5 ± 1.9 nmol/g protein.

Protein thiol content of human plasma is stable for at least 4 h at room temperature or 24 h at $-20°C$. For longer preservation plasma must be frozen at $-80°C$ (Figs 3 and 4).

Acknowledgments
This work was made possible with the technical help of M. Rual and S. Charrel.

References

[1] Lunec, J. (1990) Free radicals: Their involvement in disease processes. *Ann. Clin. Biochem.* 27: 173–182.

[2] Merry, P.G., Winyard, C.J., Morris, M., Grootveld, M. and Blake, D.R. (1989) Oxygen free radicals, inflammation and synnovitis: the current status. *Annals. Rheum. Dis.* 48: 864–870.

[3] Oliver, C.N., Ahn, B.W., Moerman, E.J., Goldstein, S. and Stadman, E.R. (1987) Age related changes in oxidized proteins. *J. Biol. Chem.* 262: 5488–5491.

[4] Davies, K.J.A. (1987) Protein damage and degradation by oxygen free radicals. II. Modification of amino-acids. *J. Biol. Chem.* 262: 9902–9907.

[5] Levine, R.L. (1983) Oxidative modification of glutamine synthetase. I. Inactivation is due to loss of one histidine residue. *J. Biol. Chem.* 258: 11823–11833.

[6] Faure, P., Lafond, J.L., Coudray, C., Rossini, E., Favier, A., Halimi, S. and Blache, D. (1994) Zinc prevents the structural and functional properties of free radical treated insulin. *Biochimica Biophysica Acta* 1209: 260–264.

[7] Wolff, S.P. and Dean, R.T. (1986) Fragmentation of proteins by free radicals and its effect on their susceptibility to enzymatic hydrolysis. *Biochem. J.* 234: 399–403.

[8] Davies, K.J.A. (1987) Protein damage and degradation by oxygen free radicals. IV. Degradation of denatured proteins. *J. Biol. Chem.* 262: 9914–9920.

[9] Davies, K.J.A. (1987) Protein damage and degradation by oxygen free radicals. I. General aspects. *J. Biol. Chem.* 262: 9895–9901.

[10] Davies, K.J.A. (1988) A secondary antioxidant defense role for proteolytic systems. *In:* G.M. Simic, K.A. Taylor, J.F. Ward and C. Von Sonnag (eds.): *Oxygen Radicals in Biology and Medicine.* Plenum Press, New York and London.

[11] Capeilere-Blandin, C., Delaveau, T. and Descamps-Latscha, B. (1991) Structural modification of human b2 microglobulin treated with oxygen-derived radicals. *Biochem. J.* 277: 175–182.

[12] Davies, K.J.A. (1987) Protein damage and degradation by oxygen free radicals. III. Modification of secondary and tertiary structure. *J. Biol. Chem.* 262: 9908–9913.

[13] Jocelyn, P.C. (1972) *Biochemistry of the SH Group.* Academic Press, London, New York.

[14] Friedman, M. (1973) *The Chemistry and Biochemistry of the Sulfhydryl Group in Amino Acids, Peptides and Proteins.* Pergamon Press, New York.

[15] Kvist, U. (1980) Sperm nuclear chromatin decondensation ability. *Acta Physiol. Scand.* (Suppl.) 486: 1–24.

[16] Held, K.D. and Hopcia, K.L. (1993) Role of protein thiols in intrinsic radiation protection of DNA and cells. *Mutation Res.* 299: 261–269.

[17] Di Monte, D., Bellomo, G., Thor, H., Nicotera, P. and Orrenius, S. (1984) Menadione-induced cytotoxicity is associated with protein thiol oxidation and alteration in intracellular Ca^{++} homeostasis. *Arch. Biochem. Biophys.* 235: 343–350.

[18] Di Simplicio, P., Cheeseman, K.H. and Slater, T.F. (1991) The reactivity of the SH group of bovine serum albumin with free radicals. *Free Rad. Res. Comms.* 14: 253–262.

[19] Abrams, J. (1991) Interactions between organic nitrates and thiol groups. *Am. J. Med.* 91 (suppl. C): 106S–112S.

[20] Sunman, W., Hughes, A.D. and Sever, P.S. (1993) Free-radical scavengers, thiol-containing reagents and endothelium-dependent relaxation in isolated rat and human resistance arteries. *Clinical Science* 84: 287–295.

[21] Radi, R., Beckman, J.S., Bush, K.M. and Freeman, B.A. (1991) Peroxynitrite oxidation of sulphydryls. *J. Biol. Chem.* 266: 4244–4250.

[22] Girard, P. and Potier, P. (1993) NO, thiols and disulfides. *FEBS Letts.* 320: 7–8.

[23] Jocelyn, P.C. (1987) Spectrophotometric assay of thiols. *Methods Enzymol.* 143: 44–67.

[24] Walmsley, T.A., Abernethy, M.H. and Fitzgerald, H.P. (1987) Effect of daylight on the reaction of thiols with Ellman's reagent. 5,5-Dithiobis(2-Nitrobenzoic acid). *Clin. Chem.* 33/10: 1928–1931.

[25] Jocelyn, P.C. (1962) The effect of glutathione on protein sulphydryl groups in rat liver homogenates. *Biochem. J.* 85: 480–485.

[26] Suzuki, Y., Lyall, V., Biber, T.U.L. and Ford, G.D. (1990) A modified technique for the measurement of sulfhydryl groups oxidized by reactive oxygen intermediates. *Free Rad. Biol. Med.* 9: 479–484.

[27] Berson, G. and Marquet, A. (1990) Dosage des groupes sulfhydryles des protéines myofibrillaires cardiaques: effet des radicaux libres oxygénés in vitro. *C.R. Soc. Biol.* 184: 31–36.

[28] Stadman, E.R. (1990) Metal ion-catalyzed oxidation of proteins: Biochemical mechanism and biological consequences. *Free Radic. Biol. Med.* 9: 315–325.

[29] Levine, R.L., Garland, D., Oliver, C.N., Amici, A., Climent, I., Lenz, A.G., Ahn, B.-W., Shaltiel, S. and Stadman, E. (1990) Determination of carbonyl content in oxidatively modified proteins. *Methods Enzymol.* 186: 464–472.

Analysis of Free Radicals in Biological Systems
Favier et al. (eds)

Measurement of oxidized nucleobases and nucleosides in human urine by using a GC/MS assay in the selective ion monitoring mode

M.-F. Incardona[1], F. Bianchini[1], A.E. Favier[2] and J. Cadet[1]

[1]CEA/Département de Recherche Fondamentale sur la Matière Condensée-SESAM/LAN, F-38054 Grenoble Cédex 9, France
[2]Laboratoire de Biochimie C, CHRU de Grenoble, Hopital Michallon, B.P. 217 X, F-38043 La Tronche, France

Summary. Urinary excretion of oxidized bases and nucleosides has been suggested as a non-invasive biomarker of oxidative damage and repair. Oxidative adducts, resulting from a variety of reactive oxygen species, are in fact excreted in urine following repair. A method for the analysis of 5-hydroxyuracil (5-OHUra), 5-hydroxy-2'-deoxyuridine (5-OHdUrd), 5-hydroxymethyluracil (5-HMUra), 5-hydroxymethyl-2'-deoxyuridine (5-HMdUrd) in urine has been developed, based on GC/MS quantification. Stable isotopically labelled (M + 4) analogues, synthesized in our laboratory, are used as internal standards, assuring appropriate quantification in complex biological fluids. Urines are first purified by HPLC on a reverse phase column and fractions corresponding to the compounds of interest derivatized for GC/MS analysis. Two different derivatives for GC/MS determination, the trimethylsylyl (TMS) and the tert-butyldimethylsylyl (t-BDMS) derivatives, have been tested. Detection limit in the selective ion monitoring (SIM) mode is generally two-fold lower with t-BDMS. Preliminary results on four healthy subjects show that the oxidized bases 5-OHUra and 5-HMUra are excreted in urine (approximately 0.7 nmol/kg/day), while the corresponding nucleosides are not detectable. The developed method could have important applications in the analysis of oxidative lesions in DNA exposed to oxidative stress or in DNA from biological samples.

Introduction

The oxidative damage to DNA appears to be involved in ageing, carcinogenesis and mutagenesis [1–3]. Oxidized purine and pyrimidine bases constitute one of the major classes of oxidative DNA lesions, together with strand breaks, crosslinks and abasic sites. Hydroxyl radical, one-electron oxidation and singlet oxygen which may be produced by physical agents such as solar light and ionizing radiations are mainly involved in the formation of oxidized nucleobases [4, 5]. In addition, pathological conditions like infection and inflammation were also shown to induce a burst of oxygen radicals.

The measurement of oxidized DNA bases and nucleosides in urine has been proposed as a non-invasive approach for the monitoring of oxidative stress in humans and animals [6–10]. The release of oxidized bases and nucleosides in urine appears to be the result of DNA repair

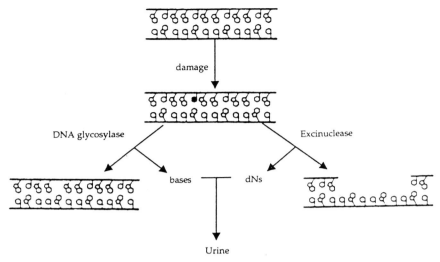

Fig. 1. Two enzymatic repair pathways of damaged DNA.

processes, namely base excision repair and nucleotide excision repair, respectively [11–14]. In the base excision repair the damaged base is first released by a glycosylase, giving rise to an apurinic site subsequently excised by endonucleases. In the nucleotide excision repair two phospho-diester bonds at both sides of the lesion are hydrolyzed, leading to the removal of the damaged nucleoside. However, the enzymatic release of oxidized nucleosides in biological fluids such as urine has been recently questioned [15] (Fig. 1). The GC/MS method associated with the selective ion monitoring (SIM) mode [16] is particularly suitable for the measurement of oxidized nucleosides and nucleobases in urine [17]. We developed a GC/MS assay for the measurement of four modified bases and nucleosides. These include 5-hydroxymethyluracil (5-HMUra), 5-hy-droxymethyl-2'-deoxyuridine (5-HMdUrd), 5-hydroxyuracil (5-OHUra) and 5-hydroxy-2'-deoxyuridine (5-OHdUrd). These compounds, whose

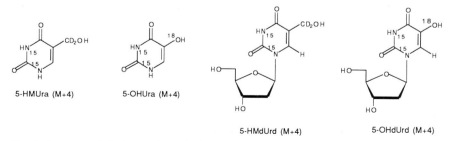

Fig. 2. Structure of the internal standards.

formation may involve ·OH reactions and/or one-electron oxidation [5, 18, 19] could be biologically important. In fact, 5-HMUra has been demonstrated to be present in the DNA of human blood cells [20]. In addition, a specific 5-HMUra glycosylase has been detected in several eukaryotes [21].

Accurate quantitative measurements require the use of stable isotopically labeled standards [22–24]. In this respect, emphasis was placed on the synthesis of enriched modified nucleobases and nucleosides. The internal standards differ from the compounds to be searched by at least three mass units due to the presence of stable isotopic atoms such as deuterium, nitrogen 15 and oxygen 18 (Fig. 2). Optimization of GC/MS analysis was also achieved in order to lower the threshold for detection of the DNA lesions described above. In the present chapter we will focus on the description of the method which includes HPLC prepurification of urine samples and GC/MS analysis.

Principle of the method

The experimental protocol utilized for the measurement of oxidized nucleobases and nucleosides in urine is summarized in Figure 3.

Materials and methods

Prepurification of oxidized bases and nucleosides in urine

Apparatus
The prepurification of the oxidized bases and nucleosides in the urine samples was achieved by HPLC separation on a 5 μm particle size Interchrom semi-preparative octadecylsilyl (ODS) silica gel column (250 × 10 mm) (Interchim, Montluçon, France) equipped with a RP-18 guard column. A Gilson model 201 controller fractions collector (Gilson International, Villiers-le-Bel, France) was programmed to collect 2 ml per fraction.

Eluents
The oxidized bases (5-OHUra and 5-HMUra) were purified on the ODS column by using 50 mM $HCOONH_4$ in water at a flow-rate of 2 ml/min. The purification of the oxidized nucleosides (5-OHdUrd and 5-HMdUrd) was achieved with a mixture of $HCOONH_4$-MeOH (95:5, v/v) at a flow-rate of 2 ml/min. After each cycle of purification, the column was washed successively with approximately 30 ml water and 30 ml methanol, prior to reconditioning with the appropriate solvent.

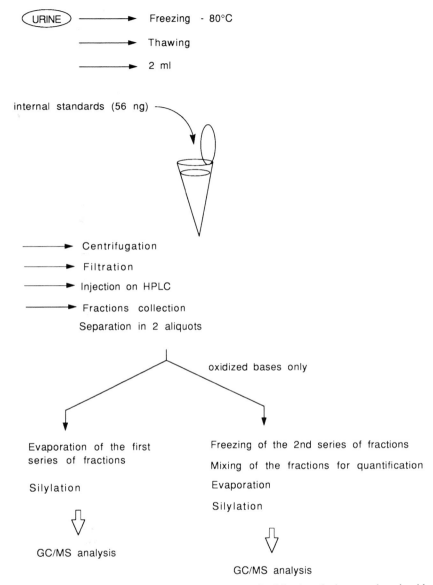

Fig. 3. Experimental procedure for the measurement of oxidized nucleobases and nucleosides in urine.

Methodology

Typically, urine samples (generally collected over 24 h) stored at −80°C, were thawed and internal standards (56 ng) were spiked in a sample size of 2 ml of urine. After centrifugation for 5 min, the supernatant was filtered and then injected twice on the ODS column. A slight isotopic effect was found to affect only the HPLC elution times of the oxidized bases. Therefore, in order to better locate the sample to be analyzed, each of the 4 ml HPLC fractions was divided into two:

The first series of fractions was used to determine where the compound of interest was eluted. Fractions were evaporated to dryness in a speedVac apparatus and the dry residues were resuspended in 500 μl acetonitrile and centrifuged. The supernatants were transferred to 2 ml glass vials (Touzard & Matignon, Paris, France) and dried under vacuum. The content of each vial was then analysed by GC/MS (vide infra). Generally, 5-OHUra was present in fractions 10 and 11 whereas 5-HMUra was in fractions 14 and 15.

The second series of fractions was frozen during the GC/MS analysis of the first series. Once the fractions containing the compound to be analyzed were determined, these were combined and analyzed as reported above.

The isotopic effect was not observed on the HPLC elution of the oxidized nucleosides. Under the eluting conditions (see above), 5-OHdUrd was present in fraction 15, whereas 5-HMdUrd was eluted in fraction 18.

GC/MS analysis

Apparatus

GC/MS analyses were performed using a Nermag, Model R10-10c mass spectrometer in the electron impact (EI) mode and a Nermag, Model DN gas chromatograph equipped with a Ros injector. The source temperature was set at 210°C. The separations were achieved on a 25 m × 0.32 mm CP Sil 5 capillary Chrompack column (Chrompack, Les Ulis, France) covered with a 1.2 μm film using helium as the carrier gas. Under these conditions, the column head pressure was 0.5 bar. The injection temperature was 250°C and the initial temperature of the column was 180°C. Immediately after the injection, the temperature was increased to 290°C at a rate of 7°C/min.

Derivatization

Two different silylating agents were used, N-(tert-butyldimethylsilyl)-N-methyl-trifluoro-acetamide (MTBSTFA) (Fluka, St. Quentin-Fallavier, France) and N-(trimethylsilyl)-N-methyl-trifluoro-acetamide (BSTFA) (Pierce, Rockford, Illinois). They form the tert-butyldimethylsilyl (t-BDMS) and the trimethylsilyl (TMS) derivatives, respectively. The detection limit of the TMS derivative was approximately two-to-five fold higher, compared to the t-BDMS derivative. In addition, with BSTFA as the derivatizing agent, results were less reproducible; an interfering peak was observed in the analysis of 5-HMUra. Therefore, t-BDMS derivatives were used in the protocol.

Typically, 100 µl of a mixture of acetonitrile-MTBSTFA (1:1, v/v) was added to the dry contents of the tubes which were then heated at 110°C for 20 min. It should be noted that each of the two nucleosides is converted into the corresponding nucleobase during derivatization.

Analysis and quantitation

Typically, 5 µl or 10 µl of each of *t*-BDMS derivatives of oxidized bases and nucleosides was injected. The quantification was obtained by monitoring the specific M-57 ions resulting from the loss of a *tert*-butyl group from the molecular ion:

— 427.4 and 431.4 for 5-HMUra or 5-HMdUrd

— 413.3 and 417.3 for 5-OHUra or 5-OHdUrd

Application of the method

Calibration curves
The calibration curves were obtained by measuring the peak area ratios of the base peak (M-57) for the four compounds of interest and the corresponding internal standards [24]. This was achieved by selecting the ions 427.4/431.4 for 5-HMUra and 5-HMdUrd (Fig. 4), while the calibration curve for 5-OHUra and 5-OHdUrd (Fig. 5) was calculated by measuring the ratio 413.3/417.3. The calibration curves were linear up to 112 ng. The detection thresholds for the four molecules are summarized

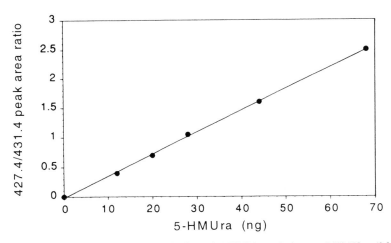

Fig. 4. Calibration curve for the quantitation of 5-HMUra relative to 5-HMUra (M + 4) (internal standard initial quantity: 28 ng; injection quantity: 140 pg).

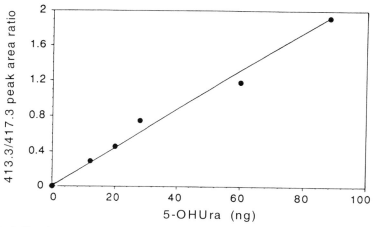

Fig. 5. Calibration curve for the quantitation of 5-OHUra relative to 5-OHUra (M + 4) (internal standard initial quantity: 28 ng; injection quantity: 140 pg)

Table 1. Detection threshold for nucleobases and nucleosides analysed

Compounds	5-HMUra	5-HMdUrd	5-OHUra	5-OHdUrd
Detection threshold (pg-pmoles)	20−0.14	60−0.23	40−0.31	120−0.49

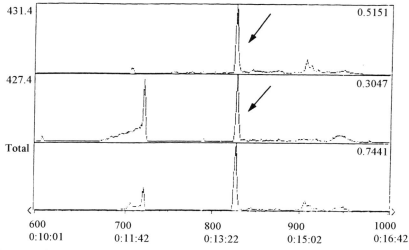

Fig. 6. GC/MS-SIM chromatogram of urinary 5-HMUra.

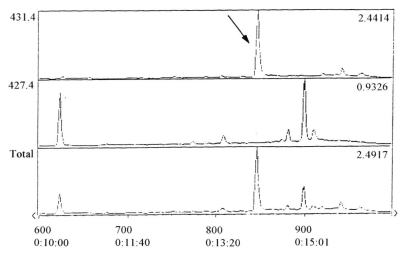

Fig. 7. GC/MS-SIM chromatogram of urinary 5-HMdUrd.

in Table 1. 5-OHUra, 5-OHdUrd, 5-HMUra, 5-HMdUrd background levels in human urine were determined using these calibration curves.

Standard recovery
The averaged recovery of 5-OHUra and 5-HMUra standards from urine was between 70 and 80%. However, the presence of the internal

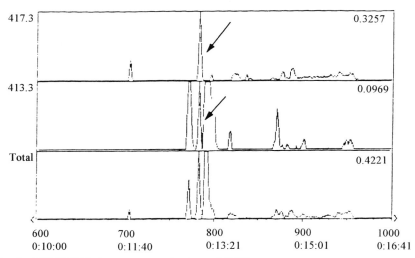

Fig. 8. GC/MS-SIM chromatogram of urinary 5-OHUra.

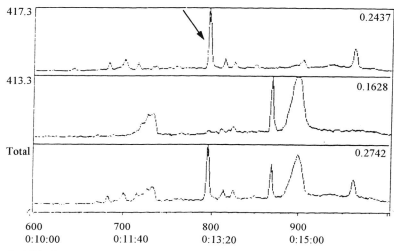

Fig. 9. GC/MS-SIM chromatogram of urinary 5-OHdUrd.

standard helped correcting for the partial loss of the compounds during analysis.

Application to human urine
The four DNA lesions were searched in 6 h urine samples of four healthy subjects. The corresponding chromatograms are shown in Figures 6–9. It was clearly shown that both 5-HMUra and 5-OHUra are present in human urine. The mean level of 5-HMUra was 10.5 nmoles/mmole of creatinine, with a variation of 1.7 nmoles/mmole between individuals. The variation due to experimental error was 6%. The presence of interfering impurities in the GC/MS elution profile led to a less accurate determination of 5-OHUra compared to 5-HMUra. Nevertheless, after averaging ten measurements, 5-OHUra level in normal urine was estimated to be similar to that of 5-HMUra. On the contrary, the corresponding nucleosides were not detectable, indicating that their levels were, at best, lower than 0.6 nmoles/mmole of creatinine for 5-HMdUrd and lower than 1.3 nmoles/mmole of creatinine for 5-OHdUrd, which represents at least ten-fold lower levels than nucleobases.

Extension of the method: Measurement within DNA

The GC/MS method is also suitable for measuring the oxidized nucleobases and nucleosides within naked DNA exposed to various conditions of oxidation. The assay may also be applicable to monitoring the levels of oxidative damage in cellular DNA exposed to oxidative stress and in DNA obtained from biological samples. However, prior hydrolysis of DNA is necessary in this case. Two approaches including acidic treatment and enzymatic digestion are usually employed to induce the release of modified bases and nucleosides, respectively [25]. The first method consists of the cleavage of the N-glycosidic bond. This can be achieved using hydrogen fluoride stabilized in pyridine [26–28]. The method is reliable to obtain a quantitative release of 5-OHUra. However, this method is not applicable to 5-HMUra as it is significantly unstable under acidic conditions. An alternative approach involves the enzymatic digestion of DNA into nucleosides. This is achieved with enzymes such as nucleases and phosphatases (Douki et al., this volume, pp. 213–224). In both cases, however, it seems necessary to separate the unmodified from the modified bases (or nucleosides). This procedure should avoid any artefacts due to oxidation of the normal constituents during the steps of silylation or GC/MS analysis. One particular approach for purification is immunoaffinity chromatography, as illustrated for the prepurification of 8-oxoguanine compounds [29].

Measurements of the oxidative lesions in DNA, i.e., from lymphocytes, and in the urine from the same individuals could provide useful information for validating the possible use of urinary modified bases and nucleosides as biomarkers of oxidative damage, and for clarifying mechanisms and extent of DNA repair.

Acknowledgments
This work was partly supported by Commission of the European Communities, "Environmental exposures: modulation of host factors and biomonitoring end points". M.-F.I. acknowledges the receipt of a PhD fellowship from l'Oreal.

References

1. Adelman, R., Saul, R.L. and Ames, B.N. (1988) Oxidative damage to DNA: relation to species metabolic rate and life span. *Proc. Natl. Acad. Sci. USA* 85: 2706–2708.
2. Simic, M.G. (1989) Mechanisms of inhibition of free radical processes in mutagenesis and carcinogenesis. *Mutat. Res.* 202: 377–386.
3. Fraga, C.G., Shinenaga, M.K., Park, J.W., Degan, P. and Ames, B.N. (1990) Oxidative damage to DNA during aging: 8-hydroxy-2'-deoxyguanosine in rat organ DNA and urine. *Proc. Natl. Acad. Sci. USA* 87: 4533–4537.
4. von Sonntag, C.V. (1987) *The Chemical Basis of Radiation Biology*. Taylor & Francis, London.
5. Cadet, J. (1994) DNA damage caused by oxidation, deamination, ultraviolet radiation and photoexcited psoralens. *In*: K. Hemminki, A. Dipple, D.E.G. Shuker, F.F. Kadlubar, D. Segerback and H. Bartsch (eds): *DNA Adducts: Idenitfication and Biological Significance*, IARC Scientific Publication No. 125, IARC Lyons, pp 245–276.
6. Bergtold, D.S., Simic, M.G, Alession, H. and Cutler, R.G. (1988) Urine biomarkers of oxidative DNA damage. *In*: M.G. Simic, K.A. Taylor, J.F. Ward and C. von Sonntag (eds): *Oxygen Radicals in Biology and Medicine*. Plenum Press, New York, pp 483–490.

7. Cundy, K.C., Kohen, R. and Ames, B.N. (1988) Determination of 8-hydroxyguanosine in human urine: a possible assay for *in vivo* DNA damage. *In:* M.G. Simic, K.A. Taylor, J.F. Ward and C. von Sonntag (eds): *Oxygen Radicals in Biology and Medicine*. Plenum Press, New York, pp 479–482.

8. Loft, S., Vistisen, K., Ewertz, M., Tjonneland A., Overvad, K. and Poulsen, E.H. (1992) Oxidative DNA damage estimated by 8-hydroxydeoxyguanosine excretion in humans: influence of smoking, gender and body mass index. *Carcinogenesis* 13: 2241–2247.

9. Shinenaga, M.K., Gimeno, C.J. and Ames, B.N. (1989) Urinary 8-hydroxy-2'-deoxyguanosine as a biological marker of *in vivo* oxidative DNA damage. *Proc. Natl. Acad. Sci. USA* 86: 9697–9701.

10. Shigenaga, M.K., Aboujaoude, E.N., Chen, Q. and Ames, B.N. (1994) Assays of oxidative DNA damage biomarker 8-oxo-2'-deoxyguanosine and 8-oxoguanine in nuclear DNA and biological fluids by high-performance liquid chromatography with electrochemical detection. *In:* L. Packer (ed.): *Oxygen Radicals in Biological Systems*, Methods in Enzymology, Vol. 234, Academic Press, San Diego, CA, pp 16–33.

11. Boiteux, S., Gajewski, E., Laval, J. and Dizdaroglu, M. (1992) Substrate specificity of the *Escherichia coli* Fpg protein: excision of purine lesions in DNA produced by ionizing radiation or photosensitization. *Biochemistry* 31: 106–110.

12. Demple, B. and Harrison, L. (1994) Repair of oxidative damage to DNA: enzymology and biology. *Ann. Rev. Biochem.* 63: 918–948.

13. Hamilton, K.K., Lee, K. and Doetsch, P.W. (1994) Detection and characterization of eukaryotic enzymes that recognize oxidative DNA damage. *In:* L. Packer (ed.): *Oxygen Radicals in Biological Systems*, Method in Enzymology, Vol. 234, Academic Press, San Diego, CA, pp 33–44.

14. Tchou, J., Bodepudi, V., Shibutani, S., Antoshechkin, I., Miller, J., Grollman, A.P. and Johnson, F. (1994) Substrate specificity of Fpg protein. *J. Biol. Chem.* 269: 15318–15324.

15. Lindhal, T. (1993) Instability and decay of the primary structure of DNA. *Nature* 362: 709–715.

16. Fuciarelli, A.F., Wegher, B.J., Gajewski, E., Dizdaroglu, M. and Blakely, W.F. (1989) Quantitative measurement of radiation-induced base products in DNA using gas-chromatography-mass spectrometry. *Radiat. Res.* 119: 219–231.

17. Djuric, Z., Luongo, D.A. and Harper, D.A. (1991) Quantitation of 5-hydroxymetlyuracil in DNA by gas chromatography with mass spectral detection. *Chem. Res. Toxicol.* 4: 687–691.

18. Wagner, J.R., van Lier, J.E., Decarnoz, C., Berger, M. and Cadet, J. (1990) Photodynamic method for oxyradical induced DNA damage. *In:* L.Packer and A.N. Glazer (eds): *Oxygen Radicals in Biological Systems*, Method in Enzymology, Vol 186, Academic Press, pp 502–511.

19. Wagner, J.R., Hu, C.-C. and Ames, B.N. (1992) Endogenous oxidative damage of deoxycytidine in DNA. *Proc. Natl. Acad. Sci. USA* 89: 3380–3384.

20. Djuric, Z., Heilburn, L.K., Reading, B.A., Booner, A., Valeriote, F.A. and Martino, S. (1991) Effects of a low-fat diet on levels of oxidative damage to DNA in human peripheral nucleated blood cells. *J. Natl. Cancer Inst.* 83: 766–769.

21. Boorstein, R.J., Levy, D.D. and Teebor, G.W. (1987) Phylogenetic evidence of a role for 5-hydroxymethyluracil-DNA glycosylase in the maintenance of 5-methylcytosine in DNA. *Nucleic Acids Res.* 17: 7653–7661.

22. Stillwell, W.G., Xu, H.-X., Adkins, J.A., Wishnok, J.S. and Tannenbaum, S.R. (1989) Analysis of methylated and oxidized purines in urine by capillary gas chromatography-mass spectrometry. *Chem. Res. Toxicol.* 2: 94–99.

23. Faure, H., Incardona, M.-F., Boujet, C., Cadet, J., Ducros, V. Favier, A. (1993) Gas chromatographic-mass spectrometry determination of 5-hydroxymethyluracil in human urine by stable isotopic dilution. *J. Chromatogr.* 613: 1–7.

24. Dizdaroglu, M. (1993) Quantitative determination of oxidative base damage in DNA by stable isotopic-dilution mass spectrometry. *FEBS Lett.* 315: 1–6.

25. Cadet, J. and Weinfeld, M. (1993) Detecting DNA damage. *Anal. Chem.* 65: 675A–682A.

26. Polverelli, M., Berger, M., Mouret, J.F., Odin, F. and Cadet, J. (1990) Acidic hydrolysis of the N-glycosidic bonds of deoxyribonucleic acid by hydrogen fluoride stabilized in pyridine. *Nucleosides Nucleotides* 9: 451–452.

27. Cadet, J., Ravanat, J.L., Buchko, G.W., Yeo, H.C. and Ames, B.N. (1994) Single oxygen DNA damage: chromatographic and mass spectrometric analysis of damage products. *In:* L. Packer (ed.): *Oxygen Radicals in Biological Systems*, Methods in Enzymology, Vol. 234, Academic Press, San Diego, CA, pp 79–88.
28. Douki, T., Voituriez, L. and Cadet, J. (1995) Measurement of pyrimidine (6-4) pyrimidone photoproducts in DNA by a mild acid hydrolysis and HPLC-fluorescence detection assay. *Chem. Res. Toxicol.* 8: 244–253.
29. Degan, P., Shigenaga, M.K., Park, E.M. Alperin, P.E. and Ames, B.N. (1991) Immunoaffinity isolation of urinary 8-hydroxy-2'-deoxyguanosine and 8-hydroxyguanine and quantitation of 8-hydroxy-2'-deoxyguanosine in DNA by polyclonal antibodies. *Carcinogenesis* 12: 865–871.

Analysis of Free Radicals in Biological Systems
Favier et al. (eds)

Methodological and practical aspects for *in vitro* studies of oxidative stress on cell culture models (toxicity and protection)

M.J. Richard and P. Guiraud

Laboratoire de Biochimie des Micronutriments et Radicaux Libres, GREPO, CHRU, F-38043, Grenoble Cédex 3, France

Summary. Cell culture systems are especially useful to study the cytotoxicity mediated by reactive oxygen species (ROS) under defined conditions, and to understand cytoprotection and cellular adaptative response. This article reviews the methodological and practical aspects for studying cell injury using chromium-release assay, loss of cellular "adhesion/proliferation" ability and inactivation of mitochondrial enzymes in different model systems subjected to intracellular or extracellular oxidative stress. Their advantages, limitations as well as their biological significance are discussed.

Introduction

Among the damaging events which may injure a cell, considerable attention is being devoted to the role of free radicals in cell injury in various pathologies and to the development of antioxidant drugs. In cells, the deleterious effects of active oxygen species may become dominant when the balance between radical formation and removal is disturbed to produce "oxidative stress". The cell culture systems are, therefore, especially useful to study the cytotoxicity mediated by reactive oxygen species (ROS) under defined conditions, and to understand cytoprotection and cellular adaptative response.

This article reviews the methodological and practical aspects for studying cell injury on fibroblasts using chromium-release assay, loss of cellular "adhesion/proliferation" ability and inactivation of mitochondrial enzymes in model systems subjected to various oxidative stress: extracellular sources of superoxide ($O_2^{\cdot -}$) or hydrogen peroxide (H_2O_2) and intracellular oxidative stress mediated by UVA exposure.

Cell culture systems for oxidative stress

To assess cytotoxicity, various cell types may be used ranging from differentiated cells with a finite *in vitro* life span to transformed cells. Depending on the cells used, endogenous antioxidant defenses differ leading to differences between cell disturbances. Although different methodologies for measuring toxic effects of oxidative stress will be discussed, attention will be focused mainly on studies involving human fibroblasts.

To induce oxidative stress, two alternative approaches may be followed to disturb the prooxidant-antioxidant balance: either by increasing free radical fluxes via extracellular $O_2^{\cdot-}$ or H_2O_2 generation, hyperoxia, free radical-generating drugs such as menadione, exposure to irradiation such as UVA or by inhibiting the extent of antioxidant defenses. For instance the use of buthionine sulfoximine which induces glutathione depletion [1] increased cell injury. Similar results were obtained with drugs like 3-aminotriazole or diethyldithiocarbamate which inhibited metalloenzymes, catalase and superoxide dismutase, respectively. This chapter will only deal with cell injuries caused by an increase in extracellular or intracellular free radical generation.

Free radical generation

Different free radical generators can be used to induce oxidative stress leading to cell injury. The most commonly used are summarized in Table 1.

Factors that modulate $O_2^{\cdot-}$ or H_2O_2 toxicity for cells

The composition of the cell culture media may modulate cellular damages. In general, the cell culture media may contain scavengers of $O_2^{\cdot-}$

Table 1. Most commonly used generators inducing oxidative stress for cells

Extracellular sources		References
H_2O_2		
Bolus	H_2O_2 stock solution	[8, 26, 29]
Continuous flux	Glucose oxidase-glucose	[30]
$O_2^{\cdot-}$		
Bolus	Potassium superoxide	[31, 32]
Continuous flux	Xanthine oxidase-xanthine	[5]
	$(O_2^{\cdot-}, H_2O_2, {}^{\cdot}OH)$	[2]
	FMN reductase $(O_2^{\cdot-})$	[33]

Intracellular sources	References
UV irradiation	[1, 34]
Hypoxia-reoxygenation	[22, 35, 36]
Gamma radiation	[37]
Chemicals (paraquat, hydralazin, menadione)	[38, 39, 40]
Hyperoxia	[41]

or H_2O_2, (phenol red, pyruvate) as well as other antioxidants such as alpha-tocopherol or ascorbic acid. On the other hand, the medium may exert a prooxidant effect. It has been reported that histidine greatly enhances the cytotoxicity of hypoxanthine-xanthine oxidase (HX-XO) system or the cytotoxic effect of H_2O_2, [2–4]. Also the presence of transition metals in the medium may stimulate hydroxyl radical ($^{\cdot}OH$) formation from H_2O_2, by the transition metal-catalyzed Haber-Weiss reaction. This phenomenon is supposed to be limited because $^{\cdot}OH$ formation can only occur when transition metals are available in their reduced form. However, the presence of ascorbic acid or other reducing agents may increase the formation of $^{\cdot}OH$ by maintaining the transition metals in their reduced forms.

Finally, in complex media many other secondary reactive species or toxic products may be formed. Therefore, the use of a simple buffer system is recommended. However, it may be important to add glucose and glutamine to the buffers in order to provide the cells with their major energy sources. For instance, the two frequently used buffer solutions, Dulbecco phosphate buffer (PBS) and Hanks' balanced salt solution (HBSS), differ only in their glucose content. The higher cytotoxicity observed with PBS as compared to HBSS [5] could be explained by the reduced glucose levels. The modified Eagle's medium (MEM) includes aminoacids and vitamins in addition to the constituents of HBSS. These compounds are able to scavenge free radicals and this may explain why the cytotoxicity of HX-XO system was lower in MEM than in HBSS [5]. In the same experimental conditions Noel-Hudson et al. [5] observed a better survival in presence of fetal calf serum (MEM + 10% FCS). This is consistent with the finding of Bishop et al. [6], who demonstrated that serum had a pronounced scavenging effect towards H_2O_2. Calcium (Ca) salts may be another important factor to consider in cytotoxicity assays. When cells are exposed to oxidative stress, intracellular Ca flux increases. It is known that Ca may physiologically activate various enzymes involved in catabolic processes, e.g., proteases, lipases and endonucleases [7]. Thus it appears possible that cellular damages may be different in presence or absence of calcium.

The exposure time, and so the conditions of incubation during the assay, must be taken into account to choose the buffer. When the cells are incubated under CO_2 it is necessary to use a bicarbonate buffer and not a phosphate buffer in order to provide an adequate pH because cytotoxicity also depends on pH [2].

To study the pure cytotoxic effect of $O_2^{\cdot -}$, catalase should be added to the culture media to remove H_2O_2 as well as to prevent $^{\cdot}OH$ formation. The purity of the catalase preparations is of utmost importance because many commercial catalase samples appear to be contaminated with superoxide dismutase (SOD).

The other factors that modulate cellular response to free radical-mediated stress are linked to the cells themselves. The cell density [5, 8] is an important factor. Toxicity varies inversely with plating density. This effect seems to be dependent on the ratio between the rate of RLO generation and cellular antioxidant content. Not only does cell density influence the toxicity of RLO, but it also affects the cell cycle [9, 10] and cell differentiation. In effect, both cell density and cell cycle should be controlled.

Methodological and practical aspects for *in vitro* cytotoxicity studies

MTT evaluation of living cells

MTT (3-(4,5-dimethylthiazol-2-yl)-2,5-diphenyltetrazolium bromide) is a tetrazolium salt whose tetrazolium ring is cleaved in active mitochondria by dehydrogenase enzymes. Thus, the yellow substrate is transformed into a dark blue formazan product only in living cells.

Chemicals and reagents

RPMI 1640, fetal calf serum (FCS), Puck's saline solution and trypsin, were purchased from Gibco (Grand Island, New York, USA). 3-(4,5-dimethylthiazol-2-yl)-2,5-diphenyltetrazolium bromide (MTT) was from Sigma Chemical Co. (Saint-Louis, Missouri, USA). Dimethyl sulfoxide (DMSO) was purchased from Merk (Nogent sur Marne, France). Other reagents were from Prolabo (Paris, France). Stock solutions were prepared as follows: Tyrode buffer [NaCl 8 g/L, KCl 0.2 g/L, $NaHCO_3$ 1 g/L, NaH_2PO_4, H_2O 5.8 mg/L, $MgCl_2$, $6H_2O$ 1 mM, $CaCl_2$, $2H_2O$ 2 mM, HEPES 0.238 g/L, D-glucose 1 g/L, pH 7.3 adjusted with HCl] was sterilized by filtration (0.2 μm) and stored at 4°C. MTT stock solution (5 g/L) was prepared in Dulbecco phosphate buffer (PBS). Before use, this solution was filtered (0.2 μm) and diluted six times in tyrode buffer or in the medium according to the procedure used to measure cytoprotection and cytotoxicity.

Cell treatment

The confluent fibroblasts were harvested, counted and seeded into 96-well microplates (Falcon Plastic, California, USA) in growth medium (RPMI 1640 + FCS 10%; 100 μl) at a final concentration of 2.5×10^4 cells per well. The controls represent wells filled with 100 μL of growth medium without cells. Multiwell plates were incubated at 37°C, in a 5% CO_2 atmosphere for 24 h.

Oxidative stress

H_2O_2 *induced oxidative stress.* Hydrogen peroxide solution was prepared in Dulbecco phosphate buffer (PBS) at a final concentration ranging from 10^{-4} to 10^{-1} M before use. The H_2O_2 concentration was verified spectrophometrically ($1 = 230$ nm; $e = 81$ M·cm^1). Cells were washed twice with PBS before being subjected to oxidative stress for 30 min at room temperature in darkness. Control cells were maintained in PBS during the same period.

UVA irradiation. Cells were rinsed twice and irradiated in PBS buffer with a Uvasun 2000 apparatus (Mutzhas, Munich, Germany). The spectrum is from 340 to 420 nm with a maximum intensity at 375 nm. A compensated Kipp and Zonen thermopile coupled to a digital voltmeter was used to measure UV-A energy effectively received by the cells through the culture dishes. Non-irradiated cells were left on the bench in PBS buffer while irradiation was being carried out.

Cell survival

Cell survival was quantified by a modified colorimetric method previously described by Mossman [11]. After treatment of cells, plates were rinsed twice ($100 \mu l$ buffer/well). Each well then received $120 \mu l$ of the diluted MTT solution and cells were incubated for $2-4$ h under a humidified atmosphere containing 5% CO_2 at 37°C. Then the dark-blue crystals formed by living cells were dissolved by adding $100 \mu l$/well of DMSO and mixing thoroughly. Alternatively, acidified isopropanol (HCl 0.04 N) could be used and mixed thoroughly overnight. The absorbance was recorded at 570 nm using a multiscan spectrophotometer (Titertek Multiskan, Labsystems Group, Les Ulis, France) interfaced with a computer. The absorbance or optical density (OD) is directly proportional to the number of viable cells.

Processing of results

The percentage of cytotoxicity (Fig. 1) is determined from the equation: [(ODcontrol − OD stressed cells)/(ODcontrol)] × 100. In order to evaluate cytoprotection, antioxidants are prepared in buffer and generally added to the cells $10-15$ min before inducing the oxidative stress. They are maintained in contact with the cells during the stress. The toxic effect of the stress is evaluated by applying the following equation [5]: X = [(Control wells − Exposed wells)/Control wells] × 100. The protective effect of an antioxidant is evaluated by the equation: Protection % = [(X − Y)/X] × 100.

The time-lag between oxidative stress and MTT evaluation of living cells is a determinant variable (Fig. 2). Indeed, it has been shown that oxidative stress (tumor necrosis factor, radiation) is a mediator of apoptosis [12].

Limitations

MTT can react with superoxide and can give a positive reaction by a non-specific pathway without the involvement of active mitochondria.

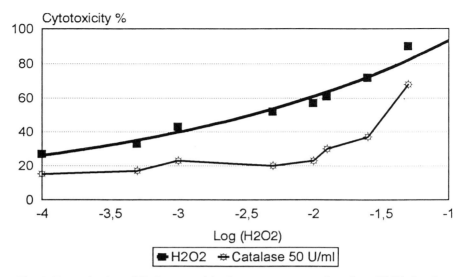

Fig. 1. Determination of H_2O_2 cytotoxicity in presence or not of catalase (50 U/ml) using MTT assay. The measures are done immediately after the oxidative stress according to the procedure described in the methodological section.

Further refinements and interpretation of tetrazolium-based assays for antioxidant drug evaluations will definitely benefit from careful attention to concepts and observations already described in the literature. For example, it is important to note that some components (ascorbic

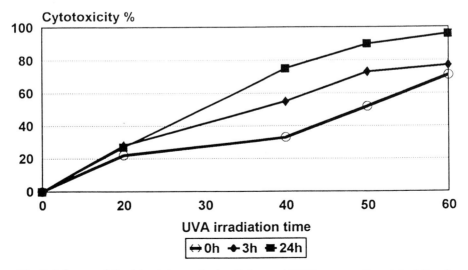

Fig. 2. Influence of the delay between the irradiation and MTT assay on the cytotoxicity of UVA. The measures are done immediately after irradiation (T0) and after 3 h or 24 h.

acid, sulfhydryl agents) are capable of reducing tetrazolium salts by direct interactions, whereas other compounds (malonate, rotenone) can block cell-mediated MTT reduction indirectly by inhibiting early steps in cellular respiration [13]. A major difficulty with this assay is the solubilization of the formazan crystals before reading the color absorption. Acidified isopropanol, dimethyl sulfoxide [14] and ethanol have been identified as suitable solvents. An ultrasound water bath can be used to dissolve the crystals after addition of acid isopropanol [15]. The use of 2,3-bis(2-methoxy-4-nitro-5-sulfophenyl)-5-[(phenylamino)carbonyl]-2H-tetrazolium hydroxide, inner salt, sodium salt (XTT) will simplify this step. In general, XTT is metabolically reduced by viable cells to a water soluble formazan product; however, numerous cell lines were not able to efficiently reduce XTT during 4 h incubation time, thus making it necessary to use longer incubation times for obtaining absorption values significantly greater than the background values. However, supplementation of the XTT incubation mixtures with an electron-coupling agent like phenazine methosulfate (PMS) results in adequate absorbance levels [16].

Cell ability to adherate and proliferate

This method has been recently developed in our laboratory [8] and has been validated only in a few laboratories [17]. It is a variant of the cell-forming colony (colony forming unit) test. It may be useful for adherent cells and reflects both the ability of a cell to adhere and proliferate after the induction of a stress.

Chemicals and reagents
RPMI 1640, fetal calf serum (FCS), Puck's saline solution and trypsin, were purchased from Gibco (Grand Island, New York, USA). Xanthine oxidase, hypoxanthine were obtained from Sigma Chemical Co (Saint-Louis, Missouri, USA).

For protein determination stock solution was prepared as follows: Na_2CO_3 40 g/l, NaK tartrate, 4 H_2O 20 g/l, $CuSO_4$, 5 H2O 10 g/l. Just before use the working solution is prepared by mixing 24 V Na_2CO_3; 0.5 V tartrate salt, 0.5 V $CuSO_4$. Folin and Ciocalteu's phenol reagent 2N is from Sigma. It is diluted by half just before analysis.

Cell treatment
Fibroblasts are placed in 35 mm diameter culture dishes and incubated at 37°C under a 5% CO_2 atmosphere in their usual growth medium (RPMI 1640 + FCS 10%). The original cell density used is 10^5-10^6 per dish.

Oxidative stress
Hydrogen peroxide and UVA irradiation are applied as described above.
   Hypoxanthine-xanthine oxydase system. Hypoxanthine solution stock
(30 mM) is prepared in Tyrode buffer and stored at 4°C whereas
xanthine-oxidase is diluted just before use to give a final concentration
20 U/l. Cells are incubated in presence of the HX-XO mixture for
90 min. Under these conditions hypoxanthine concentrations used vary
from 0.05 to 0.3 mM.

Replating assay
After oxidative stress, cells are rinsed with Puck's saline, harvested by
trypsination, replated into another dish and reincubated for 18 h in
fresh culture medium. Non-adhering cells and cell debris are removed
by rinsing vigorously three times with 0.9% NaCl. Survival levels and
proliferation capacities are determined by assaying total protein using
the method of Shopsis and Mackay [18]. Cells are digested with 0.5 ml
NaOH (0.5 N) per dish during 2 h at room temperature. The samples
are frozen at $-20°C$ until further analysis. For analysis, 150 μl of
alcalin solution of protein are added to 375 μl of the working solution.
After 15 min, 150 μl of Folin Ciocalteu's phenol reagent is added. Tubes
are mixed and left in darkness during 45 min before measuring absorbance at 630 nm. Assays are performed in duplicate. A calibration
curve is made with 2.5×10^4 to 10^6 cells per dish and the protein
concentrations measured are directly proportional to the number of
adherent cells.

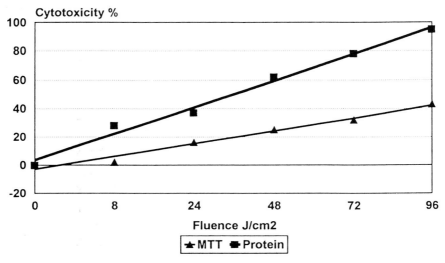

Fig. 3. Immediate (MTT assay) and delayed (replating assay) cytotoxicity of UVA irradiation.

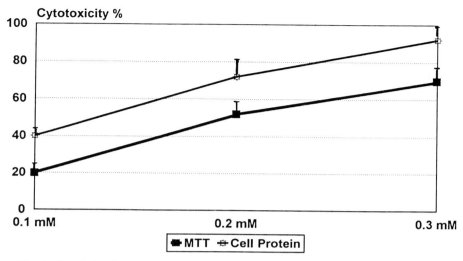

Fig. 4. Correlation between MTT and replating assays in assessement of cytotoxicity of xanthine oxidase (XO 20 U/l)-hypoxanthine mixture.

The plating efficiency of the untreated cells ranges from 90 to 95%. Cytotoxicity is determined via the percentage of cells which have re-attached *versus* control cells. This method only investigates a delayed toxicity (Fig. 3).

Limitations

It has been reported that trypsin can amplify the toxic effects of oxidative stress when it is added to the cells before oxidative stress [19]. In the present method we use proteases, after oxidative stress, while transferring treated cells to a new dish. We have often noted an effect of the lagtime between stress and trypsination using the mixture HX-XO [8]. In this system, we observe a greater susceptibility to killing, which is either all-or-nothing, if trypsination is done immediately after the oxidative stress. However, when trypsination is performed after a delay of 90 min between stress and trypsination, a good correlation between MTT and replating assay was observed (Fig. 4). Nevertheless, the cytotoxicity of the mixture HX-XO appears to be more important when we use MTT assay rather than replating assay, whereas UVA irradiation gives opposing results (Fig. 3). The hypothesis of a direct reduction of MTT salt by the superoxide anion generated by HX-XO could explain these results. Finally, a potential susceptibility of the cells to trypsin could be avoided by scrapping. On the whole, a main disadvantage of this technique is that it is time consuming and lacks automation.

Chromium release assay

The basis of the assay is the use of the radioactive isotope of chromium (^{51}Cr), which binds to the cellular proteins of cultured cells. Target cells are pre-labelled by incubation with ^{51}Cr, and incubated with oxygen free radicals generators, or UV irradiated. The amount of radioactivity which is released into the supernatant reflects the amount of cell lysis. This assay is mostly adapted to the measurement of cell killing caused by disruption of the cell membrane.

Chemicals

Cell culture reagents are as previously described. Sterile radioactive sodium chromate (Na_2 $^{51}CrO_4$), biologically tested for cytotoxicity was purchased from NEN Du Pont de Nemours (Bad Homburg, Germany). Triton X 100 was from Sigma Chemical Co (St-Louis, Missouri, USA).

Cell treatment

A monolayer of fibroblasts in 35 mm diameter dishes, grown to confluence is incubated in 1 ml complete culture medium containing 10 μCi ^{51}Cr for 16 h. Unincorporated ^{51}Cr is aspirated and cells are washed three times with the Puck solution. After hydrogen peroxide treatment, cells are incubated with 1.5 ml of fresh culture medium in the incubator for 2 to 24 h (Fig. 5). Supernatants are then removed for counting (aliquots) in an auto gamma scintillation spectrophotometer (Packard 5160). To measure maximal ^{51}Cr release, 2 × 1 ml of culture medium

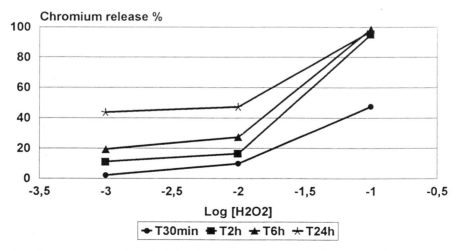

Fig. 5. Determination of cytotoxicity of H_2O_2 using chromium release assay. The measures are done 30 min, 2, 6, and 24 h after the exposition of the cells.

containing 0.67% Triton X-100 are added per dish to lyse cells, and the suspension is counted. These values are added to the supernatant values to give maximal ^{51}Cr release. Spontaneous release is obtained by counting supernatant of control cells. The amount of cellular cytotoxicity is quantified based on a simple calculation of the amount of cell bound ^{51}Cr *versus* cell free ^{51}Cr.

Oxidative stress
Hydrogen peroxide is applied as described above.

Processing of results
Specific release is calculated for all samples according to

$$\% \text{ specific release} = [^{51}\text{Cr supernatant}$$

$$/(^{51}\text{Cr supernatant} + {}^{51}\text{Cr lysat})] \times 100$$

In order to compare results from the different methods (Fig. 6), we consider

$$\% \text{ of } {}^{51}\text{Cr retention} = 100 - (\% \text{ specific } {}^{51}\text{Cr release})$$

Limitations
This method can be applied either to adherent or non adherent cells. It is adaptable either to microplates or to Petri dishes but can be automated only with specific and expensive material. It is the most sensitive

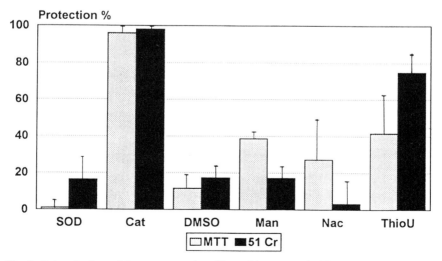

Fig. 6. Determination of the cytoprotective effects of known antioxidants (SOD: superoxide dismutase 100 μg/ml, Cat: catalase 100 U/ml, DMSO: dimethyl sulfoxide 0.1 mM, Man: mannitol 0.1 mM, Nac: N-acetylcysteine 0.05 mM, ThioU: thiourea 50 mM) using MTT assay and chromium (^{51}Cr) release in fibroblasts exposed to H_2O_2.

method to measure precisely the membrane damages but is not adequate for monitoring cellular mortality. When membrane damage is
only a secondary effect, it may be difficult to interpret the results.
Moreover, the technique is time consuming. However, Salnikow et al.
[20] showed that incorporation of ^{51}Cr into proteins of human osteosar-
coma cells did not differ significantly between a 3 h or 24 h incubation
period. Thus, the period of labelling could be reduced. Another important limitation of this technique is the requirement of a short half-life
radioactive material.

Discussion

Different methods can be used to evaluate cytotoxicity. Dye exclusion
has been used as a simple qualitative or quantitative estimation of cell
viability. Moribund cells as well as cells with leaky membranes are
stained. Various dyes may be employed: erythrosine, eosine, alcian blue,
or trypan blue. This last method may be easily automated. Another
automated colorimetric determination to score cell injury is based on
the uptake of a vital dye by viable cells. Neutral red, which accumulates
in the lysosomes, is the most commonly employed method [21].
    To determine the integrity of the cell membranes during cell damage
it is also possible to follow the release of substances. Among the
endogenous substances, the marker most commonly used is the leakage
of lactate dehydrogenase (LDH) which is expressed as a function of the
number of cells [22] or the protein content of the cells in the culture
plates [23]. It has been previously suggested that 2-deoxy-D-^{3}Hglucose
is the most sensitive label to assess cellular damage [24] but because this
isotope has a high rate of spontaneous release, it is only suitable for
short studies (i.e., 1 to 2 h incubation). For longer experiments some
elements such as indium (In) or chromium have been proposed. It has
been shown that ^{51}Cr is a more sensitive label than ^{111}In. Although ^{51}Cr
is the most widely used, it has been proposed by Andreoli et al. [24] that
^{3}H adenine could be considerably more sensitive, but this has been
refuted by Warren and Ryan [25]. Whorton et al. [26] have tested both
^{51}Cr and LDH release by endothelial cells subjected to cytolytic action
of H_2O_2. The data concerning the release of ^{51}Cr suggest that a critical
level of damage must be reached and after this threshold ^{51}Cr release
increases rapidly. Conversely, the release of LDH is a time-dependent
phenomenon. Accordingly, results on oxidative cellular damage depend
on the markers used.
    MTT assay is a fast, easy method which facilitates the screening of an
antioxidant. Very often, this technique is considered to be equivalent to
chromium release or LDH release assay. It must be pointed out that
these assays measure different end-points. When mitochondria is impli-

cated as primary targets organelles (hypoxia, toxics) reduction of MTT has been found to be a more sensitive endpoint, followed by the neutral red assay [27]. Under the same conditions the authors did not observe any LDH leakage. Twentyman and Luscombe [14] have compared MTT assay with total viable cell counts as indicators of cytotoxicity on mouse tumor cell line. A satisfactory degree of agreement was noted between the results obtained using the two assays.

It must be remembered that neither of these previously cited assays (MTT, cell count and endogenous or exogenous substances leakage) is equivalent to a clonogenic assay. The clonogenic assay determines the proportion of cells with intact proliferative capacity [28]; but conversely to thymidine incorporation, it takes no account of reduced growth rate induced by the stress or by an antioxidant drug. It is also true for the replating assay which is a less time-consuming variant of clonogenic assay [2].

Conclusion

The different tests available to measure cytotoxicity do not reflect the same cellular damage, so the cytoprotection mediated by a substance could appear to be different. In conclusion, a great deal of attention must be paid to the choice of the marker which must be defined based on the over-all aims of the proposed experiment.

References

1. Tyrrell, R. and Pidoux, M. (1986) Endogenous glutathione protects human skin fibroblasts against the cytotoxic action of UVB, UVA and near-visible radiations. *Photochem. Photobiol.* 44: 561–564.
2. Link, E.M. and Riley, P.A. (1988) Role of hydrogen peroxide in the cytotoxicity of the xanthine/xanthine oxidase system. *Biochem. J.* 249: 391–399.
3. Tachon, P. and Giacomoni, P.U. (1989) Histidine modulates the clastogenic effect of oxidative stress. *Mutation Res.* 211: 103–109.
4. Oya, Y. and Yamamoto, K. (1988) The biological activity of hydrogen peroxide. IV. Enhancement of its clastogenic action by coadministration of histidine. *Mutation Res.* 198: 33–240.
5. Noel-Hudson, M.S., de Belilovski, C., Petit, N., Lindenbaum, A. and Wepierre, J. (1989) *In vitro* cytotoxic effects of enzymatically induced oxygen radicals in human fibroblasts: experimental procedures and protection by radical scavengers. *Toxic in vitro* 3: 103–109.
6. Bishop, C.T., Mirza, Z., Crapo, J.D. and Freeman, B.A. (1985) Free radical damage to cultured porcine aortic endothelial cells and lung fibroblasts: modulation by culture conditions. *In Vitro Cell Dev. Biol.* 21: 229.
7. Orrenius, S. and Bellomo, G. (1991) Metabolic regulation in oxidative stress. *In:* K.J.A. Davies (ed.): *Oxidative damage and repair. Chemical, biological and medical aspects.* Pergamon Press, Oxford, pp 449–457.
8. Richard, M.J. Guiraud, P., Monjo, A.M. and Favier, A. (1992) Development of a simple antioxidant screening assay using human skin fibroblasts. *Free Rad. Res. Comms.* 16: 303–314.

9. Gaboriau, F., Morliere, P., Marquis, I., Moysan, A., Gèse, M. and Dubertret, L. (1993) Membrane damage induced in cultured human skin fibroblasts by UVA irradiation. *Photochem. Photobiol.* 58: 515–520.

10. Rubin, D.B., Drab, E.A. and Bauer, K.D. (1989) Endothelial cell subpopulations *in vitro*: cell volume, cell cycle, and radiosensitivity. *J. Appl. Physiol.* 67: 1585–1590.

11. Mossman, T. (1983) Rapid colorimetric assay for cellular growth and survival: application to proliferation and cytotoxicity assays. *J. Immunol. Methods* 65: 55–63.

12. Buttke, T.H. and Sandstrom, P.A. (1994) Oxidative stress as a mediator of apoptosis. *Immunol. Today* 15: 7–10.

13. Alley, M., Scudiero, D., Monks, A., Hursey, M., Czerwinski, M., Fine, D., Abbott, D., Mayo, J., Shoemaker, R. and Boyd, M. (1988) Feasibility of drug screening with panels of human tumor cell lines using a microculture tetrazolium assay. *Cancer Res.* 48: 589–601.

14. Twentyman, P.R. and Luscombe, M. (1987) A study of some variables in a tetrazolium dye (MTT) based assay for cell growth and chemosensitivity. *Br. J. Cancer* 56: 279–285.

15. Iselt, M., Holtei, W. and Hilgard, P. (1989) The tetrazolium dye assay for rapid *in vitro* assessment of cytotoxicity. *Drug Res.* 39: 747–749.

16. Scudiero, D., Shoemaker, R., Paull, K., Monks, A., Tierney, S., Nofziger, T., Currens, M., Seniff, D. and Boyd, M. (1988) Evaluation of a soluble tetrazolium/formazan assay for cell growth and drug sensitivity in culture using human and other tumor cell lines. *Cancer Res.* 48: 4827–4833.

17. Varani, J., Bendelow, M., Sealey, D., Kunkel, S., Gannon, D., Ryan, U. and Ward, P. (1988) Tumor necrosis factor enhances susceptibility of vascular endothelial cells to neutrophil-mediated killing. *Lab. Invest.* 59: 292–295.

18. Shopsis, C.H. and Mackay, G.J. (1984) Semi-automated assay for cell culture. *Anal. Biochem.* 140: 104–107.

19. Varani, J., Ginsburg, I., Schuger, L., Gibbs, D.F., Bromberg, J., Johnson, K.J., Ryan, U.S. and Ward, P.A. (1989) Endothelial cell killing by neutrophils: synergistic interaction of oxygen products and proteases. *Am. J. Pathol.* 135: 435–438.

20. Salnikov, K., Zhitkovitch, A. and Costa, M. (1992) Analysis of the binding sites of chromium to DNA and protein *in vitro* and in intact cells. *Carcinogenesis* 13: 2341–2346.

21. Borenfreund, E. and Puerner, J.A. (1985) Toxicity determined *in vitro* by morphological alterations and neutral red absorption. *Toxicol. Lett.* 24: 119–124.

22. Palluy, O., Bonne, C. and Modat, G. (1991) Hypoxia/reoxygenation alters endothelial prostacyclin synthesis-protection by superoxide-dismutase. *Free Rad. Biol. Med.* 11: 269–275.

23. Massey, K. and Burton, K. (1990) Free radical damage in neonatal rat cardiac myocyte cultures: effects of α tocopherol, trolox, and phytol. *Free Rad. Biol. Med.* 8: 449–458.

24. Andreoli, S.P., Baehner, R.L., Bergstein, J.M. (1985) *In vitro* detection of endothelial cell damage using 2-deoxy-D-3H-glucose: comparison of chromium 51, 3H-leucine, 3H-adenine and lactate dehydrogenase. *J. Lab. Clin. Med.* 106: 253–261.

25. Warren, J.B. and Ryan, U.S. (1989) Endothelial injury assessed by isotope release: 3H-adenine compared with 51 Cr. *In vitro Cell. Dev. Biol.* 25: 334–335.

26. Worton, A.R., Montgomery, M.E. and Kent, R.S. (1985) Effect of hydrogen peroxide on prostaglandin production and cellular integrity in cultured porcine aortic endothelial cells. *J. Clin. Invest.* 76: 295–302.

27. Hsieh, G., Acosta, D. (1991) Dithranol-induced cytotoxicity in primary cultures of rat epidermal keratinocytes. I. The role of reactive oxygen species. *Toxicol. Appl. Pharm.* 107: 16–26.

28. Park, Y.M., Anderson, R., Spitz, D. and Hahn, G. (1992) Hypoxia and resistance to hydrogen peroxide confer resistance to tumor necrosis factor in murine L 929 cells. *Radiat Res.* 131: 162–168.

29. Simon, R.H., Scoggin, C.H. and Patterson, D. (1981) Hydrogen peroxide causes the fatal injury to human fibroblasts exposed to oxygen radicals. *J. Biol. Chem.* 256: 7181–7186.

30. Ody, C. and Junod, A.F. (1985) Effect of variable glutathione peroxidase activity on H_2O_2-related cytotoxicity in cultured aortic endothelial cells. *Proc. Soc. Exp. Biol. Med.* 180: 103–111.

31. Gille, J.J.P. and Joenje, H. (1992) Cell culture models for oxidative stress: superoxide and hydrogen peroxide versus normobaric hyperoxia. *Mutation Research* 275: 405–441.

32. Bolann, B.J. and Ulvik, R.J. (1991) Improvement of a direct spectrophotometric assay for routine determination of superoxide assay. *Clin. Chem.* 37: 1993–1999.
33. Gaudu, P., Touati, D., Nivière, V. and Fontecave, M. (1994) The NAD(P)H: flavin oxidoreductase from *Escherichia coli* as a source of superoxide radicals. *J. Biol. Chem.* 269: 8182–8188.
34. Leccia, M.T., Richard, M.J., Beani, J.C., Faure, H., Monjo, A.M., Cadet, J., Amblard, P. and Favier, A. (1993) Protective effect of selenium and zinc on UV-A damage in human skin fibroblasts. *Photochem. Photobiol. USA* 58: 548–553.
35. Inauen, W., Payne, K., Kvietys, P. and Granger, N. (1990) Hypoxia/reoxygenation increases the permeability of endothelial cell monolayers: role of oxygen radicals. *Free Rad. Biol. Med.* 9: 219–223.
36. Zweier, J., Kuppusamy, P. and Lutty, G. (1988) Measurement of endothelial cell free generation: evidence for a central mechanism of free radical injury in postischemic tissues. *Proc. Natl. Acad. Sci. USA* 85: 4046–4050.
37. Rubin, D.B., Drab, E.A., Ward, W.F., Bauer, K.D. (1986) Cell cycle changes and cytotoxicity in irradiated culture of bovine aortic endothelial cells. *Rad. Res.* 108: 206–214.
38. Zer, H., Freedman, H., Peisach, J. and Chevion, M. (1991) Inverse correlation between resistance towards copper and towards the redox-cycling compound paraquat: a study in copper-tolerant hepatocytes in tissue culture. *Free Rad. Biol. Med.* 11: 9–16.
39. Weglarz, L. and Bartosz, G. (1991) Hydralazine stimulates production of oxygen free radicals in Eagle's medium and cultured fibroblasts. *Free Rad. Biol. Med.* 11: 149–195.
40. Rosen, G. and Freeman, B. (1984) Detection of superoxide generated by endothelial cells. *Proc. Natl. Acad. Sci. USA* 81: 7269–7273.
41. Farris, M.W. (1990) Oxygen toxicity: unique cytoprotective properties of vitamin E succinate in hepatocytes. *Free Rad. Biol. Med.* 9: 333–343.

Analysis of Free Radicals in Biological Systems
Favier et al. (eds)
© 1995 Birkhäuser Verlag Basel/Switzerland

Detection and production of nitric oxide

C. Garrel[1], J.-L. Decout[2] and M. Fontecave[2]

[1] Groupe de Recherches des Pathologies Oxydatives, Centre Hospitalier Universitaire, BP217, F-38043 Grenoble Cedex 9, France
[2] Laboratoire d'Etudes Dynamiques et Structurales de la Sélectivité, Université Joseph Fourier, BP53, F-38041 Grenoble, France

Summary. The methods most commonly used to detect or measure nitric oxide are summarized in this chapter. Measurement of nitrite produced from ˙NO by activated human monocytes or by synthetic thionitrites is described in detail. Special emphasis is made on the spectroscopic assays for ˙NO, including spectrophotometry and EPR spectroscopy. The assays are based on the reaction between a metal center and ˙NO. A very specific EPR test, based on the reversible coupling reaction between ˙NO and the tyrosyl radical of ribonucleotide reductase, is reported.

Introduction

The measurement of nitric oxide (˙NO) production in biological systems is a rather difficult task. Actually, ˙NO is a gas which is highly unstable when dissolved in buffers. As a radical, it reacts with molecular oxygen very efficiently [1]. As a consequence, its concentration is very low (less than μM) and very sensitive methods are needed to detect and quantitate it.

Despite its instability, several direct or indirect techniques have been developed for monitoring the produciton of ˙NO. Direct detection assays are essentially based on ˙NO trapping reactions. ˙NO traps can be free radicals, metal ions or ozone. Indirect detection consists of measurements of stable end breakdown products of ˙NO, such as nitrite and nitrate ions, which are considered to be reliable markers for ˙NO formation. The methods most commonly used to measure ˙NO are summarized in Table 1. Some of these methods, including nitrite assay, spectrophotometric assay with hemoglobin and EPR measurements, will be described in detail in this chapter.

Materials and methods

Nitrite measurement-Greiss reaction

This is an indirect colorimetric assay of ˙NO [2]. NO, in solution, is rapidly oxidized by O_2 yielding quantitative amounts of

Table 1. A summary of assays used to detect nitric oxide

Method	Principle – detected species	Detection threshold	Comments
Greiss reaction	Colorimetric determination of nitrite, the NO oxidation product [2]	1 μM	– simple – most widely used technique
	Nitrite reacts with sulfanilamide in an acidic solution of N-(1-naphtyl)ethylene diamine to form an azo derivative which can be monitored by spectrophotometry at 548 nm		– indirect method: cannot distinguish nitrite derived from NO or from other sources
Chemilumi-nescence	After reaction with O_3 in the gas phase, light is emitted and can be detected using a photomultiplier tube [3] $NO + O_3 \rightarrow NO_2^* \rightarrow h\nu + NO_2$	1 nM	– great sensitivity – drawback: NO must be shifted from the biological medium medium to a gas phase, a process not always quantitative
Electro-chemistry	NO binds to a nickel-porphyrin adsorbed onto an anode and is oxidized electrochemically [4] $Ni(P) + NO \rightarrow Ni(P)(NO)$ $\rightarrow Ni(P)NO^+ + e^-$ The current generated is proportional to the amount of NO oxidized	10 nM	– great sensitivity – possible chemical and electrical interferences
Spectro-photometry of hemoglobin	$HbO_2 + NO \rightarrow metHb + NO_3^-$ NO quantification is based on modification of the visible spectrum during the oxidation [5]	1 μM	– simple – limited sensitivity – possible interferences with redox compounds
Electron Spin Resonance	(1) Reaction of NO with a diamagnetic trap gives rise to an EPR active paramagnetic stable species	0.1–1 μM	
	Different traps are used: – Oxyhemoglobin [6] – Fe(II)-diethyldithiocarbamate [7] – Fe(II)-thiosulfate [7, 8]		– on-line detection – quantitative
	(2) Reaction of the tyrosyl radical of ribonucleotide reductase with NO reversibly abolishes its characteristic EPR signal [9] $TyrO^{\cdot} + {}^{\cdot}NO \leftrightarrow TyrONO$		– highly specific
Fluorimetry	Reaction of nitrite with 2,3-diaminonaphtalene forms the fluorescent product 1-(H)-naphtotriazole [10]	10 nM	– 50–100 times more sensitive than Greiss assay – fast assay

nitrite:

$$4\,NO + O_2 + 2\,H_2O \longrightarrow 4\,NO_2^- + 4\,H^+$$

Thus, measurement of nitrite may give a reliable estimate of ˙NO.

In acidic solutions, nitrite undergoes a diazotation of sulfanilamide. In a second step reaction with N-(1-naphtyl)ethylenediamine it results in the formation of a red azo derivative absorbing at 548 nm which can be easily monitored by spectrophotometry.

In order to illustrate this method, two ˙NO generating systems have been used. The first one is from the specific oxidation of L-arginine catalyzed by ˙NO synthases within cells. ˙NO is produced in macrophages, endothelial cells and neutrophils for example, in a process that is inhibited by N-substituted-L-arginine analogs such as N-monomethyl-L-arginine (NMMA).

We will describe here the detection of NO_2^- produced by human monocytes, after induction of ˙NO synthase activity.

Another source of nitric oxide is NO-donors. These are synthetic chemicals which can, either spontaneously or by activation (with light or reducing agents, etc.), decompose into ˙NO in solution.

Evaluation of ˙NO production by human monocytes

It is now well established that a variety of human cells is able to generate nitric oxide *in vitro*, through the action of inducible or constitutive NO synthases: hepatocytes, keratinocytes, vascular cells, endothelial cells, etc.

However, there is still controversy regarding ˙NO production by the human monocytes and macrophages, and the existence of ˙NO synthase within these cells [11–14].

Macrophages and monocytes are mononuclear phagocytes which play a central role in specific immunity, non specific defense against infection, regulation of cell growth and in inflammation and, as such, constitute a major host regulatory and defense system. These cells are also important in malignancy because many tumors have been shown to be heavily infiltrated by monocytes and monocyte-mediated cytotoxicity has been demonstrated against a variety of tumor types.

Mammalian monocyte-derived macrophages produce two independent classes of inorganic oxidants that can contribute to tumoricidal and microbicidal activity: reactive oxygen intermediates (ROI) and reactive nitrogen intermediates (RNI). The role of RNI in some antimicrobial and antitumor activities of rodent macrophages is well established while it is still unknown whether this is also the case for human monocytes and macrophages.

*Methodological and practical aspects for stimulating and measuring
˙NO production by peripheral blood monocytes from healthy human
volunteers*

Blood manipulation
- Anticoagulant: Acid Citrate Dextrose (ACD):
  - D-Glucose: 2 g/l
  - Trisodium citrate: 0.25 g/l
  - Citric acid: 0.14 g/l
  - pH: 4.5
- Histopaque 1077 (Sigma Chemical Co)
- RPMI 1640 with or without phenol red (Gibco)

Cell Culture
- Culture dishes: Falcon primaria 35 mm diameter
- Agents used to activate or inhibit ˙NO production by human monocytes:
  Human recombinant Interferon-gamma (IFNγ) (Boehringer-Mannheim
  Biochemicals); Lipopolysaccharide (LPS, *E. coli* serotype 055: B5, Sigma
  Chemical Co); N_{ω}-nitro-L-arginine (Sigma Chemical Co).

Nitrite assay
- Spectrophotometer Uvikon 860 (Kontron)
- Sodium nitrite, sulfanilamide and N-(1-naphtyl)ethylenediamine
  were purchased from Sigma Co
- Greiss reagent: This is a mixture in a volume ratio 1:1 of solutions A
  and B which are stable for 2 months at $-4°C$ sheltered from light
  - Solution A: Sulfanilamide dissolved in 100 ml 5% aqueous H_3PO_4
  - Solution B: 100 mg naphtylethylenediamine dihydrochloride dissolved
    in 100 ml distilled water
  Greiss reagent should be prepared freshly as required.

Isolation of human monocytes
The following procedures require fully sterile conditions.

150 ml of whole blood from one healthy volunteer was collected from a
separation chamber which was then transferred into plastic tubes containing 16% ACD and diluted two-fold with endotoxin-free RPMI 1640.

The solution was slowly loaded on top of a density gradient material
(Histopaque 1077), in 50 ml plastic tubes (histopaque: blood volume
ratio 1:2).

After centrifugation at 200 g for 30 min at 4°C, a relatively pure
peripheral blood mononuclear cells (lymphocytes, monocytes) population was concentrated as a ring at the surface of the histopaque. The
cells were sucked off with a plastic pipette and washed in 15 ml of
RPMI 1640 and centrifuged at 720 g for 10 min at 4°C. Supernatant

was removed and the pellets containing the mononuclear cells were washed twice in RPMI 1640 by two successive 10-min centrifugations respectively at 600 and 400 g.

The cell pellets were collected and resuspended in medium A: RPMI 1640, supplemented with 10% inactivated fetal calf serum, 1 mM glutamine and an antibiotic-antimycotic solution (penicilline 170 U/ml, streptomycin 0.17 mg/ml, kanamycin 54 μg/ml and fungizone 0.5 μg/ml).

Mononuclear cells were counted automatically (Coulter counter JT2 model S plus II Coultronics SA). Cells were then plated out at a concentration of $2-4.10^6$/ml in individual 30-mm culture plates falcon primaria (Boehringer).

A trypan blue exclusion analysis was performed at this stage and showed a mean viability of 95%.

Monocyte enrichment was achieved by allowing the cells to adhere to the plates for 2 h at 37°C in 5% CO_2/95% air atmosphere. Then the medium was removed by aspiration and cells were delicately washed twice with RPMI 1640 to remove lymphocytes and all other non adherent cells.

For the final incubation, cells were cultured in RPMI 1640, supplemented as above, with various combinations of inducers of 'NO production and 'NO synthase inhibitors. It is crucial to carry out blanks to evaluate the specific contribution of the RPMI and the various additives, under strictly identical conditions.

All samples and blanks were incubated for 48 h at 37°C in 5% CO_2/95% air atmosphere. The supernatants were then removed for subsequent nitrite assay and cell protein content of individual culture plate was determined according to the Lowry method.

Determination of 'NO by the Greiss reaction
Supernatant solutions were centrifugated at 800 g for 10 min at 37°C to eliminate cell debris. 300 μl supernatant was added to an equal volume of freshly prepared Greiss reagent and incubated for 15 min at room temperature. The optical density was measured with an Uvikon spectrophotometer at 548 nm. The corresponding nitrite concentration was determined from a comparison with a standard curve generated with known concentrations of sodium nitrite. Nitrite concentration was expressed as pmol NO_2^-/μg protein, after correction of the values of the corresponding blanks.

Decomposition of thionitrites

The concentrated solutions of S-nitrosocysteamine are prepared quantitatively by reaction of cysteamine with one equivalent of tert-butyl

Table 2. Stability of S-nitrosothiols, effects of reaction conditions. S-nitrosothiols (0.5–2 mM) were dissolved in 0.5 ml of 50 mM Tris buffer (pH 7.4) or acetate buffer (pH 4), in a spectroscopic cuvette. The absorbance at 333 nm (S-nitrosocysteamine) or 339 nm (SNAP) was recorded at time intervals. Results are expressed in terms of $t_{1/2}$, the half-life of the compound, at 37°C. In some experiments buffers were first treated with Chelex resin, 0.1 mM or DTT 6 mM were present during the decomposition. The effect of light was studied by illuminating the incubation mixture with a slide projector, 20 cm from the cuvette

Conditions	$t_{1/2 \text{ (min)}}$
S-nitrosocysteamine	
buffer pH 4	720
buffer pH 7.4	5
chelexed buffer pH 7.4	40
+ desferal	180
+ desferal + illumination	5
SNAP	
buffer pH 7.4	240
+ DTT	5

nitrite [15]. This compound can be stored only in acidic aqueous solutions but not as a solid. On the other hand, SNAP can be obtained, as a pure solid, after reaction of N-acetyl-D,L-penicillamine with sodium nitrite in acidic solutions [16]. S-nitrosothiols are characterized by two absorption bands, one between 330 and 350 nm and the other, much less intense, between 500 and 650 nm.

Thionitrites spontaneously decompose in neutral aqueous solutions. During the reaction, the characteristic absorption bands disappear, which allows for effective monitoring of its decomposition. The reaction yields the corresponding disulfide and NO according to the following Equation:

$$2 \text{ RSNO} \longrightarrow \text{RSSR} + 2 \text{ } ^{\bullet}\text{NO}$$

Half-lives of thionitrites greatly depend on the structure of the molecule and on several other factors, such as the pH, illumination, presence of reducing agents and trace metal contamination. This is demonstrated in Table 2.

The decomposition of S-nitrosocysteamine or SNAP is, as expected, accompanied by the formation of nitrite, resulting from the oxidation of ${}^{\bullet}$NO. However, in most experiments the reaction is not quantitative. For example, a solution of 0.1 mM S-nitrosocysteamine gives approximately 60 μM nitrite, at room temperature. A portion of the nitrogen atoms is also recovered in the form of nitrate.

Materials and methods

N-acetyl-D,L-penicillamine was purchased from Sigma. S-nitroso-N-acetyl penicillamine (SNAP) was synthesized as previously described [16].

S-nitrosocysteamine: To an aqueous solution (10 ml) of cysteamine hydrochloride (125 mg, 1.10 mmol) was added t-butyl nitrite (90%, 150 μl, 1.10 mmol) under an argon atmosphere. After 5 min, water (10 ml) was added to the red mixture, then the solution was concentrated to a final volume of 10 ml to remove t-BuOH and excess t-BuONO. UV_{max} (H_2O): 333 nm ($\varepsilon = 15$), 546 nm ($\varepsilon = 790$); ^1H NMR (Me$_2$SO-d$_6$) δ 2.99 (t, J = 7.0 Hz, 2H), 3.96 (t, J = 7.0 Hz, 2H), 8.43 (broad s, 3H). The solution has to be stored in the cold, protected from light.

Nitrite assay:

(1) Preparation of the Greiss reagent.
Solution A: 1 g sulfanilamide (5.81 mmol) dissolved in 100 ml of 1.2 N HCl.
Solution B: 300 mg of N-(1-naphtyl)ethylenediamine dihydrochloride (1.16 mmol) dissolved in 100 ml H_2O.
The Greiss reagent (solution C): mixture of 1 volume of A and 1 volume of B.
(2) The standard concentration line is obtained as follows:
A 10 mM aqueous solution of sodium nitrite was prepared (69.0 mg/ 100 ml). 5 ml of solution was diluted to obtain 100 ml of 0.5 mM solution. 200 μl samples were prepared by addition of water and sodium phosphate buffer (conditions used in the assay with thionitrite: pH 7, final concentration 2 mM) to 0–200 μl aliquots of freshly prepared 0.5 mM solution of sodium nitrite. Then 400 μl of solution C were added. After 15 min incubation at room temperature and shaking, the absorbances at 540 nm and 750 nm (blank for verification) were recorded.
(3) Decomposition of S-nitrosocysteamine: the thionitrite (25 μmol) is dissolved in 25 ml of 60 mM sodium phosphate buffer pH 7 (final concentration 1 mM). The red solution is then left for 1–2 h in the dark at room temperature, during which thionitrite decomposes, giving rise to a bleaching of the solution. After a 10-fold dilution, 0.2 ml of the solution is added to 0.4 ml of solution C. After 15 min shaking at room temperature, the absorbance at 540 nm is recorded.

Methemoglobin spectrophotometry assay

This technique is based on the rapid oxidation of reduced hemoglobin to methemoglobin metHb by $^\bullet$NO [5, 17]. The threshold for detection using $^\bullet$NO is 1 μM. The advantages of that method include the ready availability of spectrophotometers, avoidance of sample acidification, and the relative stability of methemoglobin. Furthermore, $^{\cdot}$NO synthesis can be measured continuously. However, the assay is not highly specific since oxidation of HbO_2 to metHb also takes place with NO_2^- but nitrite is much slower in promoting the reaction. HbO_2 spontaneously and slowly oxidizes to metHb, however minor amounts of metHb in the starting solution of HbO_2 do not interfere with the $^{\cdot}$NO quantification.

The commercially available hemoglobin is 50–95% methemoglobin and must be reduced with an excess of dithionite [18]. The protein is then purified by gel chromatography using a Sephadex column. $^{\cdot}$NO is detected by observing the characteristic shift in the Soret absorbance peak of hemoglobin from 433 to 406 nm. Accumulation of oxy-hemoglobin is not a problem because of the higher affinity of hemoglobin for $^{\cdot}$NO than for O_2. As a consequence, oxyhemoglobin can be used instead of reduced hemoglobin. In that case the shift is from 415 nm (Soret peak of HbO_2) to 406 nm. The reaction is as follows:

$$HbO_2 + NO \longrightarrow Hb^+ + NO_3^- \quad (Hb = metHb)$$

Another method is derived from the metHb assay described by Kaplan [17]. The proportion of metHb in a mixture of metHb and HbO_2 can be obtained, at any time, from the ratio $R = OD_{578}/OD_{525}$.

Actually, in addition to the Soret bands, the hemes have less intense absorption bands between 500 and 600 nm (α ou β bands). HbO_2 has two bands at 578 ($\varepsilon = 15.2$ mM^{-1} cm^{-1}) and 542 nm ($\varepsilon = 14.2$) while metHb has a shoulder at 578 nm ($\varepsilon = 3.52$) and a broad band at 635 nm ($\varepsilon = 4.09$). During the transformation of HbO_2 to metHb an isobestic point is seen at 525 nm. Consequently, the absorption of 525 nm is proportional to the total amount of hemoglobin in the reaction mixture. Thus there is a linear correlation between R and the proportion of metHb in solution. R_{max} is obtained for 100% HbO_2, R_{min} is obtained for 100% metHb.

The time dependence of metHb formation during decomposition of a S-nitrosoderivative can be monitored in the presence of HbO_2.

Materials and methods

Preparation of hemoglobin solutions
Oxyhemoglobin solutions may be stored under anaerobic conditions at $-80°C$ during several days without alterations of the protein. However, it is recommended to store it rather concentrated (>1 mM) at slightly alkaline pH.

Method A
5 ml of blood from a healthy person is collected in a tube containing 0.5 mg/ml heparin. Then 0.1 ml of blood is washed three times with

physiological serum. Hemolysis is achieved with 4 ml of phosphate buffer 10 mM pH 6.5 followed by centrifugation at 5000 rpm for 10 min. The oxyhemoglobin content of the supernatant is determined from the UV-visible spectrum, λ_{max}: 542 nm ($\varepsilon = 14.2$ mM^{-1} cm^{-1}); 578 nm ($\epsilon = 15.2$ mM^{-1} cm^{-1}).

Method B

Human methemoglobin was purchased from Aldrich (86% metHb, 2% desoxyHb, 7% oxyHb).

Three flasks are flushed with argon for 1 h. The first one contains 2 ml of 10 mM phosphate buffer pH 7.5, the second 4.2 mg sodium dithionite in 1 ml of the same buffer ($C = 24$ mM), the third 85 mg of metHb. Under anaerobic conditions, 0.75 ml of the buffer is transferred to the metHb flask to dissolve the protein (final conc. 1.7 mM). Then 1 ml of dithionite solution is transferred to the metHb solution and the mixture is shaked for 5 min. Desalting of the mixture is achieved by filtration on 25 ml Sephadex G 25, equilibrated with two volumes of 10 mM phosphate buffer 7.5. Elution is with the same buffer. During elution deoxyHb is transformed into HbO_2, which is collected (70 mg, yield 83%).

Nitric oxide spectrophotometric assay

Nitric oxide concentrations were determined spectrophotometrically by monitoring the oxidation of oxyhemoglobin to methemoglobin in phosphate buffer pH 6.5. In a standard reaction, S-nitrosocysteamine (82 μM) was incubated with an excess of oxyhemoglobin solution (30 μM, 4 hemes/protein). OD_{578} and OD_{525}, the absorbances at 578 nm (maximum for HbO_2) and at 525 nm (isobestic point) were recorded at various time intervals for 2 h. The percentage of remaining oxyhemoglobin was obtained from $R = OD_{578}/OD_{525}$. R reference values corresponding to 100% (R_{max}) and 0% (R_{min}) HbO_2 were obtained from pure solutions of 30 μM oxyhemoglobin and methemoglobin, respectively. A straight line can then be obtained between R_{min} and R_{max}. From that standard curve one can calculate from each couple of experimental values (OD_{578}, OD_{525}) the proportion of metHb. Total oxidation of oxyhemoglobin to metHb is achieved by addition of potassium ferricyanide (5 mg/ml).

EPR spectroscopy

$^{\cdot}NO$ is a radical with an unpaired electron in the p orbital and theoretically could be EPR-active. However, the relaxation times of the excited electron is too short and NO EPR signal cannot be detected by conventional EPR.

However, there are several tools to solve these problems and EPR spectroscopy remains an interesting technique to detect the presence of ˙NO, in aqueous media, cells and tissues. For this, ˙NO traps should be added to reaction mixtures in order to generate stable and detectable paramagnetic species:

$$\text{Trap} + \text{˙NO} \longrightarrow \text{Trap} - \text{NO˙}$$

In general, ˙NO traps are metal iron complexes since ˙NO is a very good ligand of ferrous ions and the resulting nitrosyl complexes are EPR active with characteristic EPR signals. Two ˙NO traps are widely used: the oxyhemoglobin complex [6], the Fe-diethyldithiocarbamate complex [7]. Unlike the diazotization assay, ˙NO is measured directly without acidification of ˙NO reaction mixture. Another trap which can be used easily is the iron (II)-thiosulfate complex [7, 8].

˙NO trapping by hemoglobin

Hemoglobin is also useful as a spin trap since nitrosyl-hemoglobin (HbNO) is readily detected by EPR [6]. ˙NO interacts with hemoglobin, producing HbNO as an intermediate leading to methemoglobin during reaction with O_2. The low temperature ($=110$ K) spectrum of HbNO, obtained after 30 s incubation of a S-nitrosothiol with hemoglobin, in a phosphate buffer, pH 6.5, has a characteristic three-line hyperfine signal that is not seen when hemoglobin is exposed to nitrite. The g values for HbNO are 2.060, 2.010, 2.005 with a nitrogen coupling constant of 17.2 G.

However, the assay is not highly sensitive because of the rather large instability of the HbNO complex. As a consequence, it is not quantitative and can be only used as a rapid qualitative assay for the presence of ˙NO.

˙NO trapping by Fe-diethyldithiocarbamate

During incubation with yeast, mammalian cells or tissues, DETC penetrates the cell wall and complexes the intracellular iron in the form of Fe(DETC)$_2$ in hydrophobic membrane compartments [7]. ˙NO generated, for example, from the decomposition of a S-nitrosothiol, in yeast cells suspensions is trapped as the nitrosyl [NO-Fe(DETC)$_2$] complex and its production can be quantitated by EPR following calibration with a standard. An advantage of this technique is that it measures ˙NO production directly. Moreover, it is specific for ˙NO since reaction with NO_2^- gives rise only to minor amounts of the nitrosyl complex [19]. One limitation is that ˙NO can only be trapped and stored for EPR detection in lipophilic compartments since the ferrous complexes of DETC precipitate in neutral aqueous solution. Addition of SOD or anaerobiosis greatly increases the amplitude of the signal, probably preventing the destruction of ˙NO by superoxide anions present in the incubates.

The spectrum of the NO-Fe(DETC)$_2$ complex shows an isotropic triplet signal at $g_{iso} = 2.03$ at $37°C$. The amount of ˙NO trapped can be calculated from the amplitude of one of the triplet lines calibrated by means of a dinitrosyl-Fe-thiosulfate complex standard.

˙NO trapping by Fe-thiosulfate

˙NO forms a stable $Fe^{II}(S_2O_3^{2-})_2(NO)_2$ paramagnetic complex in deoxygenated aqueous solutions of ferrous sulfate and sodium thiosulfate [7, 8]. At 77 K, a broad signal is observed at $g_\perp = 2.041$. This signal can be integrated to determine the ˙NO concentration from a standard curve obtained with pure nitric oxide.

Scavenging of the tyrosyl radical of the small subunit of ribonucleotide reductase by ˙NO

Protein R2, the small subunit of ribonucleotide reductase from *Escherichia Coli*, contains a stable tyrosyl radical, which is responsible for a characteristic EPR signal of protein solutions at 77 K [20]. The stability of this protein radical is explained by the fact that it is deeply buried in the interior of the protein and the access to the radical site is greatly constrained. However, because of the small size and the electrical neutrality of ˙NO, on one hand, and because of the general intrinsic reactivity of phenoxyl radicals towards ˙NO, on the other hand, a specific reaction between ˙NO and the protein radical takes place, which can be used as an assay for detection of ˙NO [9]. Actually, ˙NO couples to the radical giving rise to an EPR-silent nitroso adduct, a process which can be monitored by the disappearance of the tyrosyl radical EPR signal. Furthermore, the reaction is reversible, so that when ˙NO disappears from the reaction mixture, the EPR signal increases back again and is totally recovered at the end of the reaction. A further confirmation that this process was due to ˙NO can be obtained from the very efficient inhibition of the scavenging of the tyrosyl radical by oxyhemoglobin.

Material and methods

Materials

Human methemoglobin, ferrous sulfate heptahydrate and sodium thiosulfate were from Sigma. Diethyldithiocarbamic acid sodium salt trihydrate was from Janssen. Normal baking yeast was purchased in a local store. Protein R2 was prepared from overproducing strains of *E. coli* [9].

EPR spectroscopy

Detection of ·NO by paramagnetic HbNO complex. To a mixture of 200 μl of oxyhemoglobin solution (1 mM, pH 7.5, phosphate buffer 10 mM) and 100 μl of sodium acetate buffer (pH 6.5, 10 mM), 100 μl of aqueous S-nitrosocysteamine solution (0.5 mM) were added. The mixture was stirred and the EPR spectrum was recorded at 77 K as a function of time of reaction at room temperature.

Detection of ·NO by paramagnetic Fe(DETC)₂NO complex [7]. A suspension of yeast cells (200 mg/ml) in 0.1 M HEPES pH 7.5 was incubated with DETC (2.5 mg/ml) for 30 min at 37°C. The suspension (2 × 2 ml) was washed once by centrifugation and the plug were resuspended in 2 × 1 ml 0.1 M HEPES pH 7.5. To 200 ml of this suspension was added 200 μl of an aqueous solution containing S-nitrosocysteamine (0.5–1 mM). The reaction was performed during 15 min at 37°C. The samples were transferred into an EPR tube, frozen in liquid nitrogen and the EPR spectra were recorded at 110 K using a Varian E102 spectrometer. The microwave power was 40 milliwatts, the microwave frequency was 9.2 GHz, the modulation amplitude was 2 G and the time constant was 1 s. For calculation of the complex concentration, double integral of the EPR signal was compared to that of the EPR signal of a frozen solution of stable paramagnetic $Fe^{II}(S_2O_3^{2-})_2$ $(NO)_2$ complex as described recently.

Detection of ·NO by paramagnetic $Fe^{II}(S_2O_3^{2-})_2(NO)_2$ [7, 8]. Ferrous sulfate heptahydrate (20 mg) and sodium thiosulfate (356 mg) were dissolved in 20 ml of water previously deoxygenated by argon bubbling. Argon was flushed through the solution for 15 min. The resulting solution can be kept at 4°C for a week.

·NO detection: To a mixture of 200 μl of the freshly prepared ferrous sulfate and sodium thiosulfate solution, 100 μl of Hepes buffer (pH 7.5, 100 mM) and 50 μl of aqueous S-nitrosocysteamine solution (0.5 mM) were added. The mixture was stirred and the EPR spectrum was recorded at 77 K.

EPR spectroscopy with pure protein R2. The reaction was carried out at 37°C into an EPR tube containing R2 (1 mg/ml) and S-nitrosocysteamine at various concentrations, in 150 μl of 50 mM Tris-HCl buffer, pH 7.5, 10% glycerol. At time intervals, the tube was frozen in liquid nitrogen and the EPR spectrum of the solution was recorded at 110 K using a Varian E102 spectrometer. The amount of tyrosyl radical was determined from the comparison of the amplitude of the typical EPR signal at $g = 2$ to that of a pure sample of protein R2 (1 mg/ml). The microwave power was 1.5 milliwatts, the microwave frequency was

9.2 GHz, the modulation amplitude was 3.2 G and the time constant was 0.25 s.

References

1. Garrel, C. and Fontecave, M. (1995) Nitric oxide: Chemistry and biology. *This volume.*
2. Bratton, A.C., Marshall, Jr., E.K., Babitt, D. and Henrdrickson, A.R. (1989) *J. Biol. Chem.* 128: 537–550.
3. Archer, S. (1993) Measurement of nitric oxide in biological models. *FASEB J.* 7: 349–360.
4. Malinski, K. and Taha, Z. (1992) Nitric oxide release from a single cell measured *in situ* by porphyrinic-based microsensor. *Nature* 358: 676–678.
5. Kaplan, J.C. (1965) Méthode de mesure rapide du taux de la methémoglobine dans les globules rogues. *Extrait de la Revue Française d'Etudes Cliniques et Biologiques* 10: 856–859.
6. Kosaba, H., Watanabe, W., Yoshihara, H., Harada, N. and Shiga, T. (1992) Detection of nitric oxide production in lipolysaccharide-treated rats by ESR using carbon monoxide hemoglobine. *Biochem. Biophys. Res. Commun.* 184: 1119–1124.
7. Mordvintcev, P., Mülsch, A., Busse, R. and Vanin, A. (1991) On-line detection of nitric oxide formation in liquid aqueous phase by electron paramagnetic resonance spectroscopy. *Anal. Biochemistry* 199: 142–146.
8. Vedernikov, Y., Mordvintcev, P.I., Malenkova, I.V. and Vanin, A.F. (1990) *In*: S. Moncada and E.A. Higgs (eds): *Nitric Oxide from L-Arginine: A Bioregulatory System,* Elsevier, Amsterdam, pp 373–378.
9. Roy, B., Lepoivre, M., Henry, Y. and Fontecave, M. (1995) Inhibition of ribonucleotide reductase by nitric oxide derived from thionitrites: reversible modifications of both subunits. *Biochemistry* 34: 5411–5418.
10. Misko, T.P., Schilling, R.J., Salvemini, D., Moore, W.M. and Currie, M.G. (1993) A fluorometric assay for the measurement of nitrite in biological samples. *Anal. Biochemistry* 214: 11–16.
11. Condino-Neto, A., Muscara, M.N., Grumach, A.S., Carneiro-Sampaio, M.S.M. and De Nucci, G. (1993) Neutrophils and mononuclear cells from patients with chronic granulomatous disease release nitric oxide. *Brit. J. Clin. Pharmacol.* 35: 485–490.
12. Hunt, N.C.A. and Goldin, R.D. (1992) Nitric oxide production by monocytes in alcoholic liver disease. *J. Hepatol.* 14: 146–150.
13. Martin, H.J. and Edwards, S.W. (1993) Changes in mechanisms of monocytes/macrophage-mediated cytotoxicity during culture. *J. Immunol.* 150: 3478–3486.
14. Padgett, E.L. and Pruett, S.B. (1992) Evaluation of nitrite production by human monocyte-derived macrophages. *Biochem. Biophys. Res. Commun.* 186: 775–781.
15. Roy, B., Du Moulinet D'Hardemare, A. and Fontecave, M. (1994) New thionitrites: synthesis, stability and nitric oxide generation. *J. Org. Chem.* 59: 7019–7026.
16. Field, L., Dilts, R.V., Ravichandran, R., Lenhert, P.G. and Carnahan, G.E. (1978) An unusually stable thionitrite from N-acetyl-D,L-penicillamine; X-ray crystal and molecular structure of 2-(acetylamino)-2-carboxy-1,1-dimethylethylthionitrite. *J. Chem. Soc. Chem. Commun.* 249–250.
17. Ignarro, L.J., Buga, G.M., Wood, K.S., Byrns, R.E., Chaudhuri, G. (1987) Endothelium-derived relaxing factor produced and released from artery and vein is nitric oxide. *Proc. Natl. Acad. Sci. USA* 84: 9265–9269.
18. Di Orio, E.E. (1981) Preparation of derivatives of ferrous and ferric hemoglobin. *In*: E. Antonini, L. Rossi-Bernardi and E. Chiancone (eds): *Methods in Enzymology* (76), pp 57–72, Academic Press, San Diego.
19. Kirmse, R., Saluschke, S., Möller, E., Reijerse, E.J., Gelerinter, E. and Duffy, N.V. (1994) Single-crystals EPR spectroscopy of [^{57}Fe(NO)(S$_2$CNEt$_2$)$_2$]: the "triplet signal" in the EPR spectrum of [Fe(S$_2$CNEt$_2$)$_3$]**. *Angew. Chem. Int. Ed. Engl.* 33: 1797–1499.
20. Fontecave, M., Norlund, P., Eklund, H. and Reichard, P. (1992) The redox centers of ribonucleotide reductase of *Escherichia Coli. In*: A. Meister (ed.): *Advances in Enzymology and Related Areas of Molecular Biology,* J. Wiley & Sons, New-York, pp 147–183.

Analysis of Free Radicals in Biological Systems
Favier et al. (eds)
© 1995 Birkhäuser Verlag Basel/Switzerland

Salicylate hydroxylation products assay as a marker of oxidative stress in man: A sensitive HPLC-electrochemical method

C. Coudray[1], M. Talla[2], S. Martin[3], M. Fatôme and A.E. Favier[1,2]

[1]*Groupe de Recherche et d'Etude sur les Pathologies Oxidative (GREPO), Laboratoire de Biochimie C, Centre Hospitalo-Universitaire de Grenoble, BP 17X, F-38034 Grenoble, France*
[2]*Laboratoire de Biochimie Pharmaceutique, UFR de Pharmacie, Domaine de la Merci, F-38700 La Tronche, France*
[3]*Centre de Recherches du Service de Santé des Armées, Avenue des Maquis du Grésivaudan, F-38700 La Tronche, France*

Summary. The *in vivo* measurement of highly reactive free radicals, such as ˙OH radical, in humans is very difficult. For this reason, secondary products of oxidative stress are frequently measured. The most indirect methods (TBARs [thiobarbituric acid reactants] test, conjugated diene, hydroperoxides) are rather unspecific and can give conflicting results. New, more specific markers are currently under investigation (protein oxidation, DNA adducts and aromatic probes). All these methods are based on the ability of ˙OH to attack the benzene rings of aromatic molecules to produce hydroxylated compounds that can be measured directly.

In *vivo*, radical metabolism of salicylic acid produces two main hydroxylated derivatives 2,3- and 2,5-dihydroxybenzoic acid [2,3- and 2,5-DHBA]). The latter can also be produced by enzymatic pathways through the cytochrome P-450 system, while the former is reported to be solely formed by direct hydroxyl radical attack. Therefore, measurement of 2,3-DHBA, following oral administration of salicylate in its acetylated form (aspirin), could be proposed for assessment of oxidative stress *in vivo*.

In this work, evidence is presented for a sensitive method for the detection of hydroxyl free radical generation *in vivo*. The methodology employs a high pressure liquid chromatography with electrochemical detection for the identification and quantification of the hydroxylation products from the reaction of ˙OH with salicylate. A detection limit of less than 1 pmol for the hydroxylation products has been achieved with electrochemical detector responses which were linear over at least five orders of magnitude. Using this technique, we measured plasma levels of 2,3- and 2,5-DHBA and dihydroxylated derivatives/salicylic acid ratios following the administration of 1000 mg aspirin in 20 healthy subjects.

Introduction

Oxygen-derived species such as superoxide radical anion ($˙O_2$) and hydrogen peroxide (H_2O_2) have been implicated as damaging agents in the action of many toxins and various diseases and in aging [1–7]. The possibility that much of the toxicity produced by increased $˙O_2$ and H_2O_2 generation is mediated by metal-ion-dependent formation of the highly reactive ˙OH radical has also been discussed in detail by Halliwell and Gutteridge [8, 9] and others [10, 11]. Although much circumstantial evidence supports the biological relevance of the iron-ion-catalysed formation of ˙OH radical from ˙O_2 and H_2O_2, [12] there has been as yet no direct demonstration that ˙OH radical is formed *in vivo*. The inflamed rheumatoid joint, the myocardial infarction, diabetes and cancer chemotherapy, in which conditions are proposed to be ideal for ˙OH generation, seems to be good places to look.

The *in vivo* measurement of highly reactive free radicals, such as the ˙OH radical, in humans is difficult. Due to its high reactivity, ˙OH radical has a very short half-life and is therefore present in extremely low concentrations. For these reasons, secondary products of oxidative stress are frequently measured. The commonly used indirect methods (TBARs test, conjugated diene, hydroperoxides) are rather unspecific and can give conflicting data. This is partly due to the lack of direct methods for measuring oxidative stress and/or oxygen free radical production in humans. Recently, some "direct" methods have been proposed to probe the formation of ˙OH radical *in vivo*, such as electron spin resonance measurement ethylene from 2-keto 4-methylthiobutyrate [13, 14]. These methods are neither sensitive nor practical.

At the moment, more specific markers are under investigation (aromatic probes, DNA adducts and protein oxidation). All these methods are based on the ability of ˙OH to attack the benzene rings of aromatic molecules to produce hydroxylated compounds that can be measured directly. Aromatic hydroxylation has already been used for measuring *in vitro* ˙OH [12, 15, 16]. For *in vivo* studies, the scavenger molecules, and the hydroxylated products must not undergo extensive metabolisme. Aromatic compounds react with high rate constants with ˙OH, to form a specific set of hydroxylated products [16]. If an aromatic compound can be safely administered to humans in doses that produce concentrations in body fluids sufficient to scavenge ˙OH, then assaying such products would be a reasonable evidence that ˙OH is being formed *in vivo*, provided that these products are not formed by enzymatic hydroxylation. A suitable candidate for use in humans may be salicylate or its acetylated form, i.e., aspirin.

Aspirin (O-acetylsalicylic acid, ASP) is a commonly used analgesic and anti-inflammatory agent in man. After injestion, a substantial amount of ASP is hydrolysed to salicylic acid (SA) by esterases in the gastrointestinal tract, in the liver and, to a smaller extent, in the serum [17]. Salicylate reaches its peak in plasma about 0.5 to 1.5 h after the oral intake of aspirin. SA is further metabolised by conjugation to glycine (liver glycine N-acetylase) to form salicyluric acid, by hydroxylation (liver microsomial hydroxylases) to form gentisic acid, and by formation of the phenolic glucuronide (conjugation of the hydroxyl group of SA with the first, hemiacetal carbon of D-glucuronic acid to form an ethereal linkage) and the acyl glucuronide (conjugation of the carboxylic group of SA with the first, hemiacetal carbon of D-glucuronic acid to form an ester linkage) [18]. About 60% of salicylate

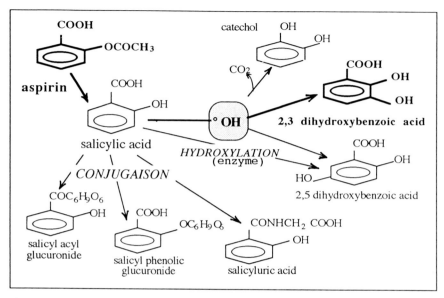

Fig. 1. Major reported enzymatic and oxidative by-products of aspirin in human.

remains unmodified and can undergo ˙OH attack to produce two products namely 2,3-dihydrobenzoate (2,3-DHBA) and, to a much smaller extent, catechol [19], which have not been reported as products of enzymatic metabolism. Thus 2,3-DHBA appears to be a useful marker of *in vivo* ˙OH production. We, therefore, propose to develop techniques to identify and quantify this derivative in human body fluids (Fig. 1).

Materials and methods

Chemicals

Aspirin (Aspégic) was obtained from Synthelabo, Le Plessis Robinson, France; sodium salicylate from Merck; 2,3 dihydroxybenzoic acid, 2,5 dihydroxybenzoic acid, 2,6 dihydroxybenzoic acid and 3,4 dihydroxy-benzoic acid were purchased from Sigma. Acetonitril, methanol, trisodium citrate, sodium acetate, ether and ethyl acetate from Prolabo. All chemicals were analytical reagent grade and used without further purification.

Instrumentation

For dihydroxybenzoic acids (DHBAS) assay, we used a Hitachi Liquid Chromatograph pump equipped with autosampler WISP (Waters) (injected volume = 100 μl). Two electrochemical amperometric detectors (Bioanalytical Systems, model LC4B or Shimadzu L-ECD-6A) with a plastic cell equipped with a glassy carbon electrode operated at + 0.7 V, and an Ag/AgCl reference electrode. The signals from the detector were acquired on Varian DS601 data system and subsequently processed.

The analytical stainless-steel column was 150 × 4.6 mm, and packed with octadecyl silane (Spherisorb ODS$_2$, C18) with average particle size of 3 μm (Alltech). A guard column packed with 10 μm spheri-10 PR18 (30 × 4.6 mm I.D., from Alltech) was also used. Peak areas at 0.7 V were recorded with a 1 mV potentiometric screen. All separations were performed at ambient temperature.

Optimisation procedures

Choice of oxidation potential
The hydrodynamic voltamograms between 0 to 1400 mV were recorded using solutions containing 100 mg/l of all studied substances in mobile phase (85% phosphate buffer 100 mM, 0.10 mM SDS, pH 3.3 and 15% methanol, V/V). A CES detector (Erosep Instrument) equipped with 16 electrodes set up at variable potentials was used. The choice of appropriate oxidation potential by measuring signal/noise ratio of concerned substances especially 2,3-DHBA.

Choice of eluent
The composition and the pH of the eluent, and the flow rate were examined. The retention time, the separation of peaks and the ratio peak/noise were recorded.

Choice of internal standard
Four potential internal standards (2,4-DHBA, 2,6-DHBA, 3,4-DHBA, and catechol) were studied.

Extraction procedure
Volume of sample, solvent type and volume and extraction duration were studied. For SA, two deproteinisation procedures were tested, i.e., with ethanol and with acetonitril.

Preparation of standard curves
Saline and plasma, with no detectable salicylate as measured by the present method, were used to make up solutions of different concentrations (10 to 200 nM, and 100 to 2000 nM, of 2,3-DHBA and 2,5-DHBA) respectively. Stock solutions (0.1 mM, 1 mM) of 2,3-DHBA, and 2,5-DHBA respectively were stable for at least 2 months when stored at 4°C in the dark. For SA assay, saline and plasma were used to

make up solutions varying in concentrations (100 to 2000 μM from a stock solution of 100 mM). The concentrations were determined by the peak-area ratio method using 3,4-DHBA as internal standard for DHBAS assay and 2,6-DHBA for SA assay.

Analytical data

The detection limit was determined according to Gatautis and Pearson [20]. A sample, with salicylate or 2,3- and 2,5-DHBA concentrations of three to five times the noise level (2 nM for 2,3- and 2,5-DHBA, and 6 μM for salicylate) was measured 30 times. The detection limit was calculated according to the formula $DL = (2 \times SD \times c)/S$ where DL is detection limit, S is the mean relative area, SD is the corresponding standard devation and c is the concentration of the "test" solution.

The linearity was established with correlation coefficients. Calibration standard solutions (20 to 4000 nM of 2,3-DHBA; 20 to 5000 nM of 2,5-DHBA; and 20 to 4000 μM of salicylate) were determined. The variation coefficients and correlation coefficients were then calculated.

The accuracy was evaluated by standard addition recoveries; known amounts of 2,3-DHBA (100 nM); 2,5-DHBA (1000 nM) and salicylic acid (300 and 600 μM), were added to saline and to pooled plasma. The preparations were then analysed. The obtained peak areas were then converted to concentration using appropriate regression equations, and the values from plasma and saline were then compared. The ratio of the concentration in the plasma to the corresponding concentration in saline was then used as an index of recovery.

The precision was determined according to the ValTec protocol [21]. Intra- and interassay reproducibility was conducted on pooled samples of human plasma with known concentrations of 2,3-DHBA (100 nM), 2,5-DHBA (1000 nM) and SA (600 μM).

The specificity of the method was studied by assaying plasma samples form subjects before and after receiving ASP. Comparison of the retention times with those of standards established the peak identity.

Salicylic acid measurement

The chromatograph was from Kontron Instruments (Rotkreus, Switzerland) and consisted of two solvent-delivery pumps (model T414), a sample injector 234 Kontron, an analytical stainless-steel column packed with ultrasphere ODS 5 μm (150 $\times$ 4.6 mm I.D., from Alltech), a guard column packed with 10 μm spheri-10 PR18 (30 $\times$ 4.6 mm I.D., from Alltech), and a muliwavelength detection system (HPLC detector 430, Kontron). The system was controlled by a computer (Data System

450, Kontron). Salicylic acid assay was performed as elsewhere indicated [22]. Briefly, aliquots (100 μl) of standard solutions or plasma samples were mixed with 100 μl of 2.5 μM 2,6-DHBA and deproteinized by 200 μl of ethanol in polypropylene conical tubes. The samples were mixed on a vortex-type mixer for exactly 2 min. The tubes were then centrifuged at 1600 $\times$ g for 15 min, and 50 μl of the supernatant was diluted with 950 μl of mobile phase. The diluted solution was then filtered on 0.45 μ (Alltech) and 50 μl were injected onto the column. Mobile phase consisted of sodium citrate and acetate 30 mM, pH 5.45/methanol (85/15). Flow rate was 1 ml/min and the detector was set up at 295 nm. Standard curves were constructed from measurements of peak-area ratios. Saline and plasma blanks were analysed with each set of standards. The used concentrations were as follows: 0, 62.5, 125 250, 500, 1000 μmoles/1 of SA.

The human study

To show the applicability of the proposed method, salicylic acid, 2,3- and 2,5-DHBA concentrations were measured in 20 human subjects after a single oral dose (1000 mg in 150 ml water, prepared just prior to administration) of soluble (Aspégic). 7 ml blood samples were drawn from the antecubital vein with a vacutainer system in heparinized tubes before and 120 min following ASP administration. The tubes were centrifuged within 30 min of blood collection at 1600 g for 10 min. Plasma was separated, quickly frozen and kept until required for analysis.

Results

Optimisation procedures

Choice of oxidation potential
The hydrodynamic voltamograms of the various monohydroxylated products and of salicylic acid are presented in Figure 2. The 2,3- 2,5- and 3,4-DHBA acid compounds produce a maximum current when the potential is about 0.7 V. However, at this voltage the other hydroxybenzoic acid products 2,4- and 2,6-DHB acids catechol and salicylic acid do not yield a detector response. At a detector potential of 1.1 V, salicylic acid and all the dihydroxybenzoic acid products are recorded. An oxidation potential of 0.7 V *versus* Ag/AgCl reference electrode was chosen so as to get maximum current response and minimum background noise for the hydroxylated products of salicylate. Because SA occurs at high concentrations in plasma, it cannot be monitored at

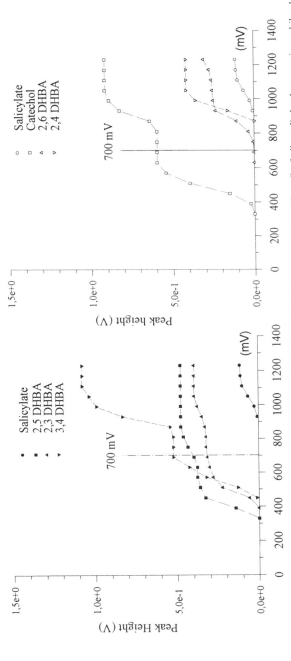

Fig. 2. Hydrodynamic voltamograms of hydroxylated products of salicylic acid. Solutions containing 100 mg/l of all studied substances in mobile phase (85% Phosphate buffer 100 mM, 0.10 mM SDS, pH 3.3 and 15% methanol, V/V) were recorded on a CES detector (Erosep Instrument) equipped with 16 electrodes set up at variable potentials.

0.7 V, and the noise average is very important at 1.1 V, therefore we decided to detect SA spectrophotometrically at 295 mm.

Choice of eluent

An isocratic delivery system was used consisting of a single eluent containing sodium acetate/trisodium citrate (30 mM/30 mM) pH 3.90. The flow rate was 0.2 ml/min. The mobile phase was carefully selected to achieve maximum separation and sensitivity. The mobile phase was sparged continuously with helium gas during elution. Moreover, the isocratic elution allows a minimum of down-time. If this eluent revealed very appropriate for DHBAS assay, the retention time of SA with this eluent was 37 min and that of 2,6 DHBA was 48 min, with a flow rate of 1 ml/min. Several eluents consisting of buffer with different percentage of methanol (0 to 30%) were then tested. We found that the eluent containing 15% of methanol constitutes a good compromise with a retention time for SA of 7 min and 11 min for the internal standard 2,6 DHBA.

Choice of an internal standard

We assayed different substances previously used by other investigators (3,4 DHBA, 2,4-DHBA, 2,6-DHBA, and catechol). Though catechol could be one of the radical metabolites of salicylate, we were then unable to detect it in our patients. Moreover, the analysis time was increased when we used catechol as internal standard. The 2,4- and 2,6-DHBA do not yield a detector response at a detector potential set at 0.7 V. The 3,4-DHBA was thus used for the measurement of DHBAs. For SA assay, the 2,6 DHBA was found to be the most appropriate. In the SA assay conditions, the retention time of 3,4 DHBA was very short and it coeluted with non identifiable peaks. 2,6 DHBA had a retention time more important than that of SA without any interfering peaks.

Extraction procedure

Extractions with ether and ethyl acetate were compared. DHBAs were better extracted with ethyl acetate. Double extraction results were not sufficiently different, and therefore one simple extraction with ethyl acetate was applied. The extraction with different volumes of ethyl acetate was then examined. We obtained satisfactory results with 3 ml of solvent when 0.4 ml of sample is used.

Optimised procedure

Aliquots of standard solutions and of plasma samples (400 μl) were mixed with 100 μl of 2.5 μM 3,4-DHBA and acidified by 75 μl of concentrated HCl in 10 × 70 mm glass tubes. The samples were mixed on a vortex-type mixer for 30 s; 3 ml of ethyl acetate were then added and mixed for exactly 2 min. The tubes were finally centrifuged at 1600 × g for 15 min, and 2 ml of ethyl acetate phase were dried under nitrogen steam. The dry residue was taken up by 200 μl of mobile phase and 100 μl were injected into the column. Mobile phase consisted of sodium citrate and acetate 30 mM, pH 3.90. Flow rate was 0.2 ml/min. Standard curves were drawn using peak-area ratios. Saline and plasma blanks were analysed with each set of standards. The used final concentrations for calibration curves were as follows: 25, 50, 100, 200, 400 nanomol/l for 2,3-DHBA 125, 250, 500, 1000, 2000 nanomol/l for 2,5-DHBA.

Analytical performance

Detection limit
The detection limit was found to be 0.37 nM for 2,3-DHBA and 0.62 nM for 2,5-DHBA (electrochemical detection) and 3.47 μM for SA (spectrophotometric detection) when calculated according to the formula of Gatautis and Pearson. These limits demonstrate the excellent sensitivity of the proposed method for both DHBAs and SA measurement. These ranges allow the determination of the three compounds after ingestion of ASP in moderate dose in all patients at risk for intensive oxidative stress.

Linearity
The concentrations and peak areas showed linear relationships for 2,3- and 2,5-DHBA in the concentration range investigated (0 to 4000 and 5000 nmol/l respectively). Figure 3 illustrates this linearity for 2,3- and 2,5-DHBA with their corresponding correlation coefficients and variation coefficients (Fig. 3).

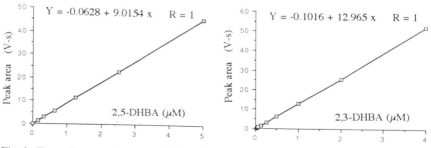

Fig. 3. The calibration plots for salicylate-derived-hyroxylated products.

Table 1. Within-run (WR) and between run (BR) precision of salicylic acids and its hydroxylated products

		2,3-DHBA	2,5-DHBA
Spiked plasma		100 nmol/l	1000 nmol/l
	WR	4.1%	2.1%
	BR	6.2%	7.4%
Standard		100 nmol/l	1000 nmol/l
	WR	3.3%	2.1%
	BR	5.4%	5.7%

Accuracy

The absolute recoveries of 2,3-DHBA added to a concentration of 25 nmol/l to six different plasma samples ranged from 86 to 102% with a mean of 94%. Similar studies on 2,5-DHBA (250 nmol/l) yielded recoveries from 94 to 103% with a mean of 98%. SA assay showed a recovery percentage with a mean of 96.2%.

Precision

The within-run and between-run coefficients of variation are indicated in Table 1. In terms of within-run precision, they ranged from 3.3 to 4.1% for 2,3-DHBA and 2.1 for 2,5-DHBA. The between-run co-efficients of variation ranged between 5.4 and 6.2% for 2.3 DHBA and from 5.7 to 7.4% for 2,5 DHBA. Our CV are better than those of previous studies conducted by others at similar concentrations. More-over, the SA method permitted very low within-run coefficient of variation less than 2%. The between-run coefficient of variation of SA assay is 3.9%. The between-run and recovery assays were conducted during a 15 day period. During this period, the stability was good for the three compounds, as shown by the correlation coefficients.

2,3-DHBA and 2,5-DHBA in healthy volunteers

Figure 4 shows a chromatogram of a blank plasma sample and a chromatogram of a plasma sample from the subject receiving ASP. Both 2,3- and 2,5-DHBAs were well separated with the following retention times: 2,3-DHBA 22.0 min; 2,5-DHBA 19.6 min and 27.0 min for the

Fig. 4. HPLC-electrochemical detection chromatogram showing separation of isomeric DHBAs. (A) Plasma sample from a normal subject before 1000 mg *per os* ASP administration (T0). (B) Plasma sample from the same subject 2 h after ASP administration (T2). (C) Elution of a standard mixture containing 125 nM 2,3-DHBA, 1000 nM 2,5-DHBA and 500 nM of internal standard 3,4-DHBA. The flow rate for the HPLC mobile phase (pH 3.85-3.90) was 0.2 ml/min and the chromatographs were recorded. The identification of peaks is as follows: (1) 2,5-DHBA; (2) 2,3-DHBA; and (3) 3,4-DHBA.

A

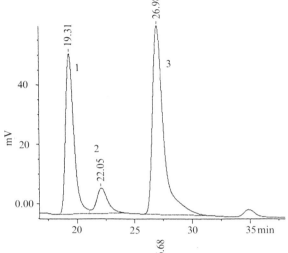

B

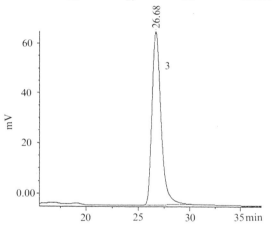

C

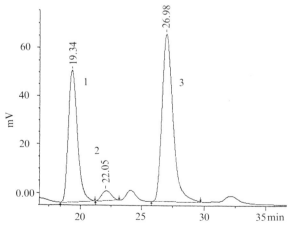

Table 2. Plasma salicylic acid and its hydroxylated products levels 2 h after ingestion of 1000 mg of aspirin in adult humans

	Total (n = 20)	Men (n = 10)	Women (n = 10)
2,3-DHBA (nmol/l)	63.2 ± 23.8	57.2 ± 21.4	69.3 ± 25.5
2,5-DHBA (nmol/l)	832 ± 309	656 ± 190	1008 ± 312
Salicylic acid (μmol/l)	487 ± 116	427 ± 62.8	548 ± 128
2,3-DHBA/SA (nM/μM)	0.129 ± 0.034	0.133 ± 0.041	0.125 ± 0.025
2,5-DHBA/SA (nM/μM)	1.69 ± 0.387	1.55 ± 0.433	1.83 ± 0.293
2,3/2,5-DHBA (nM/μM)	0.080 ± 0.031	0.092 ± 0.039	0.069 ± 0.013

internal standard (3,4 DHBA). No interfering peaks were seen in blank plasma samples. Concentrations of 2,3- and 2,5-DHBA as low as 10 nmol/l could be accurately measured. Analysis of samples drawn before aspirin administration shows that volunteers' plasma fractions were free of salicylic acid or its dihydroxybenzoic acids. The volunteers' plasma, analysed 2 h after aspirin administration, contained high concentration of salicylic acid, reaching in some subjects more than 800 μM.

Table 2 shows the results obtained from 20 laboratory volunteers ranging from 20 to 40 years with a mean age of 28.2 ± 5.22 years (10 women and 10 men). As expected, the concentration of 2,3-DHBA was low and that of 2,5-DHBA was relatively high. Two hours after ASP administration, the plasma SA level was 487 ± 116; 245–729 μmoles/l (M ± SD; M ± 2.086SD). The plasma level of 2,3-DHBA was 63.2 ± 23.8; 13.6–113 nM (M ± SD; M ± 2.086SD) and that of 2,5-DHBA was 832 ± 309; 187–1477 nM (M ± SD; M ± 2.086SD). Plasma aspirin levels were not determined in this study. Due to individual variations in the pharmacokinetics of aspirin, the plasma level of salicylic acid could vary largely from one individual to another (344 to 868 μmol/l). Consequently, the concentration of dihydroxybenzoic acids was corrected for this variation. For this reason, we reported our results as 2,3-DHBA/ salicylic acid ratios (mmoles/moles).

Discussion

Despite the emergence of electron spin resonance and spin trapping techniques, the identification and the characterisation of oxygen-derived free radicals in humans still pose a formidable task. The present paper describes a chromatographic method for the specific identification and quantification of hydroxylated salicylate products in body fluids. It is hoped that this methodology will be used by those attempting to detect and measure ·OH generation in oxidative diseases such as diabetes, myocardial infarction, rheumatoid arthritis, cancer, ageing.

Attack by ˙OH radicals, generated by the Fenton system at pH 7.4, upon salicylate, produces three products [23]. The major products are 2,3-DHBA and 2,5-DHBA. However, formation of the decarboxylation product catechol has not been reported in previous experiments in which salicylate was exposed to Fenton systems under physiological conditions [24], but its formation is not unexpected [25], especially since radioloysis of aerated salicylate solutions produces some catechol. Also, decarboxylation of benzoate has been used as assay for ˙OH radicals *in vitro* [26].

Salicylate reacts with ˙OH at a rate constant of about $5 \times 10^{+9}$ to 10^{+10} $M^{-1}S^{-1}$ [27]. The fairly low oral doses of aspirin used in our study (1000 mg per os, or approximately 15 mg/kg) gave body-fluid salicylate concentrations between 245 and 729 μM (confidence interval $= M \pm 2.086SD$), which is sufficient to trap ˙OH radicals. However, *in vivo* metabolism of salicylic acid produces two main hydroxylated derivatives (2,3- and 2,5-DHBA) which could be measured.

Many techniques have already been used to evaluate the hydroxylated compounds in *in vitro* as well as in *in vivo* studies. A colorimetric method has been initially described by Richmond et al. [16], in which the hydroxylated products phenols were treated with sodium tungstate and sodium nitrite in acid medium. The pink colour was developed in alkaline medium and measured with a spectrophotometer at 410 nm [28]. HPLC techniques were then developed to separate and quantify the hydroxylated derivatives in animal [15, 29, 30] or in humans [24, 31, 32]. The spectrophotometric and spectrofluometric methods were adequate for *in vitro* studies, where high concentrations of ˙OH are usually formed; however, these detection modes are not sensitive enough to quantify hydroxylated derivatives in general and 2,3-DHBA in particular in *in vivo* studies. Consequently, the electrochemical detection mode became popular in most laboratories interested in the quantification of oxidant stress *in vivo* and especially in man.

2,5-DHBA is usually considered a bona fide marker for detecting hydroxyl radicals. However, its level is markedly lower than that of 2,5-DHBA [29]. Here, we examined the level of both 2,3- and 2,5-DHBA in the human plasma. Our findings show a significantly higher level of 2,5-DHBA than 2,3-DHBA (three- to five-fold). The selective difference between 2,3- and 2,5-DHBA could be associated with differing metabolic rates of production of these adducts, different efficiencies of scavenging ˙OH radicals at the 3 or the 5 position of the salicylate aromatic ring, and differences in the stability of the adducts against metabolic or biochemical modifications by cellular components. Indeed, when added to a tissue homogenate, the decrease in 2,3-DHBA concentration was 2.5 to 3 fold higher than the 2,5-DHBA. This difference in the metabolic removal of 2,3-DHBA and 2,5-DHBA may be even more pronounced within the tissue where the enzymatic content is higher than the homogenate [33]. However, this difference seems to be due to the fact that 2,5-DHBA, but not 2,3-DHBA, is produced by an enzymatic hydroxylation pathway through the cytochrome P-450 microsomal system [34]. The latter seems to be formed exclusively via a radical pathway, which further substantiates the validity of the assumption that 2,3-DHBA authentically reports free radical fluxes *in vivo*.

Thus, 2,3-DHBA appears to be a sensitive and specific marker of *in vivo* oxidative stress studies. There is now increasing evidence for 2,3-DHBA production after aspirin administration in conditions of oxidative stress such as exposure to 100% oxygen [35], treatment with the redox-cycling drug Adriamycin [29], *in vitro* neutrophile activation [36], diabetes [31], or rheumatoid arthritis [24].

Recently, this technique has been used by Tubaro et al. [32], who demonstrated the formation of hydroxyl radicals in acute myocardial infarction in man. In this study, all the subjects received 100 mg aspirin p.o. daily, venous blood samples were taken 30 min after the first dose (time 0) and then at 3-, 6-, 12-, 24-, and 48 h and 5 days. Serum was analysed by HPLC-electrochemical detection for 2,3- and 2,5-DHBA contents. 2,3-DHBA was present in all subjects with acute myocardial infarction and undetectable in healthy volunteers at all time points studied. On the contrary, serum level of 2,5-DHBA did not show statistically significant differences between acute myocardial infarction patients and healthy volunteers. Their data support the hypothesis that hydroxyl radicals are formed during acute myocardial infarction in man.

In another study, Ghiselle et al. [31] have reported plasma levels of 2,3- and 2,5-DHBA following oral administration of 1000 mg of aspirin in well-controlled diabetic patients and in healthy subjects. They also determined plasma level of TBARs as an index of lipid peroxidation process. They have noted that 2,3-DHBA levels were significantly higher in diabetic patients than in controls (63.4 ± 20.0 vs 49.0 ± 6.8 nmol/l). However, TBARs levels were not significantly different in groups. They concluded that salicylate hydroxylation is useful to reveal differences *in vivo* oxidative stress.

Finally, Grootveld and Halliwell [24] reported that patients suffering from rheumatoid arthritis have 2,3-DHBA values twice as high as normal subjects. They also found increased production of 2,5-DHBA in patients than in controls. About 4.5 h after ingestion of 600 mg ASP, they found approximately 100 nM 2,3-DHBA; 600 nM 2,5-DHBA and 200 μM SA in rheumatoid arthritis patients. In control subjects, these values were approximately 50 nM 2,3-DHBA; 250 nM; 2,5-DHBA and 200 μM of SA. These results suggest an increase in ˙OH production in these patients.

Grootveld and Halliwell [19, 24] were the first to report the presence of 2,3-DHBA in blood plasma and urine of healthy adult volunteers.

They found that plasma concentration of 2,3-DHBA peaked approximately 2 h after ASP ingestion. Thus, we used this time period in our technique. The levels of 2,3-DHBA and 2.5-DHBA were very low 1 h after ASP administration, but increased to 50 and 160 nM 2,3-DHBA 2 h after the ingestion of 600 and 1200 kmg ASP in adult man. Our results are in good agreement with these findings.

The presence of 2,3-DHBA in the plasma of healthy subjects after aspirin intake could be due to the baseline rate of intracellular 'OH formation from ionising radiation and Fenton reactions *in vivo* [24]. This phenomenon is supported by the detection of baseline values of several markers of 'OH production such as allantoin [37], 8-hydroxy-deoxyguanosine, thymine and thymidine glycols in plasma or urine of healthy subjects [38].

Because the plasma level of 2,3-DHBA depends on that of salicylate, and because the latter could vary among individuals ingesting the same dose of aspirin, the concentration of 2,3-DHBA should be expressed as 2,3-DHBA/salicylate ratio (mmoles/moles). It should be noted that plasma ASP is rapidly hydrolysed (40% hydrolysis in 120 min at room temperature after sampling) and this could increase plasma level of salicylic acid [39]. Indeed, this hydrolysis is faster in whole blood than in plasma, and is attributed to plasma and red blood cell esterases. It is thus necessary to remove red blood cells from plasma as rapidly as possible. However, it appears that plasma enzymatic hydrolysis is completley inhibited by physostigmine (0.2 mM). This should prevent an increase in salicylate plasma level after blood was drawn. In our case, plasma ASP level, 2 h after ASP administration should be negligible. However, at higher doses of ASP or when samples are drawn in the early periods, inhibition of esterases should be considered.

However, the present study and the other studies on human samples [24, 31, 32] do not prove that the 2,3-DHBA detected in body fluids originate exclusively from radical attack on salicylate. The presence of low concentrations of this product in the plasma of healthy human volunteers after aspirin ingestion might be related to the baseline rate of intracellular 'OH formation from ionising radiation and Fenton reactions or, it might be also generated by an as yet unreported minor metabolic pathway [40, 41].

Thus, many aspects still remain to be investigated. First, larger doses of ASP should be examined. In fact, in humans, ASP is generally not acutely toxic until the plasma concentration of salicylic acid reach the millimolar range [42]. Thus, the ASP dose used in the present study was well below toxic concentrations. Secondly, salicylic acid has less pharmacological effects than ASP. For example, inhibition of platelet function by ASP could underestimate 'OH production by the cyclooxygenase pathway. The direct administration of sodium salicylate could overcome this concern. Third, because 2,3- and 2,5-DHBA could be metabolised and excreted in the urine, their determination in the urine could be of great interest. Finally, evaluation of the relationships between 2,3-DHBA levels and the antioxidant defence systems in different oxidative stress should now be undertaken.

In conclusion, direct evidence for the formation of free radicals in oxidative processes and the causal relationship between free radicals and the damage are still lacking. The major reasons for this are that free radicals are present in very low amounts in the tissues, are highly reactive, and are short lived. A newly developed methodology is now available, where a specific hydroxyl radical trap produces stable hydroxylation products that can be specifically identified and quantified with high sensitivity using HPLC coupled with electrochemical detection. Plasma 2,3- and 2,5-DHBA are specifically quantified (electrochemical detection, 0.7 V) with an octadecyl silane reversed-phase chromatographic column by peak-area ratio (internal standards 3,4-DHBA). The mobile phase was carefully selected to achieve a maximum separation rapidly. The isocratic elution allows a minimum of down-time. The improved salicylate hydroxylation products assay, as described in the present paper, provides a simple and convenient method by which ·OH radicals may be detected and quantified *in vivo*. The small plasma requirement (0.4 ml) permits determinations of ·OH from infants and young children. Finally, it should be noted that the principle behind our methodology can be applied to other aromatic compounds such as phenylalanine or benzoic acid [43–46].

Acknowledgements
This work was supported by a French Ministry of Research and Technology grant Nr. 88-C-0852 Paris, France. The authors wish to thank Dr. H. Faure for reading the manuscript. The authors also acknowledge the excellent technical assistance of Colette Augert, Sophia Bouhadjeb and Catherine Mangournet. The oxidation potential study was performed in collaboration with Mr. L. Martinez from Eurosep society (France) and Mr. G. Achilli from Euroservice (Italy).

References

1. Di Guiseppi, J. and Fridovich, I. (1984) The toxicity of molecular oxygen. *CRC Crit. Rev. Toxicol.* 12: 315–342.
2. Halliwell, B. (1989) Free radicals, reactive oxygen species and human disease: a critical evaluation with special reference to atheroscelerosis (review). *Br. J. Exp. Path.* 70: 737–757.
3. Halliwell, B. and Cross, C.E. (1991) Reactive oxygen species, antioxidants, and acquired immunodeficiency syndrome. *Arch. Intern. Med.* 151: 29–31.
4. Ames, B.N. (1989) Endogeneous oxidative DNA damage, aging and cancer. *Free Rad. Biol. Med.* 7: 121–128.
5. McCord, J.M. (1985) Oxygen-derived free radicals in postischemic tissue. *N. Engl. J. Med.* 312: 159–163.
6. Flohé, L. (1988) Superoxide dismutase for therapeutic use: clinical experience, dead ends and hops. *Mol. Cell. Biochem.* 84: 123–131.
7. Harman, D. (1988) Free radicals in aging. *Mol. Cell. Biochem.* 84: 155–161.

8. Haliwell, B. and Gutteridge, J.M.C. (1984) Oxygen toxicity, oxygen radicals, transition metals and disease. *Biochem. J.* 219: 1–14.
9. Halliwell, B., Gutteridge, J.M.C. and Blake, D.R. (1985) Metal ions and oxygen radical reactions in human inflammatory joint disease. *Philos. Trans. R. Soc. Lond.* 311: 659–671.
10. Singh, S. and Hider, R.C. (1988) Colorimetric detection of hydroxyl radical: comparison of the hydroxyl-radical-generating ability of various iron complexes. *Anal. Biochem.* 171: 47–54.
11. Obata, T., Hosokawa, H. and Yamanaka, Y. (1993) Effect of ferrous iron on the generation of hydroxyl free radicals by liver microdialysis perfusion of salicylate. *Comp. Biochem. Physiol.* 106C: 629–634.
12. Sagone, A.L., Husney, R.M. and Davis, B. (1993) Biotransformation of paraaminobenzoic acid and salicylic acid by PMN. *Free Rad. Biol. Med.* 14: 27–35.
13. Floyd, R.A. (1983) Hydroxyl-free radical spin-adduct in rat brain synaptosomes. Observations on the reduction of the nitroxide. *Biochem. Biophys. Acta.* 756: 204–216.
14. Lawrence, G.D. and Cohen, G. (1985) *In vivo* production of ethylene from 2-keto 4-thiobutyrate in mice. *Biochem. Pharmacol.* 34: 3231–3236.
15. Floyd, R.A., Watson, J.J. and Wong, P.K. (1984) Sensitive assay of hydroxyl free radical formation utilizing high pressure liquid chromatography with electrochemical detection of phenol and salicylate hydroxylation products. *J. Biochem. Biophys. Meth.* 10: 221–235.
16. Anbar, M., Meyersein, D. and Neta, P. (1966) The reactivity of aromatic compounds toward hydroxyl radicals. *J. Phys. Chem.* 70: 2660-2662.
17. Leonards, J.R. (1962) Presence of acetylsalicylic acid in plasma following oral ingestion of aspirin. *Proc. Soc. Exp. Biol. Med.* 110: 304–308.
18. Schachter, D. (1957) The chemical estimation of acyl glucuronide and its application to studies on the metabolism of benzoate and salicylate in man. *J. Clin. Invest.* 36: 297–302.
19. Grootveld, M. and Halliwell, B. (1988) 2,3 dihydroxybenzoic acid is a product of aspirin metabolism. *Biochem. Pharmacol.* 37: 271–280.
20. Gatautis, V.J. and Pearson, K.H. (1987) Separation of plasma carotenoids and quantification of B-carotene using HPLC. *Clin. Chim. Acta.* 116: 195–206.
21. Vassault, A., Grafmeyer, D., Naudin, Cl., Dumont, G., Bailly, M., Henny, J., Gerhardt, M.F. and Georges, P. (1986) Protocole de validation de techniques. (Document B, stade 3). *Ann. Biol. Clin.* 44: 686-715.
22. Coudray, C., Mangournet, C., Bouhadjeb, S., Faure, H. and Favier, A. (1995) A simple HPLC-spectrophotomeric assay of salicylic acid in biological fluids. *Journal of Chromatographic Science*; *submitted.*
23. Maskos, Z., Rush, J.D. and Koppenol, W.H. (1990) The hydroxylation of the salicylate anion by a Fenton reaction and gamma radiolysis: a consideration of the respective mechanisms. *Free Rad. Biol. Med.* 8: 153–162.
24. Grootveld, M. and Halliwell, B. (1986) Aromatic hydroxylation as a potential measure of hydroxyl radical formation *in vivo*. Identification of hydroxylated derivatives of salicylate in human body fluids. *Biochem. J.* 237: 499–504.
25. Hamzah, R.Y. and Tu, S.-C. (1981) Determination of the position of monooxygenation in the formation of catechol catalyzed by salicylate hydrolase. *J. Biol. Chem.* 256: 6392–6394.
26. Winston, G.W. and Cederbaum, A.I. (1982) Oxidative decarboxylation of benzoate to carbon dioxide by rat liver microsomes: a probe for oxygen radical production during microsomal electron transfer. *Biochemistry* 21: 4265–4270.
27. Hiller, K.O., Hodd, P.L. and Willson, R.L. (1983) Antiinflammatory drugs: protection of a bacterial virus as an *in vitro* biological measure of free radical activity. *Chem-Biol. Interact.* 47: 293–305.
28. Singh, S. and Hider R.C. (1988) Colorimetric detection of the hydroxyl radical. Comparison of the hydroxyl radical-generating ability of various iron complexes. *Anal. Biochem.* 171: 47–54.
29. Floyd, R.A., Henderson, R., Watson, J.J. and Wong, P.K. (1986) Use of salicylate with high pressure liquid chromatography and electrochemical detection (LEC) as a sensitive measure of hydroxyl free radicals in adriamycin-treated rats. *Free. Rad. Biol. Med.* 2: 13–18.
30. Powell, S.R. and Hall, D. (1990) Use of salicylate as a probe for ·OH formation in isolated ischemic rat hearts. *Free Rad. Biol. Med.* 9: 133–141.

31. Ghiselli, A., Lauranti, O., De Mattia, G., Maiani, G. and Ferro-Luzzi, A. (1992) Salicyalte hydroxylation as an early marker of *in vivo* oxidative stress in diabetic patients. *Free Rad. Biol. Med.* 13: 621–626.

32. Tubaro, M., Cavallo, G., Pensa, M.A., Natale, E., Ricci, R., Milazzotto, F. and Tubaro, E. (1992) Demonstration of the formation of hydroxyl radicals in acute myocardial infarction in man using salicylate as probe. *Cardiology* 80: 246–251.

33. Udassin, R., Ariel, I., Haskel, Y., Kitrossky, N. and Chevion, M. (1991) Salicylate as an *in vivo* free radical trap: studies on ischemic insult to the rat intestine. *Free Rad. Biol. Med.* 10: 1–6.

34. Ingelman-Sundberg, M., Kaur, H., Terelius, Y., Pearson, J.O. and Halliwell, B. (1991) Hydroxylation of salicylate by microsomal fractions and cytochrome P450. Lack of production of 2,3 dihydroxybezoate unless hydroxyl radical formation is permitted. *Biochem. J.* 276: 753–757.

35. O'Connell, M.J. and Webster, N.R. (1990) Hyperoxia and salicylate metabolism in rats. *J. Pharm. Phamacol.* 42: 205–206.

36. Davis, W.B., Mohammed, B.S., Mays, D.C. et al. (1989) Hydroxylation of salicylate by activated neutrophils. *Biochem. Pharmacol.* 38: 4013-4019.

37. Grootveld, M. and Halliwell, B. (1987) Measurement of allantoin and uric acid in human body fluids. A potential index of free radical reactions *in vivo*? *Biochem. J.* 243: 803–808.

38. Cathcart, R., Schwiers, E., Saul, R.L. and Ames, B.N. (1984) Thymine and thymidine glycol in human and rat urine. A possible assay for oxidative DNA damage. *Proc. Natl. Acad. Sci. USA* 81: 5633–5637.

39. Cham, B.E., Ross-Lee, L., Bochner, F. and Imhoff, D.M. (1980) Measurement and pharmacokinetics of acetylsalicylic acid by a novel high performance liquid chromatographic assay. *Therap. Drug. Monit.* 2: 365–372.

40. Halliwell, B. and Grootveld, M. (1987) The measurement of free radical reactions in humans: some thoughts for future experimentation. *FEBS Lett.* 213: 9–14.

41. Halliwell, B., Kaur, H. and Ingelman-Sunberg, M. (1991) Hydroxylation of salicylate as an early assay for hydroxyl radicals: a cautionary note. *Free Rad. Biol. Med.* 10: 439–441.

42. Woodbury, D.M. and Fingl, E. (1975) Analgesic-antipyretics, anti-inflammatory agents and drugs employed in the therapy of gout. *In*: L.S. Goodman and A. Gilman (eds): *The pharmacological basis of therapeutics*, Fifth Edition, MacMillan Publishing Co., Inc., New York, pp 325-358.

43. Ishimitsu, S., Fujimoto, S. and Ohara, A. (1980) Quantitative analysis of the isomers of hydroxyphenylalanine by high-performance liquid chromatography using fluorimetic detector. *Chem. Pham. Bull.* 28: 992–994.

44. Ishimitsu, S., Fujimoto, S. and Ohara, A. (1984) Hydroxylation of phenylalanine by the hypoxanthine-xanthine oxidase system. *Chem. Pham. Bull.* 32: 4645–4649.

45. Kaur, H. and Halliwell, B. (1984) Aromatic hydroxylation of phenylalanine as an assay for hydroxyl radicals. Measurement of hydroxyl radical formation from ozone and in blood from premature babies using improved HPLC methodology. *Anal. Biochem.* 220: 11–15.

46. Lamrini, R., Crouzet, J.M., Francina, A., Guilluy, R., Steghens, J.P. and Brazier, J.L. (1994) Evaluation of hydroxyl radicals production using $^{13}CO_2$ gas chromatography-isotope ratio mass spectrometry. *Anal. Biochem.* 220: 129–136.

Subject index

(The page number refers to the first page of the chapter in which the keyword occurs)

312

Bioradicals

Detected by ESR Spectroscopy

Edited by
H. Ohya-Nishiguchi / L. Packer
University of California, Berkeley, CA, USA

1995. Approx. 340 pages. Hardcover
ISBN 3-7643-5077-6 (MCBU)

This book presents the historical background as well as the up-to-date developments in, and guidelines for, bioradicals and ESR research.

Bioradicals Detected by ESR Spectroscopy is a newly coined term which encompasses paramagnetic species in biological systems, such as active oxygen radicals and transition metal ions.

Research on the structures and functions of bioradicals have been attracting growing attention in the biological sciences, resulting in an increasing demand for comprehensive studies which allow researchers from many fields to understand the true importance of these species.

ESR spectroscopy is of great interest to interdisciplinary research and is applied in many fields, ranging from physics and chemistry to biology and medicine. New ESR technologies of multiquantum ESR, STM-ESR and open-space ESR, several ESR imaging techniques, spin trapping and new methods in *in vivo* spin trapping are all described and dicussed in this volume. In addition, it demonstrates the applications of ESR in food and medical sciences, i.e., the estimation and characterization of the antioxidant ability of foods and food components and the elucidation of the underlying chemical mechanisms.

Birkhäuser Verlag • Basel • Boston • Berlin

Oxygen Free Radicals in Tissue Damage

Edited by
M. Tarr / F. Samson
University of Kansas Medical Center, Kansas City, KS, USA

1993. 296 pages. Hardcover • ISBN 3-7643-3609-9

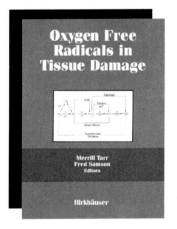

This important volume provides insights into current thinking and controversies regarding oxygen free radicals and reactive oxygen species in tissue dysfunction and pathology. Bringing together contributions from scientists actively involved in reactive oxygen species research, this volume presents these contributors' ideas and concepts regarding new advances in this rapidly expanding field of research. To readers new to this research field, this volume provides an introduction and a state-of-the-art overview of important topics relevant to understanding oxygen free radicals. Readers knowledgeable in oxygen free radicals should gain new insights.

Included in this authoritative and timely text are chapters detailing the chemistry of oxygen free radicals as well as discussing methods for detecting and generating these highly reactive and short-lived chemicals. Chapters also discuss the roles played by oxygen free radicals in damage to the heart, the brain, the microvascular system, and the immune system.

A discussion of the properties of various chemicals which can provide protection against these potentially destructive substances is also presented.

**This book will be valuable to all biomedical researchers ~
especially to those investigating the mechanisms of tissue damage
~ and to libraries that serve basic medical sciences.**

With contributions by:
K.L. Audus, J.S. Beckman, R. Bolli, D.C. Borg, J. Chen, K.A. Conger, R.A. Floyd, I. Fridovich, J.I. Goldhaber, D.N. Granger, E.D. Hall, J.H. Halsey Jr., N. R. Harris, C.J. Hartley, H. Ischiropoulos, M.O. Jeroudi, S. Ji, J.R. Kanofsky, M. E. Layton, X.Y. Li, T.L. Pazdernik, F. Samson, M. Tarr, J. P. Uetrecht, D.P. Valenzeno, J.N. Weiss, B.J. Zimmerman, L. Zu, M. Zughaib.

Birkhäuser Verlag • Basel • Boston • Berlin

Free Radicals:

From Basic Science to Medicine

Edited by
G. Poli / E. Albano / M.U. Dianzani
Univ. di Torino, Italy

1993. 528 pages. Hardcover
ISBN 3-7643-2763-4 (MCBU)

This book is a compilation of reviews covering the major areas of free radical science. Physiological and pathological aspects of free radical reactions are considered. While the contributions reflect the differing backgrounds and interests of investigators involved in basic or applied studies, all take a „vertical" approach to the main fields in which free radicals are assumed to play an important role, e.g. aging, cancer, metabolic disorders, inflammation, radiation, and mutagenesis. Special emphasis is placed on the medical aspects of free radicals, an area which until recently has been accorded only scant attention by the medical profession. Finally, the book reports comprehensively on prevention and therapy of free radical based pathology by means of various antioxidants.

Free Radical Research • Biochemistry • Pathology

Birkhäuser Verlag • Basel • Boston • Berlin

DATE DUE

MAY 0 6 2001	
	OCT 2 4 2010
MAR 3 0 2005	
AUG 0 7 2010	
MAY 2 2 2012	
SEP 0 9 2012	